Another
Look at
Evaluating
Training
Programs

Fifty articles from

Training & Development

and *Technical Training*

magazines cover the

essentials of evaluation

and return-on-investment.

Compiled by

Donald L. Kirkpatrick

ASTD

Library of Congress Catalog Card Number: 98-71680
ISBN: 1-56286-088-7

Ordering information: Books published by the American Society for Training & Development can be ordered by calling 800.628.2783 or 703.683.8100.

Printed in the United States of America.

ASTD
1640 King Street
Box 1443
Alexandria, VA 22313-2043
PH 703.683.8100, FX 703.683.8103
www.astd.org

Table of Contents

Preface ix

Part 1. Kirkpatrick Articles

Great Ideas Revisited 3
Donald L. Kirkpatrick

Evaluating Training Programs: Evidence vs. Proof 9
Donald L. Kirkpatrick

Part 2. Philosophy and Concepts

Making a Play for Training Evaluation 15
Theodore J. Krein and Katharine C. Weldon

Training 101: Navigating the Evaluation Rapids 21
Mary Lynn Pulley

The Basics: Evaluation 26
Angus Reynolds

Measurement Made Simple 28
Leigh Ann Williams

You Ask/Trainers Answer 31
Various

Testing Employee Performance: A Review of Key Milestones 33
Rob Schriver

Evaluating Training Results 36
Paul R. Erickson

Input, Process, Output: a Model for Evaluating Training 39
David S. Bushnell

HRD's Failure To Sell Itself 42
Richard P. Lookatch

Five Uneasy Pieces in the Training Evaluation Puzzle 46
Glenn M. McEvoy and Paul F. Buller

Evaluation That Goes the Distance 50
Paul R. Bernthal

Involving Managers in Training Evaluation 55
Linn Coffman

Ten Evaluation Instruments for Technical Training 58
Anne F. Marrelli

Training Investments: The Global Perspective 66
Edward E. Gordon

The Team Approach to Formative Evaluation 70
James Heideman

How To Ensure Transfer of Training **74**
Paul L. Garavaglia

An Integrated Evaluation Model for HRD **78**
Robert O. Brinkerhoff

Evaluation Techniques that Work **81**
Herman Birnbrauer

Your New Role in the Organizational Drama:
Measuring Effectiveness **84**
Neal E. Chalofsky and *Carlene Reinhart*

Management Training Evaluation: An Update **91**
William H. Clegg

Measuring Training Results: Key to Managerial Commitment **97**
James D. Bell and *Deborah L. Kerr*

Part 3. Reaction

Keeping Your Pilots on Course **103**
Jan Chernick

Part 4. Learning

Evaluating Test Questions: More Than Meets the Eye **111**
David Blair and *Steve Giles*

Analyzing Knowledge-Based Tests **113**
Richard L. Sullivan, Jerry L. Wircenski, and *Mary Jo Major*

Does the Trainee Know Best? **119**
Jerry L. Harbour

IBM Takes the Guessing Out of Testing **123**
George M. Alliger and *Harold M. Horowitz*

Using Evaluation to Improve Performance **127**
Vivian Marshall and *Rob Schriver*

Using Competency Exams for Evaluating Training **131**
Jack E. Smith and *Sharon Merchant*

Part 5. Behavior

Training 101: Another Look at Employee Surveys **139**
H. John Johnson

Collecting Data the E-Mail Way **142**
Lorraine Parker

Getting Them Out and Getting Them Back **145**
Kenneth M. Nowack

Everything You Always Wanted To Know About Employee Surveys. . **149**
Karen B. Paul and *David W. Bracken*

Measuring Skills and Behavior **154**
Kate Ludeman

Putting Values Into Evaluation **159**
Roger Poulet and *Gerry Moult*

How Effective Is Outdoor Training? **164**
Richard J. Wagner and *Christopher C. Roland*

Part 6. Results

Who's Afraid of Level 4 Evaluation? A Practical Approach **171**
Sandra Shelton and *George Alliger*

The Bottom Line **175**
Basil Paquet, Elizabeth Kasl, Laurence Weinstein, and *William Waite*

Zooming in on Training Goals. **182**
James Heideman and *Bruce Sanderson*

Cost-Benefit Analysis Techniques for Training Investments **185**
Werner J. Schmidt

Part 7. Return-on-Investment

Is There an ROI in ROI? **191**
Tom Barron

Costing Out the Value of Training. **195**
Paul E. Brauchle

Measuring Training's ROI **200**
Scott B. Parry

Demonstrating ROI of Training **205**
Eric A. Davidove and *Peggy A. Schroeder*

Selling Technical Training to Top Management **207**
Mary Coeli Meyer

ROI: The Search for Best Practices. **210**
Jack J. Phillips

Was it the Training? **216**
Jack J. Phillips

How Much Is the Training Worth? **220**
Jack J. Phillips

Return on Investment: Accounting for Training **224**
Anthony P. Carnevale and *Eric R. Schulz*

About the Author **257**

Listing by
Publication

Note: The information in the author biographies following each article was correct at the time of the original printing of the magazines. For current author contact information, please call the ASTD Magazines and Periodicals department at 703.683.7250. For reprint or subscription information regarding *Technical Training* or *Training & Development*, please call 800.628.2783 or 703.683.8100.

Technical Training

The Basics: Evaluation (August/September 1994) 26
Angus Reynolds

You Ask/Trainers Answer (April 1994) 31
Various

Testing Employee Performance: A Review of Key Milestones
(April 1997) 33
Rob Schriver

Ten Evaluation Instruments for Technical Training (July 1993) 58
Anne F. Marrelli

Training Investments: The Global Perspective (July 1996) 66
Edward E. Gordon

The Team Approach to Formative Evaluation (April 1993) 70
James Heideman

Evaluating Test Questions: More Than Meets the Eye
(May/June 1996) 111
David Blair and *Steve Giles*

Analyzing Knowledge-Based Tests (January 1993) 113
Richard L. Sullivan, Jerry L. Wircenski, and *Mary Jo Major*

Using Evaluation to Improve Performance (January 1994) 127
Vivian Marshall and *Rob Schriver*

The Bottom Line (May 1987) 175
Basil Paquet, Elizabeth Kasl, Laurence Weinstein, and *William Waite*

Zooming in on Training Goals (July 1997) 182
James Heideman and *Bruce Sanderson*

Cost-Benefit Analysis Techniques for Training Investments
(April 1997) 185
Werner J. Schmidt

Is There an ROI in ROI? (January 1997) 191
Tom Barron

Costing Out the Value of Training (May/June 1992) 195
Paul E. Brauchle

Selling Technical Training to Top Management
(February/March 1993) 207
Mary Coeli Meyer

Training & Development

Great Ideas Revisited (January 1996) **3**
Donald L. Kirkpatrick

Evaluating Training Programs: Evidence vs. Proof
(November 1977) **9**
Donald L. Kirkpatrick

Making a Play for Training Evaluation (April 1994) **15**
Theodore J. Krein and Katharine C. Weldon

Training 101: Navigating the Evaluation Rapids (September 1994) . . **21**
Mary Lynn Pulley

Measurement Made Simple (July 1996) **28**
Leigh Ann Williams

Evaluating Training Results (January 1990) **36**
Paul R. Erickson

Input, Process, Output: a Model for Evaluating Training
(March 1990) **39**
David S. Bushnell

HRD's Failure To Sell Itself (July 1991) **42**
Richard P. Lookatch

Five Uneasy Pieces in the Training Evaluation Puzzle
(August 1990) **46**
Glenn M. McEvoy and Paul F. Buller

Evaluation That Goes the Distance (September 1995) **50**
Paul R. Bernthal

Involving Managers in Training Evaluation (June 1990) **55**
Linn Coffman

How To Ensure Transfer of Training (October 1993) **74**
Paul L. Garavaglia

An Integrated Evaluation Model for HRD (February 1988) **78**
Robert O. Brinkerhoff

Evaluation Techniques that Work (July 1987) **81**
Herman Birnbrauer

Your New Role in the Organizational Drama:
Measuring Effectiveness (August 1988) **84**
Neal E. Chalofsky and Carlene Reinhart

Management Training Evaluation: An Update (February 1987) **91**
William H. Clegg

Measuring Training Results: Key to Managerial Commitment
(January 1987) **97**
James D. Bell and Deborah L. Kerr

Keeping Your Pilots on Course (April 1992) **103**
Jan Chernick

Does the Trainee Know Best? (June 1988) **119**
Jerry L. Harbour

IBM Takes the Guessing Out of Testing (April 1989) **123**
George M. Alliger and *Harold M. Horowitz*

Using Competency Exams for Evaluating Training (August 1990) . . **131**
Jack E. Smith and *Sharon Merchant*

Training 101: Another Look at Employee Surveys (July 1993) **139**
H. John Johnson

Collecting Data the E-Mail Way (July 1992) **142**
Lorraine Parker

Getting Them Out and Getting Them Back (April 1990) **145**
Kenneth M. Nowack

Everything You Always Wanted To Know About
Employee Surveys (January 1995) **149**
Karen B. Paul and *David W. Bracken*

Measuring Skills and Behavior (November 1991) **154**
Kate Ludeman

Putting Values Into Evaluation (July 1987) **159**
Roger Poulet and *Gerry Moult*

How Effective Is Outdoor Training? (July 1992) **164**
Richard J. Wagner and *Christopher C. Roland*

Who's Afraid of Level 4 Evaluation? A Practical Approach
(June 1993) **171**
Sandra Shelton and *George Alliger*

Measuring Training's ROI (May 1996) **200**
Scott B. Parry

Demonstrating ROI of Training (August 1992) **205**
Eric A. Davidove and *Peggy A. Schroeder*

ROI: The Search for Best Practices (February 1996) **210**
Jack J. Phillips

Was it the Training? (March 1996) **216**
Jack J. Phillips

How Much Is the Training Worth? (April 1996) **220**
Jack J. Phillips

Return on Investment: Accounting for Training
(July 1990 Supplement) **224**
Anthony P. Carnevale and *Eric R. Schulz*

Another Look at Evaluating Training Programs
Preface

When Mark Twain made his famous statement, "everybody talks about it but nobody does anything about it," he, of course, was referring to the weather. He could also have been referring to evaluation of training programs—well, not exactly, but almost.

The first part of the statement—almost everyone in training and development talks about it—is almost as true as the second part. I became interested in the subject of evaluation in 1952 when I selected "Evaluating a Human Relations Training Program for Foremen and Supervisors" as my Ph.D. dissertation topic. At that time, hardly anyone in training was talking about evaluation and almost no one was doing anything about it. At present, most human resource development (HRD) professionals are talking about evaluation, and an increasing number are doing something about it.

Over the past 45 years, I have communicated my ideas to training and development professionals and encouraged them to evaluate their programs. In 1959 I published a series of articles describing a four-stage model—REACTION, LEARNING, BEHAVIOR, and RESULTS. Since then, I have conducted numerous evaluation workshops at national, regional, and chapter conferences of the American Society for Training & Development (ASTD). My objective has been to provide trainers with a practical approach to evaluating their programs. I have also written other articles and the chapter on evaluation in the *Training & Development Handbook*. The four-stage model has influenced subsequent work on the topic. For example, this volume's article by Bell and Kerr titled "Measuring Training Results: Key to Managerial Commitment" (page 97, this volume) refers to Del Grazio's four-level model. Note the parallels between the Kirkpatrick and Del Grazio models:

Level	Kirkpatrick (1959)	Del Grazio (1984)
One	Reaction	Happiness
Two	Learning	Learning
Three	Behavior	Practical Application
Four	Results	Bottom Line

My main contribution to the field is the 1994 book, *Evaluating Training Programs: The Four Levels,* published by Berrett-Koehler Publishers of San Francisco. The second edition of this popular book was published in July 1998. Part 1 of the book describes, in detail, the guidelines, forms, and procedures for each level. Part 2 provides 12 actual case studies of application by such companies as Motorola, Intel, Kemper Insurance, CISCO, GAP, St. Luke's Hospital, Duke Energy, Arthur Andersen, University of Wisconsin Management Institute, and First Union National Bank.

Another of my contributions to the field is the 313-page book *Evaluating Training Programs,* published by ASTD in 1975. Collected in this book are all of the evaluation articles that had been published in ASTD's *Training & Development Journal* from 1965 to 1975. A total of 46 articles are included. In order to help readers gain maximum benefit in the minimum time, I organized the articles into sections.

In 1986 we repeated the process for all of the evaluation articles published in the *Training & Development Journal* from 1976 to 1986. That ASTD publication is titled *More Evaluating Training Programs.* This publication, *Another Look at Evaluating Training Programs,* includes all evaluation articles published from 1987 to 1998 in *Training & Development.*

The majority of the articles in this volume deal with philosophy and concepts regarding evaluation. I have included all of these articles in the section titled "Philosophy and Concepts." Other articles deal with specific approaches and techniques for evaluating training. I have included these in one of the following sections: Reaction, Learning, Behavior, or Results. If, for example, the article addresses reaction and learning, it is included in the Learning section. If it addresses learning and behavior, it is included in the Behavior section, and if it describes an approach that measures results, it is included in the Results section, even though it might include ways of measuring one or more of the other three stages. Because of the heightened interest in the issue of return-on-investment (ROI), I have included a special section that includes articles dealing with ROI.

How to Use

Two evaluation principles are important to understand:
1.   We *can* borrow evaluation forms, procedures, designs, approaches, techniques, and methods from other people.
2.   We *can't* borrow evaluation results, even if the program evaluated by someone else is the same program that we are conducting. Although the content, audiovisual aids, workbooks, cases, and exercises are identical, the organization, culture, job climate, attitudes, and instructors are different.

Therefore, as you read the articles, look for the *way* the program is evaluated and not the results of the evaluation.

Unfortunately, few articles were published in this 10-year period that provide specific help for HRD practitioners. Therefore, take advantage of those that can be helpful. Read them with a yellow pen in your hand and mark those ideas that are useful. Review the book periodically to see whether you need to be reminded of some of the ideas that you thought were relevant.

A Final Comment

I continually ask training people the following questions: "What kind of pressure are you under to evaluate training programs?" "Is your management asking you to *prove* in dollars and cents that benefits exceed the cost of training?"

I'm not sure whether to say "fortunately" or "unfortunately," but in the last 45 years there has been only a small increase in the pressure on trainers to prove that what they are doing is worth the money spent. There are still only a few organizations that ask trainers for bottom-line results. My advice to trainers is to anticipate that this may happen and get a head start on it. Refer to my article titled "Evaluating Training Programs: Evidence vs. Proof." Start now to look for evidence. Begin by doing a good job of evaluating reaction, which is a measure of customer satisfaction.

Positive reaction is all that many managers need to satisfy them that training is worthwhile. Start measuring learning on certain programs. This is a higher form of evidence that shows that training is accomplishing what it intends. Begin to look at changes in job behavior. This is even better evidence. Finally, see whether or not you can show evidence that final results have been achieved in terms of such factors as more productivity, lower costs, more sales, fewer accidents, less turnover, and few absences. If and when your management asks for evidence, you will have something to show them. Or better yet, show them this evidence before they ask for it. Evidence of positive results will not only increase your status and security in the organization, but it will also improve your chances of increasing your budget to do the things you think should be done!

Donald L. Kirkpatrick

PART 1.
Kirkpatrick Articles

Great Ideas Revisited

THEN
▾▾▾▾▾▾▾▾

Techniques for Evaluating Training Programs

BY DONALD L. KIRKPATRICK

Here are the original four articles (condensed) from November through February 1959, published in T&D *when it was the* Journal for the American Society of Training Directors. *The series introduced Kirkpatrick's four-level model of evaluation.*

These articles are designed to stimulate training directors to increase their efforts in evaluating training programs.

Step 1: Reaction

Reaction may best be defined as how well trainees like a particular training program. Evaluating in terms of reaction is the same as measuring trainees' feelings. It doesn't measure any learning that takes place. Because reaction is easy to measure, nearly all training directors do it. But in this writer's opinion, many of their attempts don't meet the following standards:

▸ Determine what you want to find out.

▸ Use a written comment sheet with the items determined in the task above.

NOW
▾▾▾▾▾▾▾▾

Revisiting Kirkpatrick's Four-Level Model

BY DONALD L. KIRKPATRICK

It has been more than 37 years since Kirkpatrick's classic four-level model was first published in the Journal for the American Society of Training Directors. *Here, Kirkpatrick takes another look at his creation.*

Beginning with the November 1959 issue of *Training & Development* (then called the *Journal of the American Society of Training Directors*), I published a series of four articles, "Techniques for Evaluating Training Programs." Since then, I've written many articles and book chapters on evaluation and compiled 20 years' worth of evaluation material in *Evaluating Training Programs* (American Society for Training and Development, 1975) and *More Evaluating Training Programs* (ASTD, 1986).

Over the years, a lot of things have happened in writing about and teaching evaluation. But the content has remained basically the same. I've made a few modifications in the guidelines for each of the four levels, as well as provided more and different forms and examples in my books. But the levels: reaction, learning,

CELEBRATING
50
YEARS

▶ Design the sheet so that reactions can be tabulated and quantified.

▶ Obtain honest reactions by making the sheet anonymous.

▶ Allow trainees to write additional comments not covered by the questions designed to be tabulated and quantified.

It's important to determine how people feel about a program because training decisions by top management are frequently made on the basis of one or two comments from participants. For example, a supervisory training program may be canceled just because one supervisor told the plant manager that the program was "for the birds."

People must like a training program to obtain the most benefit. Cloyd Steinmetz, past president of ASTD, says, "It's not enough to say, 'Here's the information, take it.' We must make it interesting and motivate people to want to take it."

It's important to measure participants' reactions in an organized fashion using written comment sheets that have been designed to obtain the desired reactions. The comments should also be designed so that they can be tabulated and quantified. The training coordinator, director, or other trained observer should make his own appraisal of the training in order to supplement participants' reactions. The combination of two evaluations is more meaningful than either one by itself.

When training directors effectively measure participants' reactions and find them favorable, they can feel proud. But they should also feel humble; the evaluation has only just begun. Even though a training director may have done a masterful job measuring trainees' reactions, that's no assurance that any learning has taken place. Nor is that an indication that participants' behavior will change because of training. And still further away is any indication of results that can be attributed to the training.

The comment sheet in the box was used to measure conferees' reactions at the 1959 American Society of Training Directors Summer Institute.

The subsequent steps in the evaluation process—learning, behavior, and results—will be discussed in the next three articles in the series.

Step 2: Learning

From an analysis of reactions, training directors can determine how well a program was accepted. They can also obtain comments and suggestions that will be helpful in improving future programs. It's important to obtain favorable reactions because decisions on future training activities are frequently based on the reactions of one or more key persons. In addition, the more favorable the reactions to a program, the more likely trainees are to pay attention and learn the principles, facts, and techniques discussed.

But favorable reactions don't assure learning. Most of us have attended meetings in which the speaker used enthusiasm, showmanship, visual aids, and illustrations to make his presentation well-accepted. But a careful analysis on the content would reveal that he said practically nothing of value, though he did it very well.

It's important to determine objectively the amount of learning that takes place. For the purpose of the article,

behavior, and results have remained constant.

It all started in 1952, when I decided to write my dissertation on "evaluating a supervisory training program." In analyzing my goals for the paper, I considered measuring participants' reaction to the program, the amount of learning that took place, the extent of their change in behavior after they returned to their jobs, and any final results that were achieved by participants after they returned to work. I realized that the scope of the research should be restricted to reaction and learning and that behavior and results would have to wait. Thus, the concept of four levels was born.

In the November 1959 article, I used the term "four steps." But someone, I don't know who, referred to the steps as "levels." The next thing I knew, articles and books were referring to the four levels as the Kirkpatrick model.

Defining the four levels

In 1993, my friend and colleague Jane Halcomb urged me to write a book describing the model. She said that many people were interested in it but had trouble finding details. The book, *Evaluating Training Programs: The Four Levels* (Berrett-Koehler, San Francisco, California, 1994), uses case studies to show how the four levels can be implemented—from such companies as Motorola, Arthur Andersen, and Intel.

Some articles have said that the four-level model is too simple. "The Flawed Four-Level Evaluation Model," written by Elwood F. Holton of Louisiana State University, will be published in *Human Resource Development Quarterly* this spring. Holton says that the model isn't a model at all but a taxonomy, a classification. Perhaps he is correct. I don't care whether it's a model or taxonomy as long as training professionals find it useful in evaluating training programs.

People have asked me why the model is widely used. My answer: It's simple and practical. Many trainers aren't much interested in a scholarly, complex approach. They want something they can understand and use. The model doesn't provide details on how to implement all four levels. Its chief purpose is to clarify the meaning of evaluation and offer guidelines on how to get started and proceed.

For those of you who are unfamiliar with the four levels, it's time to describe them.

Level 1: Reaction. This is a measure of how participants feel about the various aspects of a training program, including the topic, speaker, schedule, and so forth. Reaction is basically a measure of customer satisfaction. It's important because management often makes decisions about training based on participants' comments. Asking for participants' reactions tells them, "We're trying to help you become more effective, so we need to know whether we're helping you."

Another reason for measuring reaction is to ensure that participants are motivated and interested in learning. If they don't like a program, there's little chance that they'll put forth an effort to learn.

Level 2: Learning. This is a measure of the knowledge acquired, skills improved, or attitudes changed due to training. Generally, a training course accomplishes one or more of those three things. Some programs aim to improve trainees' knowledge of concepts, principles, or techniques. Others aim to teach new skills or improve old ones. And some programs, such as those on diversity, try to change attitudes.

learning is defined in a rather limited way: What principles, facts, and techniques were understood and absorbed by trainees? We're not concerned with on-the-job use of the principles, facts, and techniques.

Here are some guideposts for measuring learning:

▶ Measure the learning of each trainee so that quantitative results can be determined.

▶ Use a before-and-after approach so that learning can be related to the program.

▶ As much as possible, the learning should be measured on an objective basis.

▶ Where possible, use a control group (not receiving the training) to compare with the experimental group that receives the training.

▶ Where possible, analyze the evaluation results statistically so that learning can be proven in terms of correlation or level of confidence.

Level 3: Behavior. This is a measure of the extent to which participants change their on-the-job behavior because of training. It's commonly referred to as transfer of training.

Level 4: Results. This is a measure of the final results that occur due to training, including increased sales, higher productivity, bigger profits, reduced costs, less employee turnover, and improved quality.

Evaluation becomes more difficult, complicated, and expensive as it progresses from level 1 to level 4—and more important and meaningful. Some trainers bypass levels 1, 2, and 3 and go directly to level 4. Recently, I was asked by trainers in a consulting organization to skip a discussion of the first three levels and tell them how to do level 4 because that's what their customers want to know. I replied that understanding all four levels is necessary and that there are no easy answers for knowing how to measure results.

The guidelines (see box) were never intended to describe exactly what to do and how to do it. But they do provide an overview of the four levels and how to proceed.

Whether it's called "Techniques for Evaluating Training Programs" or *Evaluating Training Programs: The Four Levels,* it's essentially the same story. Each source describes the following reasons for evaluating training programs:

▶ to decide whether to continue offering a particular training program

▶ to improve future programs

▶ to validate your existence and job as a training professional.

If the time, money, and expertise are available, it's important to proceed through all four levels without skipping any. In some organizations, senior managers pay little attention to the training function. As long as they don't get negative vibes, they tend not to interfere or ask questions. But during times of downsizing, management must terminate people. Sometimes, the trainers are deemed expendable. The benefits from training may outweigh the costs, but unfortunately, proof can be difficult, if not impossible, to get.

Despite whether management seems to care or demand proof of training's value, training professionals should evaluate their programs at as many of the four levels as possible. In order to do that, they must learn as much as they can about evaluation. Understanding the four levels is a good start.

Donald L. Kirkpatrick *is professor emeritus of the University of Wisconsin. You can reach him at 1920 Hawthorne Drive, Elm Grove, WI 53122. Phone: 414/784-8348.*

Comment Sheet

The following evaluation was used to measure conferees' reactions at the 1959 American Society of Training Directors Summer Institute.

Leader _____

Subject _____

Date _____

1. Was the subject pertinent to your needs and interests?
☐ no
☐ to some extent
☐ very much so

2. How was the ratio of lecture to discussion?
☐ too much lecture
☐ ok
☐ too much discussion

3. How about the leader?
   a. How well did he state objectives?
   ☐ excellent   ☐ very good   ☐ good   ☐ fair   ☐ poor
   b. How well did he keep the session alive and interesting?
   ☐ excellent   ☐ very good   ☐ good   ☐ fair   ☐ poor
   c. How well did he use the blackboard, charts, and other aids?
   ☐ excellent   ☐ very good   ☐ good   ☐ fair   ☐ poor
   d. How well did he summarize during the session?
   ☐ excellent   ☐ very good   ☐ good   ☐ fair   ☐ poor
   e. How well did he maintain a friendly and helpful manner?
   ☐ excellent   ☐ very good   ☐ good   ☐ fair   ☐ poor
   f. How well did he illustrate and clarify the points?
   ☐ excellent   ☐ very good   ☐ good   ☐ fair   ☐ poor
   g. How was his summary at the close of the session?
   ☐ excellent   ☐ very good   ☐ good   ☐ fair   ☐ poor

What is your overall rating of the leader?
☐ excellent   ☐ very good   ☐ good   ☐ fair   ☐ poor

4. What would have made the session more effective? _____
_____
_____

The guideposts indicate that evaluation of learning is more difficult than evaluation of reaction. A knowledge of statistics is necessary. In many cases, the training department will have to call on a statistician to plan the evaluation procedures, analyze the data, and interpret the results.

It's relatively easy to measure learning that takes place in training on skills, such as job instruction, work simplification, and effective speaking. An evaluation of learning can be built into the training by setting up before-and-after situations in which trainees demonstrate whether they know the principles or techniques being taught. Classroom activities serve as evaluation techniques. A comparison of before-and-after scores and responses shows the amount of learning that occurred.

For example, in a course on job instruction for foremen, they will take turns demonstrating JIT skills in front of the class. From their performance, a training director can tell whether they've learned the principles of JIT and can use them, at least in a classroom situation.

When principles and facts rather than techniques are taught, it's more difficult to evaluate learning. In such cases, the most common measurement tool is a paper-and-pencil test. In some cases, standardized tests can be purchased. Or, training directors can construct their own. Unless the test accurately covers the material presented, it won't be a valid measure of the effectiveness of the learning. Frequently, standardized tests cover only part of the training material. So, only part of the course is being evaluated.

A great deal of work is required in planning the evaluation procedure, analyzing the data, and interpreting the results. When possible, training directors should devise their own methods and techniques.

See the box for samples from the Supervisory Inventory on Human Relations, devised by Earl Planty and myself.

When training directors can prove that their programs have been effective in terms of learning as well as reaction, they will have objective data for selling future programs and increasing their status in the company.

Step 3: Behavior

When I joined the Management Institute of the University of Wisconsin in 1949, one of my first assignments was to observe a one-week course on human relations for foremen and supervisors. I was particularly impressed by Herman, a foreman from a Milwaukee company. Whenever the instructor asked a question about human relations, Herman raised his hand. He had all the answers. I thought, "If I were in industry, I'd like to work for someone like Herman."

It so happened that my cousin, Jim, worked at Herman's company and Herman was his boss. Jim told me that Herman may know all about the principles of human relations but he didn't practice them on the job. He performed as a typical "bull-of-the-woods" who had little consideration for subordinates' feelings and ideas. I realized that there was a big difference between knowing principles and techniques and using them on the job.

In the article "Human Relations Skills Can Be Sharpened" (*Harvard Business Review,* July/August 1956), Dartmouth professor Robert Katz says that if people are going to change their job behavior, the following requirements

Implementation Guidelines

Here are guidelines for measuring each of the levels.

Reaction
▶ Determine what you want to find out.
▶ Design a form that will quantify reactions.
▶ Encourage written comments and suggestions.
▶ Attain an immediate response rate of 100 percent.
▶ Seek honest reactions.
▶ Develop acceptable standards.
▶ Measure reactions against the standards and take appropriate action.
▶ Communicate the reactions as appropriate.

Learning
▶ Use a control group, if feasible.
▶ Evaluate knowledge, skills, or attitudes both before and after the training. For example, use a paper-and-pencil test to measure knowledge and attitudes and a performance test to measure skills.
▶ Attain a response rate of 100 percent.
▶ Use the results of the evaluation to take appropriate action.

Behavior
▶ Use a control group, if feasible.
▶ Allow enough time for a change in behavior to take place.
▶ Survey or interview one or more of the following groups: trainees, their bosses, their subordinates, and others who often observe trainees' behavior on the job.
▶ Choose 100 trainees or an appropriate sampling.
▶ Repeat the evaluation at appropriate times.
▶ Consider the cost of evaluation versus the potential benefits.

Results
▶ Use a control group, if feasible.
▶ Allow enough time for results to be achieved.
▶ Measure both before and after training, if feasible.
▶ Repeat the measurement at appropriate times.
▶ Consider the cost of evaluation versus the potential benefits.
▶ Be satisfied with the evidence if absolute proof isn't possible to attain.

Supervisory Inventory on Human Relations

Circle "A" for agree or "DA" for disagree. Please answer all statements even if you're not sure.

1. Anyone is able to do almost any job if he tries hard enough.
A   DA

2. Intelligence consists of what we've learned since we were born.
A   DA

3. If supervisors know all about the work to be done, they are qualified to teach others how to do it.
A   DA

4. We're born with certain attitudes and there's little we can do to change them.
A   DA

5. Supervisors shouldn't praise members of their departments when they do a good job because the members will ask for a raise.
A   DA

6. A well-trained workforce is a result of maintaining a large training department.
A   DA

7. Supervisors lose respect when they ask employees to help solve problems.
A   DA

8. In making a decision, good supervisors are concerned with employees' feelings.
A   DA

9. Supervisors are closer to subordinates than to management.
A   DA

must exist:

▸ They must want to improve.
▸ They must recognize their own weaknesses.
▸ They must work in a permissive climate.
▸ They must have help from someone who is interested and skilled.
▸ They must have an opportunity to try out new ideas.

Katz put his finger on the problems in a transition between learning and changes in behavior on the job.

Evaluation of training in terms of on-the-job behavior is more difficult than reaction and learning evaluations. It requires a more scientific approach and the consideration of many factors. Here are several guideposts for evaluating training in terms of behavioral changes:

▸ Conduct a systematic appraisal of on-the-job performance on a before-and-after basis.
▸ The appraisal of performance should be made by one or more of the following groups (the more the better): trainees; trainees' supervisors, subordinates, and peers; and others familiar with trainees' on-the-job performance.

▸ Conduct a statistical analysis to compare before-and-after performance and to relate changes to the training.
▸ Conduct a post-training appraisal three months or more after training so that trainees have an opportunity to put into practice what they learned. Subsequent appraisals may add to validity of the study.
▸ Use a control group.

In the article, "Evaluate Your Training Program" (*Journal of the American Society of Training Directors, March/April 1957*), Lester Tarnopol suggests using an employee attitude survey on a before-and-after basis with control and experimental groups. He says, "Five employees is a good minimum for measuring the behavior of their supervisor." Tarnopol recommends using measurement instruments specifically suited to the requirements of one's company and training program. He also suggests inserting some neutral questions that don't relate to the training.

The most comprehensive research on evaluation in terms of on-the-job behavior is Olav Sorensen's study, "Observed Changes Inquiry," designed to answer the following questions:

▸ Have graduates of General Electric's advanced management courses of 1956 been observed to have changed their manner of managing?
▸ What inferences can be made from similarities and differences in the changes observed among graduates and nongraduates of the management courses?

First, GE's managers (graduates and nongraduates) were asked to indicate changes they'd observed in their own managing. Next, subordinates were asked to describe the changes they observed in the managers during the same period. Then, managers' peers were asked to describe changes in the managers' behavior. Last, the supervisors of the control and the experimental groups were asked to describe the managers' changes in behavior. Sorensen didn't use a before-and-after measure. Instead, he asked each participant to indicate what changes, if any, took place.

The future of training directors and programs depend to a large extent on their effectiveness. To determine effectiveness, attempts must be made to measure training in scientific and statistical terms. But few training directors have the background, skill, and time to engage in extensive evaluations. Sometimes, it's necessary to call on statisticians, research people, and consultants for advice and help.

> ■ *Workers must recognize their own weaknesses* ■

Step 4: Results

The objectives of most training programs can be stated in terms of the desired results, such as reduced costs, higher quality, increased production, and lower rates of employee turnover and absenteeism. It's best to evaluate training programs directly in terms of desired results. But complicating factors can make it difficult, if not impossible, to evaluate certain kinds of programs in terms of results. It's recommended that training directors begin to evaluate us-

ing the criteria in the first three steps: reaction, learning, and behavior.

Some training programs are relatively easy to evaluate in terms of results. For example, in teaching typing, one can measure the number of words per minute on a before-and-after basis. But in a recent issue of the journal, E. C. Keachie warns, "Difficulties in the evaluation of training are evident at the outset in the problem technically called, 'the separation of variables.' That is, how much of the improvement is due to training as compared to other factors?"

In a sophisticated article in the March/April 1958 issue of the *Harvard Business Review,* Rensis Likert says that changes in productivity can be measured on a before-and-after basis. A group of supervisors was trained in using democratic leadership in which decision making involved the use of a participative technique. Another group of supervisors was trained to make their own decisions and not to ask subordinates for suggestions.

Such factors as productivity, loyalty, attitudes, interest, and work involvement were measured. Both training programs resulted in positive changes in productivity. But the participative approach resulted in better feelings, attitudes, and other human relations factors. Likert concludes, "Industry needs more adequate measures of organizational performance than it is now getting."

Evaluation in terms of results is proceeding at a slow pace. In a few attempts, researchers have tried to segregate factors other than training that might have had an effect. In most cases, before-and-after measures have been attributed directly to training even though other factors might have been influential.

Studies like Likert's attempt to penetrate the difficulties of measuring such programs as human relations, decision making, and so forth. In the years to come, we'll see more efforts in this direction. Eventually, we may be able to measure human relations training in terms of dollars and cents. But at the present time, our research techniques are not adequate.

It's hoped that the training directors who have read and studied these articles are now clearly oriented on the problems and approaches in evaluating training. We training people should carefully analyze future articles to see whether we can borrow the techniques and procedures described.

EVALUATING TRAINING PROGRAMS:

EVIDENCE VS. PROOF

BY DONALD L. KIRKPATRICK

"As we look at the evaluation of reaction, learning, behavior, and results, we can see that **evidence** *is much easier to obtain than* **proof.** *In some cases,* **proof** *is impractical and almost impossible to get!"*

From the courtroom we can take a lesson regarding *evidence vs. proof.* A defendant is not judged "guilty" unless the evidence is so strong that there is no "reasonable" doubt. The word "reasonable," of course, still creates a problem. What is a "reasonable doubt"? Is it the same as "a shadow of a doubt"? Probably not. But the intent is that a person is innocent until proven guilty. And attorneys, juries, judges, reporters, witnesses, court observers, and newspaper readers will continue to argue about the "guilt" or "innocence" of a defendant.

Let's apply these words, *"evidence" and "proof"* to the evaluation of training programs. In previous articles I have divided the evaluation process into four segments or stages as follows:

1. *Reaction:* How do the participants feel about the program they attended? To what extent are they "satisfied customers"?

2. *Learning:* To what extent have the trainees learned the information and skills? To what extent have their attitudes been changed?

3. *Behavior:* To what extent has their job behavior changed as a result of attending the training program?

4. *Results:* To what extent have results been affected by the training program? (Results would include such factors as profits, return-on-investment, sales, production quality, quantity, schedules being met, costs, safety record, absenteeism, turnover, grievances, and morale.)

Let's analyze each one in terms of *evidence vs. proof.*

Reaction

In measuring reaction, we can ask the participants what they thought of the program (subject, leaders, schedules, facilities, meals, etc). The key question here is whether people give *honest* reactions. If, for example, they must sign their reaction forms and they have a fear of being critical, then the reactions are only *evidence* of the feelings of the trainees. If, on the other hand, they are completely candid and honest, then the reaction sheets are *proof* of the feelings and satisfaction of the participants. This factor of honesty is one that is readily controllable and therefore *proof* of reaction is relatively easy to get. For example, forms should not be signed and should be handed in or collected in such a way that there is no way to identify the person who completed them. If this is made clear to participants, honest reactions can be obtained and we do have *proof* of the reaction.

In measuring the *learning* of knowledge, skills, and attitudes, it is relatively easy to obtain *evidence* but more difficult to obtain *proof.* For example, if we teach a class in "motivation," we can ask the participants *"What did you learn?"* The responses might vary all the way from "nothing" to "I learned how to motivate my employees." These kinds of data, whether we get them orally or in writing, are strictly *evidence.* If we are going to get *proof,* we must measure learning on an objective basis by comparing their knowledge, skills, and attitudes before the program with their knowledge, skills, and attitudes after the program. Therefore, we need some kind of a pretest and posttest to obtain these data.

At first look, this would seem to provide *proof* except for one factor. We aren't quite sure whether the change in knowledge, skill, or attitudes came from the training program or from some other source. For example, a higher score on the posttests may have been due to such factors as the very process of taking the test twice or things that were learned outside the training program. Therefore, the difference between pretest and posttest scores provide *evidence* and not *proof.*

In order to get *proof,* we need to eliminate all the other factors that could have caused changes in posttest versus pretest scores. This can be done by using experimental and control groups. The experimental group is the group that attends the training program. The control group is a like group that does not attend the training program.

For example, we may have 50 production supervisors in our organization. These could be divided on a random basis by alphabetical order with every other one in the experimental group. Or, they could be divided on a systematic basis so there are 25 in each group with similar education, experience, jobs, age, and any other factors that would affect their test scores. Both groups would complete the pretest and posttest at the same time. The average scores would be compared.

As an example, a course on "Human Relations" is to be presented to supervisors. Half of them (25) will form the experimental group to participate in the program. The other half will be given the course at a later date, so they become the control group. The "Supervisory Inventory on Human Relations"[1] (80 items) will be used as the pretest and posttest.

Let's suppose that the average scores were

PRETEST
Experimental Group 61.4
Control Group . . . . 62.7.

The course was then given to the experimental group. At the completion of the course, both groups again completed the test with scores as follows:

POSTTEST
Experimental Group. 69.9
Control Group 69.8

In this example, the gain for the experimental group was 8.5 (69.9–61.4). The gain for the control group was 7.1 (69.8–62.7). The difference between the gains was 1.4 (8.5–7.1) in favor of the experimental group. This is the net score attributable to the course because all other factors were the same for the experimental and the control group. Therefore, the gain due to the program was only 1.4, which is not significant. Therefore, we don't have *proof* that significant learning has taken place.

If we hadn't used the control group, the gain of 8.5 for the experimental group was highly significant. This would have provided good *evidence* that learning had taken place. But we had to use a control group in order to eliminate the other factors

IT IS RELATIVELY EASY TO OBTAIN EVIDENCE OF BEHAVIOR CHANGES THAT OCCUR IN PARTICIPANTS BECAUSE THEY ATTENDED A PROGRAM.

that could have contributed to the gain. If we had discovered that the posttest and pretest scores for the control group were 63.5 and 62.7 (gain of .8) instead of 69.8 and 62.7 (gain of 7.1), then the difference between the gain of the experimental and control groups would have been 7.7 (8.5–.8) which would have been highly significant. Then we would have had *proof* of the learning.

Behavior

It is relatively easy to obtain *evidence* of behavior changes that occur in participants because they attended a program. All we have to do is to ask them. Three months or three days or three weeks after a program we can interview participants and ask "What are you doing differently than what you were doing before you attended the program on human relations?" The answers might be

"I'm telling people when they do a good job."

"I'm listening to them and getting to know them better."

"I'm letting them know what I expect."

"I'm rewarding them for performance."

"I'm asking them for their ideas."

"I'm making their jobs more interesting and challenging."

We might even go to their bosses and ask, "Have you noticed any change in the behavior of your subordinates since they attended the human relations program?" The answers might vary from

"No, I haven't seen any change" to *"Yes, he's sure treating his subordinates a lot better."*

The responses from the participants and their bosses would provide some evidence of the effectiveness of the program in terms of changes in job behavior. However, these kinds of responses would not provide proof.

How could we get *proof?* This requires a research approach that would do the following:

1. Measure behavior before the training program was given.

2. Measure behavior after the training program was completed.

3. Prove that any changes in behavior were due to the program and not to other factors such as salary adjustments, coaching from the boss, influence from other people inside or outside the organization, reading of books or articles, or an experience the participant had.

It's difficult to do the first one, measure the behavior before the program. We would have to do it enough times to be sure the behavior is typical. We'd have to do it in such a way as to be sure the measurement is accurate.

The second step is more difficult. For one thing, we'd have to wait long enough to measure the behavior to be sure that the participant has had an opportunity to change behavior. This may be one week, one month, or three months and it would vary for different people. And we might have to measure it several times (one month, three months, and six months) to be sure that any change in behavior is permanent.

When we compare the posttest behavior with the pretest behavior and find a change, we must be sure that the program caused the change. In order to eliminate other factors that could have caused the change, we must use a control group. This means that the control group must be equal to the experimental group regarding any factors that could cause changes in behavior. And we must measure the control group as well as the experimental group on a preprogram and postprogram basis to determine behavior changes that are caused by factors other than the training program. These changes must be subtracted from changes in the experimental group to determine the behavior changes that resulted from the program.

This 3-step process becomes complicated, time-consuming, and expensive. But it must be done to produce *proof* instead of *evidence*.

Results

The evaluation of results is similar to the evaluation of behavior. We can quite readily obtain *evidence* regarding the effectiveness of training on results. For example, if we instigate a new orientation and training program for new employees, we can measure turnover on a preprogram and postprogram basis.

We could probably obtain *evidence* that the program is effective because turnover has been reduced. However, in order to *prove* that the training program reduced turnover, we'd have to eliminate other factors that could have caused reduction in turnover. Some of these factors are increase in pay; better selection methods; fewer jobs available in the community; a seasonal situation; improved working conditions; change in management; improved

security in the company; better working conditions; and improved benefits. Again, we need a control group to eliminate these other factors.

As we look at the evaluation of reaction, learning, behavior, and results, we can see that *evidence* is much easier to obtain than *proof*. In some cases, *proof* is impractical and almost impossible to get. Therefore, what do we do?

Let's shoot for *proof* but be satisfied with *evidence*. In most cases, our superiors would be more than satisfied with *evidence*, particularly in terms of behavior or results. It's certainly a lot more than most of them are getting now.

If you have a boss who insists on *proof* of what you are doing, you must do one of the following:

1. Provide the *proof* at all costs.
2. Convince your boss that *evidence* is good enough and that *proof* is either impossible or at least impractical.

Recently, a friend of mine called me to New York to help him convince his boss that it was impractical (if not impossible) to prove in dollars and cents that a certain leadership training program was achieving more benefits than it was costing. Both my friend and I were convinced that there was *no way* to prove this. So we proceeded to try to educate the boss regarding *evidence* versus *proof*.

Learn all you can about evaluation. Next, gather *evidence* that your programs are effective. If they are not, don't broadcast the *evidence*. Just work at improving your programs so the *evidence* will be positive and then communicate it to your superiors. Most superiors will be most happy and will not ask for *proof*. However, as time and money permit, gather *proof* of the effectiveness of your programs. And if the *proof* shows an effective program, broadcast it around the world.

References

1. *Supervisory Inventory on Human Relations* by Dr. Donald L. Kirkpatrick, 1080 Lower Ridgeway, Elm Grove, WI 53122. (Sample copy available on request.)

2. *Evaluating Training Programs.* A collection of articles from the *Training & Development Journal,* compiled by Dr. Donald L. Kirkpatrick, ASTD, Box 5307, Madison, WI 53705.

Donald L. Kirkpatrick is professor emeritus of the University of Wisconsin. You can reach him at 1920 Hawthorne Drive, Elm Grove, WI 53122. Phone: 414.784.8348.

PART 2.
Philosophy and Concepts

IN THIS ARTICLE
Evaluation Design,
Instructional Evaluation

Making a Play for Training Evaluation

By Theodore J. Krein and Katharine C. Weldon

ONLY THROUGH THOROUGH EVALUATION CAN WE GAUGE A TRAINING PROGRAM'S SUCCESS. IN THIS DIALOGUE, A TRAINING DIRECTOR EXPLAINS EVALUATION BASICS—INCLUDING DONALD KIRKPATRICK'S FOUR LEVELS—TO A NEW TRAINER.

The setting is the corporate headquarters of Montac, a hypothetical diversified company with many divisions. Mary Hoskins is Montac's training director. She is based at corporate headquarters, but her staff provides support to all divisions.

In addition to developing and administering some programs that are available to employees throughout the corporation, Mary's staff provides internal consulting on improving performance. Often the consulting leads to the development of training programs specific to the needs of a division. The consulting may also result in recommendations for non-training solutions to performance problems.

Pete Elston is a training specialist on Mary's staff. He joined the company six months ago, after graduating with a liberal-arts degree from the local university. Mary hired Pete because she saw real potential in him. His education did not include courses in learning design, but Mary was convinced that she could develop Pete into a competent training professional—sort of a Pygmalion story. In addition to encouraging Pete to take courses in instructional technology at night, Mary has made it her personal mission to coach Pete.

Under Mary's guidance, Pete has been working on a performance-improvement project for the Granville Division of Montac. He has completed the front-end analysis and has prepared a set of potential performance objectives to be achieved by a training program, which has yet to be developed. We enter the scene as Pete enters Mary's office.

Pete Elston reporting

Pete Mary, you asked me to check back with you after the design meeting with the client, once we had the objectives pretty well developed. As you know, we determined that the problems at Granville are caused by a lack of knowledge and skill. So training seems to be an appropriate way to overcome the performance gap.

Mary You've done a great job of expressing most of your objectives in performance terms. That will be important as we move ahead with this project. When we state objectives in performance language, it's easier to know for sure whether we've achieved them. The reason I wanted to meet with you today is to begin discussing the evaluation plan for your program.

Pete Isn't it premature to talk about evaluation so early? I haven't even begun developing the program.

Mary I know what you mean. We usually think of evaluation happening *after* we hold a training program, so it must seem as if we're jumping the gun by talking about evaluation now. But here's why we're not.

The jargon of our profession includes two general terms—summative evaluation and formative evaluation. In today's meeting we'll concentrate on summative evaluation. This is evaluation of a program after it's been offered, to learn whether the training achieved what we wanted it to achieve. But I also want you to be aware of formative evaluation.

Formative evaluation involves checks we make during the program-development process to make sure that the program will meet our criteria of excellence. One formative evaluation technique is to have another experienced training professional go over the initial draft of a training design. That reviewer can consider a lot of different criteria, including the design's appropriateness for achieving the program's objectives.

We can take several other steps during development that increase the likelihood that participants will learn what we want them to learn. For example, we might test a case study on some experienced staff to see if it makes sense to them. Is it relevant to their jobs? Is it consistent with the company's culture? Do the instructions make sense?

When we get into program development, we will talk about the kinds of formative evaluation that might be appropriate for our program. For now, let's get back to the topic at hand—evaluation after the fact.

Program objectives versus the ultimate objectives

Pete You said we needed to develop an evaluation plan. What did you have in mind?

Mary What outcomes do you want from your program?

Pete I want the program to achieve the objectives we have on my list. For example, here's one of them: "At the end of the program, the new market-research analyst will be able to conduct an interview to determine what the client wants to learn from the market study."

Mary At what point do you want participants to show they can do these things? I mean, is this an objective you can check on at the end of the course, or does it apply to what these analysts will do when they return to their jobs?

Pete I guess I want it to be both. I'd like to know at the end of the course that they have met the objectives and can perform the skills. If they can, I can assume they'll be able to do them on the job.

Mary I know that seems logical. But there's a catch. It's true that your ultimate objective is to influence job performance. That's always our intent—usually assumed, but seldom stated—when we create a training program. At the moment the program is finished, we can't know whether it really will be successful in improving job performance. The best we can tell at the end of the program is whether the participants learned what they were supposed to learn.

We use our best judgment and experience to design training that will have an on-the-job impact. We want transfer of learning to the job. But at times, even our best judgment fails us, and we don't get the ultimate outcomes we want.

Pete So you're telling me that I can't know whether my training really has an impact?

Four levels of evaluation

Mary What I'm saying is you can't know at the end of the program whether it will have the desired on-the-job effects. Let me give you a scheme for evaluating programs. Donald Kirkpatrick developed it at the University of Wisconsin. He described four levels for evaluating training:

- Level 1—how participants reacted to the program
- Level 2—what participants learned from the program
- Level 3—whether what was learned is being applied on the job
- Level 4—whether that application is achieving results.

Pete I didn't know there could be so much to evaluation. To be honest, I didn't know there was so much involved in putting together training programs. It wasn't all that long ago that I assumed someone would just sit down and figure out all the topics a program should cover, develop a bunch of lectures to cover the topics, and then go ahead and put on the course. It was a new experience for me when you spent all that time showing me how to arrive at performance objectives. And we still haven't begun developing the program.

Mary Billions of dollars are spent each year in the United States on training. Much of that money is wasted because many trainers don't know the appropriate steps to take to ensure that their training addresses a real business need. And they don't know how to check to see whether that business need has been satisfied. In between those two are many other steps that are required if we are to do a truly professional job in training.

Pete You mentioned the four levels. I recently spoke with somebody who said that Level 1 evaluation really isn't worth much. Now that you've described the other levels, I can see why he might feel that way.

Mary Each of the four levels has value. The mistake many people make about Level 1 is to assume that if they get favorable participant reactions to their program, then that's all they need to conclude it's effective. It's not that Level 1 evaluation does not have value. It's a matter of recognizing what the value is—and what the limitations are.

Level 1: reacting to training

Mary First, let's see what Level 1 evaluation can do for us. We'll look at five possible advantages:

- Level 1 can tell us how relevant participants thought the training was.
- It can tell us whether they were confused by any of the training.
- It can point out any areas in which trainees thought information was missing.
- It can give us an idea of how engaged the trainees felt by the training.
- It can tell us how favorable overall participant reactions were.

By the way, Pete, I don't mean to imply that this is all we can get from Level 1 evaluation. But these five information items should make my point.

First, relevance. We know that adults learn better when they can relate a presentation to their previous experience and when they can see the relevance of the program to their jobs. Most adults are serious enough about their jobs that they do not welcome attending a program they perceive to be a waste of time. If the program will not help them to do their jobs better, then they probably won't be pleased with it. We shouldn't be pleased with it, either.

Second is potential participant confusion. Are participants having trouble understanding the concepts we're trying to teach them? If there are places in the program where that's a problem, then we need to know about it. If, in their evaluation comments, several participants report similar points of confusion, then we have gained valuable information to help us make corrections—either in the program design, or in its delivery.

Summative evaluations—those that come at the end of a program—can be used to make improvements in the program. But when we use summative evaluation to improve a program, the evaluation also is formative. Some people argue that all evaluation is formative evaluation, since it's hard to imagine not making changes to a program when we discover deficiencies in midstream.

The third benefit of Level 1 evaluation is its potential for pointing out missing content areas. Participants are usually painfully aware of problems they have when they try to perform their jobs. Many expect the training to provide solutions to those problems. It is useful for us to know whether the training has failed to

provide those solutions.

Don't let me mislead you. I'm not necessarily talking about missing instructional content in a program. Problems in this area might simply mean that we failed to clearly communicate what participants should expect from the program. It might also tell us that there are participant needs that can be addressed in some other way.

Fourth, did participants feel engaged? We know that adults learn better when they are involved in the learning process than when they feel like passive targets of information dumping. If they did not feel involved in the training program, then their learning probably wasn't what we'd like it to be. We'd want to take a look at why. Is it a fault of the design, a fault of the presentation, or a combination of the two?

The fifth factor is how favorable participants felt toward the program as a whole. Often we can infer this from their answers to other questions. Merely learning that participants did not like the program isn't very helpful to us. It doesn't tell us what we might do about it. So we need participants to give us some specifics.

Participants' favorable feelings don't ensure learning. But they can influence the chances that there will be a market for the program in the future. Bad press can scuttle a program, even one that teaches valuable knowledge and skills.

So you see, it is useful to gather Level 1 evaluation information.

Level 2: learning from training

Pete What happens in Level 2 evaluation?

Mary This is where we check to see whether participants can perform according to the course objectives. And an important key to measuring the performance after training is stating the desired performance properly, during training.

Let's look at the objective you chose earlier: "The new market-research analyst will be able to conduct an interview to determine what the client wants to learn from the market study."

How might we be able to tell—

during the training program—whether the new market-research analyst can conduct that interview?

Pete It seems to me we could create a simulation role play—in which we ask the participant to conduct a simulated interview. We could watch the participant and judge whether she or he performed well.

Mary Good. But I want to make sure there's no confusion here between on-the-job performance—ultimate performance—and performance at the end of the training program. So let's add some information on measurement to the description of that objective. We might restate the objective to incorporate measurement like this:

"Given a role play that simulates an interview with a client, the new market-research analyst will be able to conduct the interview to determine what the client wants to learn from the market study."

Now we have spelled out the conditions under which we can check to see whether the participant can satisfy the objective.

Pete We might even create a checklist to use in evaluating the role-play performance.

Mary That would be a great idea. We could include in that checklist all the specific interviewing principles we want the participant to demonstrate in the interview. Then we could add to our performance objective a criterion of satisfactory performance in the role play. Then the objective might look like this:

"Given a role play that simulates an interview with a client, the new market-research analyst will be able to conduct the interview to determine what the client wants to learn from the market study, demonstrating at least eight of the 10 interviewing principles presented in the training program."

Pete As I recall, that would be a test to learn whether the participant could apply what he or she learned in class. And I guess the checklist would help guarantee that the same evaluation standard would be used to evaluate the interview performance of each program participant.

Mary You're grasping this stuff very quickly, Pete. Now, you have some other objectives for your program that refer to the participant's knowledge of the subject. Let's look at a couple of them:

◗ "The participant will be able to explain the difference between a product-centered approach to selling and a client-centered approach to selling."

◗ "The participant will be able to recognize the five indicators of client resistance and the principal response to each."

Remember, we want to be able to tell, either during the training program or at the very end of the program, whether we have achieved these objectives. Look carefully at each and tell me what you would do to determine whether we have achieved them.

Pete For the first one, we could ask the participant to explain to the program leader the difference between a product-centered approach and a client-centered approach.

Mary Good. We see objectives like this for many classroom courses. Many program leaders never conduct that actual test. Instead, a leader judges the class's performance from his or her impressions of participants' understanding in class discussion.

That may be OK if the objective isn't crucial. But if we've spent the time analyzing what participants need in order to do their jobs, and we've included this objective in our program because it is important, then it's worth the effort to find out whether individual participants achieved the objective.

A key point to remember is that the verb we choose for a performance objective usually makes clear how we can test the performance. In this case, the verb is *explain*. So the participant must explain. It's clear that a true/false or multiple-choice question would not be adequate for measuring that performance. Instead, the participant would need to explain the differences between the two approaches—verbally, or in a short written paragraph.

Pete The second objective we listed is about being able to recognize the five indicators of resistance and the responses to them. That one looks trickier, because it involves two different performances.

Mary That's right. It is best to state each objective so that it contains only a single performance. So we probably should break this objective in two. But just for fun, let's see whether we could create a test item that would satisfy both parts of this objective. Start with the verb—*to recognize*. Now, what would—or wouldn't—we have participants do if we want them to recognize something?

Pete Well, the objective suggests that they should know it when they see it. It doesn't require them to write a sentence or even to drag something out of their memories with any clues other than the key words in the objective.

So what would a test item look like that asks for recognition? The question would contain the items we want trainees to recognize. So if it's a paper-and-pencil test, the five indicators would appear on the printed page. I suppose we could present the five indicators and ask, "Are these the five indicators?" But that makes the question require only a simple yes-no response. And half the participants could get it right by guessing. That wouldn't give us much comfort that participants had learned what we want them to learn about the indicators.

What if we listed 10 items, including the five indicators, and asked the participants to pick out the five? I think that would be a better question.

Mary I agree. And if that were all we wanted to know, that question would do the job. But we also said we'd test for recognition of the appropriate responses to the five types of resistance. How might you modify the question to test for both dimensions of the objective?

Pete It looks to me as though we could do it with a matching test item. On the left we'd have the 10 items, only five of which would be indicators. On the right we'd have descriptions of responses to resistance.

You said something the other day about one question giving away the answer to another question—something to avoid. It seems that if we only put five response descriptions on the right, they might provide clues to the five indicators on the left. So I think I'd put more than five response descriptions on the right—add some distracters. Then the likelihood of participants getting the question right by chance is reduced a lot.

Mary I think that would do it. You analyzed that very well. And I'm pleased that you remembered our discussion about the challenge of developing effective test questions. That's a task you'll soon need to undertake for your program. Because you need to develop the test for Level 2 evaluation after you develop your objectives—but before you develop the training program.

There's another important point. We want to know that it is the training program that's responsible for participants achieving the objectives. So we need a before-program measure. If we don't administer a pretest, we will miss the possibility that participants could have achieved the objectives before training.

Pete Now I understand why you were saying that we need to plan for program evaluation while we are planning for program development.

Mary That's right. Now, to help you get ready for developing the test for learning, I'm going to give you this book, *Criterion-Referenced Test Development,* by Sharon A. Shrock and William C.C. Coscarelli. It's one of the better ones I know of. It's scholarly without being complicated, and it will give you the right foundation.

Level 3: applying learning

Pete So what about Level 3 evaluation—applying learning to the job?

Mary Now we're getting to your ultimate objectives—ensuring that the training has had a positive influence on job performance. Unfortunately, many trainers completely ignore this evaluation level. They seem to assume that the logic that led to their training design is good enough to ensure that the desired results are happening on the job.

But that's a shaky assumption. And it's even worse than it sounds, because many trainers never bother with Level 2 measurement, either. The only data they have are from their Level 1 evaluations. So the trainers can't even tell whether participants actually learned what they were supposed to—let alone whether they are applying what they learned on the job.

So what would you do? How would you evaluate your training program at Kirkpatrick's third level?

Pete I guess I'd ask the training participants whether they're using what they learned. I'd ask them to describe to me what they do differently now, compared to what they did before they went through the training. I'd ask for specific examples of how they are applying the knowledge and skills they learned in the training program.

Mary Do you see any limitations in asking the participants themselves how they are using what they learned?

Pete I suppose they might be biased. They might even want to look good or make me or the interviewer feel good. That could bias the data.

To overcome that bias, I could talk to their managers. I suppose I could also watch the training participants doing their jobs. But that might be harder to pull off.

Mary One thing to remember is that you're trying to get a handle on changes in performance that might be attributable to the training. Unless you have a "before" performance measure, sometimes referred to as a baseline measure, it would be difficult to find out whether the "after" performance you observe is different.

Pete So if we want to do the Level 3 evaluation properly, we should gather baseline measures before people attend the training program—just like giving those pretests in order to make Level 2 evaluation meaningful.

Mary Right. And that brings us back to the importance of planning the evaluation of a program during the program-development process.

Your idea of talking to those who have participated in the training and to their managers is a good one. One-on-one interviews—face-to-face—probably give us the most useful information. But sometimes they aren't practical. The people we want to talk to may be spread out geographically. Or individual interviews may take too much time. One reason Level 3 evaluation is performed so seldom is that it can be costly to gather the information.

Questionnaires can help us reduce the cost of Level 3 evaluation. They require careful design. If you choose the questionnaire approach, you'll have to get someone involved who is experienced in questionnaire design. You may have to make some follow-up telephone calls to participants, as well. Telephone interviews aren't as effective as face-to-face interviews, but they can be more cost-effective.

Focus groups are also useful for gathering information. They can provide information more efficiently than individual interviews. We won't explore at this time all the factors to consider in using focus groups. Let me just say that this technique also requires specialized experience.

Some people think that Level 2 evaluation is unnecessary. They say we need only to focus on whether we are getting on-the-job results—Level 3. But that idea overlooks a crucial fact: If we didn't have a Level 2 measure, we wouldn't know the reason behind a lack of results on the job. It could result from trainees' failure to learn what was intended. But it could also result from something going on in the work environment.

When you plan your information-gathering activities for Level 3 evaluation, you'll need to take into account possible factors in the job environment that could prevent the application of newly learned knowledge and skill. All too frequently, we hear comments like this from the supervisors of newly trained employees: "I know that's what they taught you. Now I'm going to show you how we really do it."

If trainees don't receive proper, on-the-job reinforcement of what they learned in the formal training program, then we've wasted most or all of our investment in training.

Level 4: measuring results from training

Mary A lot of specialized knowledge and skill goes into developing evaluation strategies. And the greatest challenge of all is figuring out how to perform Level 4 evaluation. In Level 4 evaluation, we're interested in business results. What effect does the program have on measures that are important to the business? For example:

▶ reduced employee turnover
▶ reduced costs
▶ improved quality
▶ increases in favorable comments from customers
▶ increased sales
▶ fewer grievances filed; for example, harassment complaints
▶ increased profitability.

Pete It seems like a pretty big leap from learning in the classroom to results in the real world. Can we really tell whether it was training that succeeded—or failed—when a change in results occurs?

Mary That's the big question. In many cases, we can't tell without a great deal of research. That may involve a considerable investment of time and money. So many training directors judge that the benefits of Level 4 evaluation are not great enough to justify the investment required.

Pete So Level 4 evaluation isn't practical?

Mary It can be. In some settings, the business results may be simple to measure. For example, say that we observe a significant increase in sales and profitability after sales training. If other factors have been held constant, then we have strong evidence that the training was responsible for the gains.

Or say that we provide advanced training to some machine operators in how to minimize scrap. If we then observe a reduction in the amount and cost of scrap from those operators, then we have good evidence that the training was at least a major contributor to the result.

Pete Why the hedge? Isn't it obvious that training was the factor that caused the result?

Mary Unless we conduct a carefully designed study, using experimental and control groups, we have to recognize the possibility that other variables could have contributed to the result. We need to think in terms of evidence, not proof. We have evidence that training influenced the result. But we cannot state positively that training was the sole cause.

The examples I've given are simple ones. Measuring the effects of, say, supervisory or management training becomes much more difficult. Frequently, the skills participants learn in supervisory training

For Further Reading on Training Evaluation
Donald L. Kirkpatrick. "Techniques for Evaluating Training Programs. *Training & Development Journal,* June 1979.

Jack J. Phillips. *Handbook of Training Evaluation and Measurement Methods,* second edition. Houston, TX: Gulf, 1991.

Sharon A. Shrock and William C.C. Coscarelli. *Criterion-Referenced Test Development: Technical and Legal Guidelines for Corporate Training.* Reading MA: Addison-Wesley, 1989.

programs are not reinforced in the workplace. After training, a supervisor may return to a job in which she or he works for a manager who does not have the skills.

In such instances we could say there's a lack of congruence between formal training and the workplace. It is somewhat like taking golf lessons when the game to be played is baseball. In general, organizations need to do a better job of connecting formal training to the work environment.

Pete Mary, I appreciate the short course in evaluation. I know I have a lot to learn. But I think I have a better understanding now—at least of evaluation fundamentals. I'm looking forward to talking with you about how we can apply formative evaluation to the program as we develop it.

Theodore Krein *directs the instructional design department for Ernst & Young, 2000 National City Center, Cleveland, OH 44114.* **Katharine Weldon** *directs the instructional-design department for Ernst & Young, 8075 Leesburg Pike, Vienna, VA 22182.*

Training 101

NAVIGATING THE EVALUATION RAPIDS

By Mary Lynn Pulley, a management and organizational consultant located at 2505-A Belmont Boulevard, Nashville, TN 37212.

I stood barefoot, paddle in hand and life preserver securely fastened, listening to our guide explain how to navigate the class-four rapids we'd be encountering around the bend of the canyon. This was my first "serious" rafting experience. I was on a three-day rafting trip through Brown Canyon on the Colorado River, along with five other relative novices and one experienced guide. We were about to encounter some rapids known as Washing Machine.

Our guide explained that to get through these rapids, we would have to steer our raft between a large rock on one side and a whirlpool on the other. If we were too intent on avoiding the rock, we were likely to get sucked into the whirlpool. But if we skirted the whirlpool by too wide a margin, we could easily hit the rock.

Largely because of the guide's skill, we managed to stick to the middle course and successfully navigate the treacherous waters.

At the time, I was focused on following the shouted commands of our guide, on paddling, and on staying in the raft. Now, as I reflect on this experience from the comfort of my office, I realize that those rapids provide a useful metaphor for my work in organizations.

The twin perils on each side of us had contrasting natures. The rock was an object—solid, static, visible, and hard. The whirlpool was liquid—dynamic, hidden, indefinite, and soft. Focusing too strongly on avoiding either peril put us in danger of the other. Successfully navigating the river involved steering a middle course between "hard" and "soft" options on either side.

What does that have to do with training?

We can think of the rapids as representing the turbulence that many organizations are now experiencing because of economic pressures. Think of training programs as the raft. We know that keeping them afloat and moving forward requires flexibility and responsiveness to environmental perils.

The rock represents perils that are concrete and definite, such as budget cuts. The whirlpool represents environmental dangers that are just as serious but harder to see, such as organizational politics and individuals' perceptions. Steering the middle course means paying attention to both so that one or the other does not take you by surprise. Because the rock is more visible, I think that we tend to focus on avoiding it, only to get caught in a spinning whirlpool.

How can you evaluate training programs by steering the middle course? This type of evaluation is called responsive evaluation, because the emphasis is on responding to the organizational environment in which a training program exists.

What Is Responsive Evaluation?

In evaluating training, we tend to emphasize numerical data to demonstrate their value. The most common method uses the BIS-NOC approach ("butts in seats/number of courses"). Another method is to track participants' reactions to courses, on a numerical scale.

When training evaluation does escape the classroom, then the ultimate

In today's whitewater business environment, responsive evaluation can keep training afloat and moving forward.

credibility for justifying its effectiveness is to arrive at a number that is supposed to represent the program's return on investment. In other words, we come up with a dollar amount of profit that a training program has yielded for the business.

The problem is in finding a way to prove that a particular training program is directly responsible for profits, such as a $500,000 increase in sales revenue. Often, that proof is impossible to find; too many other variables can affect profits. Certainly, linking training to business results is essential. But the process of creating such a linkage is ambiguous and highly interpretive, especially in complex business environments.

When evaluating training, your real task is to collect evidence that suggests a causal relationship between training and business results. The creativity is in developing that linkage.

Responsive evaluation pays attention to both hard and soft issues. It does not minimize the importance of showing outcome-oriented results of training. But it also recognizes that people's perceptions toward a training program are shaped by many nonnumerical factors.

Responsive evaluation is a "both/and" approach, rather than an "either/or" approach. It is both quantitative and qualitative; it is both summative and formative. It deals with process as well as outcome.

The box, "Steering a Middle Course," shows how responsive evaluation runs a middle course between different dimensions of evaluation. Note that running a middle course does not mean steering a straight path down the middle, between options and challenges.

When river rafting, you need to make continual adjustments, sometimes veering the raft toward one side of the river or the other to skirt obstacles. Likewise, when using responsive evaluation, you'll face times when quantitative, outcome-oriented information is more effective—and you'll face other times when qualitative, process-oriented information is more effective. The key is paying attention to your environment and skillfully using the appropriate techniques at the right times.

Responsive evaluation assumes that

the purpose of evaluation is to relay information to decision makers in an organization so that they feel prepared to act. In order to be useful, any information must be translated so that it is meaningful to the people who are being informed. Organizations don't consume information; people do. Decisions are made by individuals, each of whom has his or her own perceptual filters. Because people receive and respond to information so differently, it's hard to know what will be meaningful to them.

Responsive evaluation recognizes the personal and political aspects of decision making. In a sense, it encourages you to cover your bases so that you get the message across to more people—including more of the

---

Steering a Middle Course

Like a raft navigating the rapids between a rock and a whirlpool, responsive training evaluation steers a middle course between various dimensions:

Whirlpool	Rock
Soft	Hard
Qualitative	Quantitative
Descriptive	Predictive
How	What
Formative	Summative
Process	Outcome
Holistic	Analytic
Present	Past

---

people who are responsible for funding training programs.

How Does Responsive Evaluation Work?

Back in Brown Canyon, our rafting guide was familiar with the river. He knew where the canyon narrowed and where it widened. He could anticipate the times when we would come upon particularly turbulent rapids. Before we steered the raft into Washing Machine, we took the time to stop, pull over, and survey what was around the bend. We then planned our strategy based on what we could foresee.

Similarly, when using responsive evaluation, you start by scouting

your organization's environment. Pay particular attention to the people who will be making decisions about your training program. Don't place all your emphasis on generating reams of data. Think in terms of shaping information that decision makers can use. Know who will be using the information and how it will influence their decisions.

The steps that follow will help you to keep this focus in mind. The box, "Five Steps to Responsive Evaluation," summarizes the steps and questions used in responsive evaluation.

1. Identify decision makers. This first important step of scouting the organizational environment can easily be overlooked.

I learned of its importance through a disappointing experience. I once worked with a large aerospace company on developing a large-scale management development program for its midlevel managers.

At its inception, the program received an enthusiastic go-ahead and a budget from the company's executives during their semiannual retreat. But over the year-and-a-half that it took to develop the program, the aerospace industry took a downward turn. The company began to tighten its belt.

The executives who were in key decision-making roles regarding the program had heard little about it during the design phase. Then, a year-and-a-half after approval of the program's development, its implementation came up as a budget item at the spring retreat. Executives voted to cut it.

As external consultants, my partner and I had been careful to design the program with input from internal people in various parts of the company. We had assembled a design team made up of training managers from throughout the organization, and we met with them on a monthly basis to keep them apprised of our progress and to incorporate their suggestions. Our problem was that the people on our design team were not decision makers.

It's safest to identify key decision makers during the earliest conceptual phases of a program. Be clear during your needs assessment about exactly who is in a decision-making role regarding the start-up and continuance of your program.

Ask yourself several questions about those decision makers:

▶ Why is this training program important to them?

▶ What is their stake in it?

▶ What values, biases, or experiences might influence their judgment about the program?

Ultimately, this scouting phase of your evaluation strategy will help you create a communication bridge between the program and the people who might be making critical decisions about it.

2. Identify the information needs of decision makers. The next step is to find out what decision makers want to know about a program. How are they going to use the information they receive? What decisions will they make?

The most direct way to get this information from decision makers is simply to ask them for it. If possible, interview decision makers or bring them together as an evaluation advisory board. An advisory board has the added advantage of creating a sense of shared needs and perceptions among different stakeholders. Find out what questions those stakeholders have. If decision makers don't have questions about a program, it's unlikely that you will be able to generate evaluative information that has

interest or value for them.

I have used a simple but revealing exercise to help identify information needs among a group of decision makers. See the instructions in the box, "Discovering What They Want To Know". The exercise helps you view a training program through the eyes of decision makers so that you can focus your evaluation on information they care about.

Some companies are so large that it's difficult to have direct access to decision makers. In such cases, the evaluation strategy is more political than personal. Still, the principle is the same—you still must identify informational needs of decision makers. It also helps if you develop a sense of decision makers' personal priorities, concerns, and preferred communication styles.

In river rafting, a "sleeper" is a rock or other obstacle hidden just below the surface of the water. These are difficult to see until you're on top of them, but going over one can capsize or tear your raft. To avoid sleepers, you must pay close attention to the movements of water currents.

If your organization is experiencing particularly turbulent times, then you must have direct access to at least one key decision maker. Otherwise, you are likely to run into sleepers—political currents or hidden

agendas that you are unaware of until it's too late to respond.

3. Systematically collect both quantitative and qualitative data. When we stopped in Brown Canyon and mapped out our strategy for our approach to Washing Machine, we also reviewed the paddling techniques that would be most appropriate under the circumstances. We knew that we could not anticipate every shift of the river. We knew that, to some extent, we would have to get into each part of the rapids before we would know how to respond to it. But being prepared with a variety of response techniques was absolutely critical.

Likewise, knowing how to collect and use both quantitative and qualitative methods to evaluate your program will improve your ability to navigate your training programs through a turbulent business environment.

In his book, *How To Use Qualitative Methods in Evaluation*, Michael Quinn Patton recommends thinking of evaluation outcomes in terms of both the quantity and the quality of change. (Patton's book is volume 4 in the "Program Evaluation Kit" series, Sage Publications, 1987.)

Quantity usually involves a number. Many people prefer this kind of information because they believe that it represents hard, objective data. But there is no reliable way to gather quantitative measures on some important aspects of training. For instance, quantity does not capture issues of meaning, attitude, or morale.

Patton describes quality as relating to the nuance, detail, and unique characteristics that make a difference, beyond the numerical scale. What does the training program mean to participants? What does it mean to the business? Quality is usually descriptive; we tend to relay it in the form of stories or anecdotes. It adds depth and detail, and it gives life to the numbers.

Responsive evaluation also tracks the quantitative features of a program. It may seem to place a greater emphasis on the qualitative features, but that is only because so many people tend to overlook those aspects.

"Numbers convey a sense of precision and accuracy," Patton says, "even if the measurements [that]

Five Steps to Responsive Evaluation

Negotiate the evaluation rapids by stopping before you hit whitewater, in order to scout out the environment before moving on. This five-step process can help.

1. Identify the decision makers.

▶ Who is in a decision-making role?

▶ Why is this training program important to them?

▶ What is their stake in it?

▶ What values, biases, or experiences might influence their judgment about the training?

2. Identify the information needs of decision makers.

▶ What questions do they have about the training program?

▶ How are they going to use the information they receive?

▶ What decisions might be forthcoming about the training?

3. Systematically collect both quantitative and qualitative data.

▶ What kinds of information will speak to decision makers?

▶ What data do you need to collect to create a meaningful story about the program?

4. Translate data into meaningful information.

▶ What are the information needs of each group of stakeholders?

▶ How can you frame the information so that it matters to them?

5. Involve and inform decision makers on a continuous basis.

▶ How can you involve decision makers without taking up too much of their time?

yielded the numbers are relatively unreliable, invalid, and meaningless. The point, however, is not to be anti-numbers. The point is to be 'pro-meaningfulness.' Thus by knowing the strengths and weaknesses of both quantitative and qualitative data, the evaluator can help stakeholders focus on really important questions rather than, as sometimes happens, focusing primarily on how to generate numbers."

We gather quantitative and qualitative data in different ways.

Numerical data that are tied to business results are often tracked through such methods as surveys, productivity measures (such as the sales volume, the size of an average sale, or the number of incentive bonuses), and quality measures (such as reductions in error, waste, rework, or customer complaints).

We gather qualitative data through such methods as interviews, focus groups, observations, or open-ended questionnaires. Through these methods, you are essentially gathering anecdotes that help provide a more complete picture of your program, such as examples of how the training has been used back on the job.

Relying too heavily on either type of data can result in misleading conclusions. Some people criticize qualitative data because they consider them to be subjective. People believe that you cannot draw conclusions about a program based on several participants' descriptions of their personal experiences.

But tracking only numbers can be just as misleading. Quantitative measures tend to be outcome oriented, and sometimes they do not provide the information you need to know.

For example, a program evaluator was assessing the effectiveness of a welding course for unemployed people. The evaluator chose to evaluate the course's success by tracking the number of participants who made it through the training. After a year, the program appeared to be totally ineffective, because no participants had completed it!

Fortunately, the program evaluator decided to look more closely at the training to determine why it was not working. The evaluator discovered the real reason why no participants

---

Discovering What They Want To Know

Here is a simple exercise for identifying the kinds of information that key decision makers want to know about your training program.

▶ First, ask each decision maker to write five different responses to the question, "I would like to know *(fill in the blank)* about the program."

▶ Then, have each decision maker rank the five statements on her or his list in order of their usefulness.

▶ Ask the decision makers to come together as a group to share their lists and to generate a single list of five basic things that they, as a group, would like to know—that is, information that might make a difference in what they are doing.

---

graduated from the course: The skills they were learning were in such high demand that the trainees were actually being hired before they completed the program!

In fact, the program was enormously successful at reaching its real goal, which was to provide participants with skills so that they could get jobs.

Research shows that people's beliefs and actions tend to be more affected by metaphors, stories, and anecdotes than by statistical results. Yet, no matter how much decision makers are influenced by stories, they tend to feel obligated to explain their decisions with hard evidence.

Combining quantitative and qualitative evidence helps you cover all bases. The combination creates a more complete picture for both trainers and decision makers than either kind of evidence could provide alone.

4. Translate data into meaningful information. Individual pieces of information, by themselves, have little meaning. Only when the data are put together do they begin to be truly informative. When evaluating a training program, you need to put the data together so as to construct a meaningful story for decision makers.

You can make the whole story

meaningful by paying close attention to how you "frame" the information. Frames are boundaries we place around things to set them off from the background and make them stand out. Different kinds and different sizes of frames can make the same information look different.

Generally, when you communicate with decision makers, you should frame the information in terms of the larger goals of the organization. But you might also want to adjust the "size" of the frame in order to highlight certain decision makers' real concerns.

For instance, a group of first-line managers and a group of executives would have different concerns about a training program. Managers might be concerned with the effects of training on an employee's job performance and on a work unit's productivity. Executives would probably want "bigger-picture" information, such as how a training program relates to the company's strategic plans and initiatives, or how it affects overall company performance.

When you present information about your training program to different groups, you must know your audience. Then package the information so that it speaks to that particular group's concerns.

5. Involve and inform decision makers on an ongoing basis. Successful river rafting is an interesting combination of fluidity and forward motion. Navigating a turbulent river involves being responsive, rather than reactive, to conditions.

To understand the difference between being responsive and being reactive, think of the notion in physics that for every action there's an equal and opposite reaction. A mechanical reaction causes an object to be forcefully propelled in a particular direction—as with the propulsion of a jet engine.

River rafting is quite a different way to travel. It requires a responsiveness to the environment. The word responsive means to answer, correspond, or reciprocate. It implies constant checking and adapting, in response to cues from the environment. In other words, responsiveness suggests an interactive process—an ongoing dialogue with the environment.

Responsive training evaluation also creates an ongoing dialogue—a dialogue with people who have a stake in a training program.

Remember that the information you share with decision makers feeds into the organization's decision-making process. Decisions are seldom influenced by one formal presentation. Creating an ongoing dialogue helps you shape impressions and decisions over time.

Steve Bistritz, an HRD manager for a division of IBM, conducts such a dialogue with stakeholders of the company's sales-training program. He does it by electronic mail. After each class of up to 100 people completes the training, Bistritz sends evaluation data to the trainees through e-mail. Bistritz's messages are more than just numerical data describing reactions or outcomes. They also include summaries of participants' comments, as well as major patterns and important nuances that give the program its character.

Bistritz doesn't stop with training participants. He also sends his e-mail messages to people at the three levels above his—including the division president. When Bistritz began to include the higher-ups in this information loop, the division president was so impressed that he sent a message back: "This is the best note I've gotten!" Bistritz had captured his attention; now, the president continues to be keenly interested in the training program.

Bistritz plans to stay in touch with participants so that he can collect examples of how they are using the training back on the job. He will then share this information with them via e-mail and continue to feed it to decision makers above him.

Major decisions seldom occur at a single point in time. A decision might appear to be the outcome of one important meeting. But it is really a collection of events, percep-tions, and smaller decisions that have created a direction. The best way to influence that direction is to share information with decision makers continuously. Provide decision makers with an ongoing story about a program, rather than showing them one snapshot at the end of it. The story you help shape should reduce their uncertainty about the effectiveness of a program so that they feel prepared to defend it.

How Can You Use Responsive Evaluation?

A project I once worked on provides a helpful illustration of how to use responsive evaluation to create a more complete picture of training.

Managers at a medium-size engineering company had asked me to design a training program in communication skills. The program would be part of a larger effort to move the company toward a management style that promoted continuous improvement.

The communication-skills training brought together managers from different functions of the organization. Throughout the three-day course, we encouraged a great deal of interaction and networking among participants.

Six months after the training, we sent out questionnaires to determine whether participants were using the communication-skills training back on the job. The results were not very impressive. When measuring outcomes against our program objectives, we found that few of the managers or their subordinates had seen dramatic changes in communication skills.

We then pulled together several group interviews of five or six participants each. In the interviews, we learned of some significant results from our training program.

During the training, the managers had begun to develop relationships with each other, across functional areas. Many of those relationships were carried beyond the three-day course. Back on the job, the participants were using each other for networking and problem solving. That interaction was helping many of the managers develop a broader view of the company, which enhanced their ability to make strategic decisions.

That residual effect related not to the specific goals of the training program, but to the larger organizational effort toward continuous improvement. Because we gathered some qualitative data about the participants' experiences, we were able to relay a more complete story to decision makers about the program's full value.

Navigating safely. Demonstrating the value of training is as important as ever in today's business climate. Navigating these waters safely involves being responsive to your organization's environment. It means finding a balance between attending to concrete and visible factors that might obstruct your training program, and paying attention to the more subtle processes at work in the organization.

Responsive evaluation helps you direct people's perceptions of a program. It pays particular attention to decision makers, and it attempts to show them how a program relates to the larger goals of the business.

Your objective, as a training evaluator, is to provide enough evidence to suggest that linkage. That means correctly identifying the decision makers and determining what they want to know. Then, you must navigate a course that strikes a balance between quantitative and qualitative evidence.

With the right mix of data, you can tell decision makers a meaningful and complete story about the effects of a training program. And with their support, you can keep your training programs afloat and moving forward—no matter what lies around the next bend in the river.

By Angus Reynolds

THE BASICS: *EVALUATION*

Evaluation measures the effectiveness of instructional materials in solving a performance problem identified during an analysis. It is the collection and interpretation of data to determine the value of training. Evaluation can help you decide whether to continue offering a program, get ideas for improvement, or justify the use of training to solve a performance problem. This process is called summative evaluation.

Donald Kirkpatrick, working in management training, developed an effective process for evaluating all training. His four (summative) evaluation levels are widely used and extremely effective in technical and skills training. You'd be well advised to implement all four levels in your own training:

▼ **Reaction**. Level one evaluation, reaction, gathers the trainee's opinion about the instruction, often on written forms. These evaluation sheets are known by many names and nearly all trainers use them. Although highly popular, evaluation sheets have the least value in establishing and maintaining good training.

▼ **Learning**. The learning level is usually measured by some sort of test: a written test or, better still, a performance test. The test measures how well our instruction succeeded. Unfortunately, many organizations don't bother to test the knowledge and skills employees should have gained from the training. And there is an unfortunate tendency to omit tests for supervisory training.

▼ **Behavior**. The name of the third level, behavior, is sometimes confusing. We want to ensure that the learning is translated into on-the-job performance—the behavior. This evaluation usually involves sending a survey to the trainee's manager 60 days or 90 days after training. The survey asks how well the training has enabled the employee to do what's needed to get the job done. Sometimes we also ask the former trainee how effective the training was in preparing him or her for the job.

Behavior evaluation helps maintain the quality of training programs by helping ensure that the training produces graduates who measure up in today's changing environment. Fewer organizations measure behavior than measure learning, but those that do will find it worth the effort.

▼ **Results**. The highest level of evaluation, results, gets to the "bottom line," or return on investment (ROI). The results level measures management's original goal for the training program, such as expecting less equipment downtime, reduced scrap, or improved quality or productivity. Such a measure of value is easily determined for technical and skills training.

First, we examine the known value to the organization of reduced equipment downtime, reduced scrap, or improved quality or productivity. Compared to the cost of training to make these improvements, the return should be manyfold. Most organizations don't bother with this step, but trainers who do can show how much their training is saving the organization. Remember: Management understands results and ROI, so give it to them.

Trainers do each higher level of evaluation less often than the one below it. Smart trainers know that the steps most commonly used provide the least important information. Why then don't we see more higher-level evaluation? I think we are starting to. The word is finally getting around, and increasing numbers of organizations are doing a better job of evaluation. To improve the quality of your organization's training, try to establish and maintain a complete summative evaluation system. Make it a point to use the more valuable techniques in your organization.

TAKE ACTION

Many companies take considerable pains to avoid wasting money. Their managers have the right to expect that the resources expended for training will benefit the organization. So we want to do a top-notch evaluation of our training programs. This evaluation will provide timely, accurate information to support training program decision making.

To learn more about summative evaluation, read Donald Kirkpatrick's chapter in the *Training and Development Handbook: A Guide to Human Resource Development* (3d edition), edited by Robert L. Craig (New York: McGraw-Hill Book Company, 1987).

COURSE DEVELOPERS' EVALUATION

If you are a course developer, you must consider more than summative evaluation. You also must be skilled at conducting formative evaluation, which occurs during the development phase of instructional systems development. In comparison, summative evaluation is conducted during and after instruction. Think of formative evaluation as an instructional quality process activity. Its purpose is to revise materials before widespread use.

Formative evaluation consists of several steps suggested by Robert Gagné, often called the father of instructional systems development:

▼One-on-one trials make the most sense when the instruction itself is to be individualized. The developer goes over the draft instructional materials with a typical learner while the learner thinks aloud. Lack of instructional clarity jumps out, and help is provided by modifying the materials on the spot.

▼ Small group trials usually involve three to five typical learners, although as many as 10 learners could participate if the eventual number of course participants will be large. Instruction is carried out in the normal way, but the developer takes careful notes and may question the learners after the session. The materials are revised wherever needed.

▼ The pilot test is a full-scale implementation of what should be fully developed materials. But the materials remain under the developer's eye. Although a problem may emerge that was not detected in the previous small group trial, we expect the pilot to be a complete success.

PRECAUTIONS

Have a content expert review your draft materials. You may also want to show the materials to a fellow course developer. Although these are wise and important steps, you must remember that they are not evaluation. Real evaluation must involve members of the target population for the eventual training.

It is also difficult to know when to stop formative evaluation. Any course can be constantly "tweaked" to make it even better, but at some point the cost of collecting more formative data exceeds its benefits.

Most evaluation activities are relatively simple in concept but require skillful application. Creating an evaluation plan helps detail the whys and hows of your evaluation effort, although the planning itself often becomes the major part of that effort. Still, an evaluation plan that clients can review and approve also helps you improve the evaluation process.

Finally, watch out for poorly written tests. Consult a good test construction manual to check such concepts as reliability and validity. And remember that performance tests are preferable to written tests because they help evaluate the ultimate performance you wish to enable.

To learn more about formative and summative evaluation, check out the *Handbook of Training Evaluation and Measurement Methods* (2d edition), by Jack Phillips (Houston: Gulf Publishing, 1991).

Angus Reynolds is the Advanced Training Technologies group's instructional technologist for EG&G Energy Measurements (Box 4339, Station A, Albuquerque, NM 87196; 505/845-1713).

Measurement Made Simple

If you've ever felt ill-equipped to measure training results, here's good news. If you can do a traditional needs-assessment, you already have the skills you need.

BY LEIGH ANN WILLIAMS

THERE IS A COMMON BELIEF that training design should come before a measurement discussion. In reality, I've found that working on training design and measurement planning simultaneously is more effective. That's because the information required for training design is exactly the same information needed for solid measurement. So don't wait until after the training design to decide what you want to measure, because your design may not enable the right data collection.

Clarify business goals

The first step in measuring training's effectiveness is assessing a company's business goals. This can be difficult if

WHO CAN HELP?

There are many resources available to you. A local university can be a great source of expertise on measurement. Once you know exactly what needs to be measured, a professor or student can help you figure out how. The advantage of using a university is that the cost for measurement is typically much lower than the fees charged by consulting firms. Here are some issues to keep in mind when working with universities:

▶ Does the university have the computer resources needed for your particular measurement activities? If you need to crunch data for 10,000 employees, the university may not have a way to load the data. And if it does, you may need both a computer expert and measurement expert. If they are from two different departments, how will budgets work?

▶ If the organization commissioning a measurement study is also supporting a university financially, the university may be concerned about loss of funding if the measurement study does not produce the desired results. All parties should discuss this issue prior to the study to ensure objectivity in the research.

▶ Is the university suggesting shortcuts? I'm aware of several situations where a university did not have the computer resources for the requested measurement activities. In each case, the university wanted the project badly and suggested an alternative measurement method. Sometimes that's fine. In other instances, the results are less useful.

Consulting firms are another good source of measurement assistance. Although cost can sometimes be a drawback, it may be reasonable if a consulting firm already houses the organization's data. It's helpful to ask the consulting firm for a choice of ways to do the measurement. What are the strengths, limitations, and costs of each approach?

an organization's goals are vague. For example, a company might say, "It is our goal to deliver the best customer service in the industry." What does that really mean?

The following questions clarify this business goal and make training design and measurement easier:

▶ What elements are included under the umbrella description of customer service?
▶ What does "in the industry" mean? Which organizations are included? What are the sizes and geographic locations of these organizations?
▶ How will you know when your

company has achieved the distinction of "best"? What will customers, shareholders, managers, and employees see, say, and do? What auditory, visual, and kinesthetic "evidence" will let you know the goal has been reached?
▶ Where is the organization today? Is it close to its desired goals, or does it

---

SIX MEASUREMENT TOOLS

There are hundreds of ways to measure change and results. Here are six examples to show what's possible.

1. **Surveys:** Written or telephone questionnaires used for quantitative or qualitative measurement or both.
▶ Example: One company used a survey to test how well employees understood the business strategy; it learned there were many different interpretations. In a follow-up survey, the company can test for improved understanding. They can do additional analysis to determine the impact of specific training activities on improved understanding.
▶ Note: It is not advisable to use a single survey across cultures. One company translated a U.S. survey directly into other languages without considering cultural differences and translation issues. Among other difficulties, one Asian language has no equivalent for the "yes/no" distinction. So, the translation did not make sense. Employees answered all the questions yes. And to add insult to injury, this U.S. company merged the worldwide survey responses into a single data set for use in business planning.

2. **Focus groups or interviews:** Face-to-face meetings to collect qualitative information.
▶ Example: An organization might use these to determine whether employees feel they have the right amount of latitude to make decisions about their work.
▶ Note: Focus groups require some degree of participant trust. In organizations where trust has eroded or in cultures where a misstep can hinder one's rise in the corporate hierarchy, focus groups may not be useful. People may withhold infor-

mation or say what they think others want to hear.

3. **Performance management:** The method an employer uses to assess how well each employee is doing his or her job.
▶ Example: It is possible to create a performance management process using the goal and evidence structure described on page 30. If a manager and employee define results along with measurable evidence, then it is possible to assign a numeric coding to progress (i.e., 1=achieved; 2=progress made; 3=no progress made/not achieved). The employer can then load these codes into its human resources information system to do various types of analysis.
▶ Note: Most U.S. employers do not have this type of "rigor" in their performance management systems. The systems are looser, less formal, and often less demanding. To use performance management as a results measurement tool may require substantial revision of an organization's performance management process.

4. **Case group or control group studies:** A method in which one group participates in an initiative and a second group does not. The researcher then compares the results of the two groups.
▶ Example: Division A participates in creativity training and division B does not. We study the difference between the business results of A and B after one year. A outperforms B.
▶ Note: For a good study, it's best to match the case group and control group person for person. For every college-educated, 45-year old male in the case group, it is ideal to have a person of the same profile in the control group. This is very difficult

to achieve in an employment setting. Also, it is almost impossible to keep the case group and control group intact in the workplace. Normal attrition will affect the composition of both groups. If attrition is significant, the results of the study may be suspect or totally unusable.

5. **Analysis of raw data:** Looking at a data set and drawing conclusions directly from the data set.
▶ Example: Last year, 60 percent of the employees achieved their performance goals. This year, 75 percent did.
▶ Note: If an employer only cares that things are moving in the right direction, then this type of analysis is often enough. And it tends to be inexpensive, compared with other options.

6. **Multivariate analysis:** Statistical analysis that lets the researcher remove variables that have nothing to do with the issue under study.
▶ Example: When measuring the effectiveness of health education, we need to "level the playing field." Some participants may have high blood pressure; some may not. Some may have a grade school education only. Others may have college degrees. We can use multivariate analysis to level the influence of all these factors so that we can isolate the effectiveness of the health education.
▶ Note: Multivariate analysis can be extremely expensive. If an organization wants to publish its results in a technical journal, it may want to consider this type of analysis. One organization chose this type of analysis because it needed defensible results for union negotiations. In most cases, however, organizations do not need analysis of this rigor.

---

have a long way to go?

Once you define what tangible evidence you are looking for, you have the foundation for training design and measurement activities. Then, training design becomes a vehicle for achieving goals and attaining evidence. And, measurement confirms if you have attained the evidence.

Stairstep connection

Now that you know what you're looking for, you will need ways of attaining the goals. Showing a direct connection between a training program and business results can be difficult. However, showing a "stairstep connection" is easier and is generally acceptable to senior management. A stairstep connection is a more gradual path, with more steps in-between. Hewitt Associates, an international human resources consulting firm, offers the following stairstep questions that are useful for training design and measurement conversations:

▶ What business results is the organization trying to achieve? For example, is the organization trying to increase profitability by 10 percent? Is it trying to increase market share in China by 25 percent?

▶ What business strategy is the organization using to deliver those business results (such as reengineering, new product launches, downsizing)?

▶ What do employees need to do or do differently to execute the business strategy?

▶ What are the employees' needs? Employees may need training or other resources to meet employer requirements.

▶ What is the human resources strategy? Does the organization have a strategy to help employees do what's needed?

▶ What HR initiatives support that strategy (such as training programs, performance management process, compensation, benefits)?

While it is difficult to link a training program, or other HR initiatives, directly to a business result, it is possible to measure the effectiveness of a training program by assessing a behavioral shift. Take a look at the following statements, which combine the evidence discussion with the Hewitt Associates' questions:

▶ Desired business results need specific evidence; business strategies are vehicles for reaching these goals.

▶ Demands on employees and employee needs require specific evidence; HR strategy and initiatives help attain these goals.

While this framework may seem simple, it is difficult to elicit the goals and the evidence in many organizations. Many companies go straight to designing the vehicle without defining what the evidence is. This can make measurement discussions difficult, if not impossible. It can also make training design tough. In many situations, management tells trainers it wants a creativity training program or a diversity training program. It's easy to say yes and proceed with the request. However, without information on the goals and evidence, it's difficult to know whether the organization really needs these programs—or whether management just thinks so.

Even in the best case, obtaining goals and concrete evidence can be challenging. There are many variables to consider. For example, in a large organization the business results, strategy, and people requirements may differ among divisions or units. Additionally, within a division, what's asked of individual employees may be different—so HR initiatives may need to vary by audience.

The measurement discussion

After information is collected, the following questions will help you shift from a business results discussion to a measurement discussion:

▶ How are you measuring your business results?

▶ How are you measuring the effectiveness of your business strategy in delivering those results?

If the organization can't answer these, here are some additional considerations:

▶ Is measurement really important to this organization? Does it really care about measuring training? If so, why does it care?

■ Companies often design the vehicle without defining the evidence ■

▶ If the organization wants to measure, your measurement methods will need careful documentation.

▶ What are the politics of measurement? Is there anyone who could suffer if the results do not turn out as expected?

▶ Let's say a training program costs $100,000. What percentage of this is the organization willing to spend on measurement? Ten percent? Fifty percent? Money is no object?

▶ Do you need qualitative or quantitative measures? Or both?

▶ How scientific does the measurement need to be? If we show that things are "going in the right direction," is that enough? Or do you need something you can publish in a technical journal?

Up to this point, your needs-assessment skills should yield a good measurement discussion. However, for the remaining three questions, you may want assistance from a consulting firm, university, or other measurement expert. The next few questions are the bridge between information gathering and beginning an actual measurement study:

▶ What kinds of data are needed to measure the evidence?

▶ Is there enough data to create some sort of baseline measure? If not, what can be done to capture the needed data?

▶ What measurement method will you use? If applicable, what statistical technique?

You can use these questions to learn the current position of the organization, the results the organization wants to achieve, and the specific evidence that will let the organization know it has succeeded.

So, next time senior management wants to see training's impact on results, you can step up to the challenge with confidence.

Leigh Ann Williams *is coowner of NLP Comprehensive, 4895 Riverbend Road, Suite A, Boulder, CO 80301. Phone: 303/442-1102; fax: 303/442-0609. E-mail: leighnlp@aol.com.*

" What evaluation methods or strategies give you the best feedback on training's effectiveness? "

ED STRENK, *Senior Training Specialist, UniFirst, Wilmington, Massachusetts.*

Here is the best strategy to evaluate training's effectiveness:

▼ **Needs Analysis.** Complete a needs analysis of your audience to establish a baseline of skills and behaviors.

▼ **Developmental Testing.** Evaluate the training before it goes out the door. Make use of extensive developmental testing and piloting.

Ask: "Does the training elevate baseline skills and behaviors?"

▼ **Follow up.** Evaluate the training after it goes out the door. Some effective tools include surveys, observation, cost-benefit analysis, and cost-effectiveness analysis.

Ask: "Did the training elevate baseline skills and behaviors?"

WAYNE E. BAUGHMAN, *Certified Trainer, Assistant Dean, Business-Industry Training, Cincinnati Technical College, Cincinnati, Ohio.*

I give the evaluation form to the participants before the session. I explain to them that this is my contract with them to provide an outstanding training course. This permits them to know what is expected of me. It also gives them an opportunity to improve the training as it is happening—not after they have gone home.

ROBERT A. SANREGRET, *Program Manager, Learning Tree International, Reston, Virginia.*

We find pretests and posttests the most powerful and accurate tool for determining how much information the students understood and retained. These tests involve 20 to 30 questions per day of training in a multiple-choice format. Questions on the pre- and posttests are identical to maintain statistical relevance in the delta between pre- and posttest data.

TODD EICHAS, *Skilled Trades Training Coordinator, UAW-GM, Rochester, New York.*

For physical skills, a performance evaluation provides excellent feedback on the training's effectiveness in the training environment. It does not, however, guarantee that these skills will transfer to the shop floor.

Ideally, observation of a trainee's performance in the "pressure cooker" of a real-world situation gives the most credible feedback. This can be accomplished if a strong rapport is maintained between your training department and the customers you train.

JOHN J. GRIER, *President, Management Education Center, Inc., Bristol, Pennsylvania.*

On-the-job performance monitored by job training instructors *and* employers. This is not generally happening.

RUSSELL F. WATKINS, JR., *Training Specialist, Unisun Training Center, Charleston, South Carolina.*

In a lengthy class, I assign each person to lead a morning review session of the previous day's materials. They create a flipchart sheet pertaining to key ideas. I then review the sheets for accuracy and to evaluate what I've presented.

R. BRIAN LONG, *Training & Development Manager, Spartan Mills, Startex Division, Spartanburg, South Carolina.*

Our feedback arrives from two points: the receiver's post-training work methods and the supervisor's comments. Employees are audited two weeks after the training ends to insure that the training stuck. Supervisors file evaluations on their employees.

JAMES S. PEPITONE, *Principal, Pepitone Berkshire Piaget, Dallas, Texas.*

The appropriate training evaluation methodology is best determined by the training objective. This is because the principal goal of evaluation is to indicate how well an objective is achieved. The correct process for designing appropriate training evaluation methodology begins with determining the "objective level" for the training. Let's use Dr. Donald Kirkpatrick's often-cited "four levels of training evaluation" methodology to clarify this approach.

Here are Kirkpatrick's four levels of evaluation as I see them:

▼ Level 1. Participant reaction (such as determining the approval rating)

▼ Level 2. Examination score (such as answering 15 out of 20 questions correctly)

▼ Level 3. Behavior change (such as increasing sales calls 20 percent)

▼ Level 4. Results achieved (such as increasing sales 10 percent).

These levels of evaluation provide pertinent feedback regarding the achievement of training objectives that we can categorize at four equivalent levels. The four levels of objectives are the following:

▼ Level 1. Attendance (attract people to the class)

▼ Level 2. Learning (increase knowledge and skills)

▼ Level 3. Behavior change (alter specific behavior)

▼ Level 4. Performance improvement (produce results).

The rule is to evaluate training at the same level as the objective. Evaluate level 2 training (learning) with a level 2 evaluation method (examination score) for an accurate indication of how well the objective was achieved.

For example, it is pointless to use a level 4 evaluation (results achieved) for a level 2 training objective (learning), because the results measured will be attributed to many factors other than training.

Conversely, you can't rely on a level 1 evaluation (participant reaction) for a level 3 training objective (behavior change), as you will still not know if your objective was achieved.

Note, however, that an objective can involve lower-level subobjectives. For example, training to increase product knowledge (level 2 objective) may include the subobjective to attract people to the class (level 1 objective).

ROBERTO MARIO MEDICI, *Quality Assurance Manager, Kobe Argentina, Mendoza, Argentina.*

1. Participants' evaluation of each course.

2. Improvement of trainees' operating efficiency.

Testing Employee Performance: A Review of Key Milestones

By Rob Schriver

Just as evaluation has become an essential component of any training initiative, employee testing has become an integral part of many an evaluation methodology. The author examines the recent history of employee testing and its contribution to the evaluation toolkit.

In This Story

▼ evaluation
▼ instructional systems design
▼ transfer of training

Evaluation is the cornerstone of training. Before a program participant completes a performance-based training module, he or she should be evaluated carefully in the instructional setting. Such on-the-spot training evaluation is the key checkpoint in the training program evaluation process. The evaluation may be in the form of a cognitive test, a performance skill check, or both, but some type of test is necessary as a means of evaluating the success of the training.

Employee testing, as a means of generating evaluative data, has not always been looked upon kindly by employers, social scientists, and others. That has changed in recent years as the value of testing as an objective evaluation method—as opposed to subjective assessments of employee performance—has gained credence in the training and development community. This article traces the recent history of employee testing and explores the reasons why testing is becoming the centerpiece of training evaluation.

In discussing employee testing, this article refers to achievement tests as opposed to aptitude tests. Also, it assumes that the best medium to deliver the test, whether computer-based, paper-and-pencil, or other has been appropriately selected and the test design itself is based on the taxonomy or level to which the instructional objective is written.

Recent History of Testing

Until recently, strong resistance to employee testing was the norm. Employers were concerned that standardized tests might be unfair to minorities, and were unsure whether or not testing could accurately measure individual traits and attributes.

Today, according to Tom Kubiszyn and Gray Borich, authors of *Educational Testing and Measurement,* attitudes toward testing have come full circle. Test advocates have been convincing in arguing the usefulness of tests in workplace training; certainly they have proven that tests provide the most objective, valid, and reliable information about an individual. Even Columbia University professor Robert Thorndike's work encouraging the use of more testing in schools and the workplace—considered controversial three decades ago—is well accepted in the 1990s.

With or without test data, decisions about employees will be made. Nevertheless, it is known that non-test data are more subjective and biased than test results, and although tests themselves do not discriminate, it is possible the people who use them might. (In this case, the term "discriminate" refers to a racial or ethnic bias, as opposed to the "differentiation" of learning gain that occurs as a result of instruction or study.) Most training professionals agree—because it is to their advantage to use the most measurable, objective, and reliable data possible there is sure to be more testing in the future. In fact, a 1994 *Human Resource Executive* report

confirmed a significant increase in test use during the 1980s, which is expected to continue throughout the 1990s.

Why Use Tests?

Jack Phillips, a renowned human resources practitioner, sees accountability in training as a key issue and a driving force in the trend toward testing. According to Phillips, training organizations are facing the same pressure to show the

Testing generates among the most measurable, objective, and reliable data possible for evaluating employee performance

return-on-investment previously felt exclusively by production and financial departments—and testing is one way to show the value of training programs.

Testing is, however, only one of the many elements of training program evaluation. So, why does it deserve emphasis? Following are some reasons that testing is so important:

1. Assessing trainees or course participants is what our jobs are about. When it is done correctly, testing provides a valid, reliable performance indicator of the student's work. It does not matter

whether the review is written, oral, or demonstrated; or even if it is conducted by an observation interview or applied research. With a test, learning is evaluated against the program objectives.

2. Selection and placement decisions are based in part on test data. An entrance test may be useful in a hiring decision; to possibly exempt a new hire from a certain training requirement; or to assess the need for remedial training. An entrance exam sometimes can help to simultaneously fulfill a company training requirement and assist in a placement decision.

This area of testing remains a sensitive one. Companies must be able to connect the need for a test, as well as the content of the test, to job relevance. Otherwise, the legality of the company's hiring practices may be questioned.

3. Motivation to learn is affected by tests. Research shows that anticipation of a test may cause students to study more. Most students are also motivated by the feedback test scores provide. Compare, for example, student motivation to study in the following two groups: One will have a weekly pop quiz; the other will have midterm and final exams. How frequently do you think each of the groups will study? Instructors and students alike can relate to this example—and know the answer.

4. Instructional improvement is often overlooked as a benefit to testing employees. Low test scores can show that training material was not covered well; high scores might indicate the need to move ahead to advanced material more rapidly. Robert Mager supports this theory in his Criterion-Referenced Instruction (CRI) program. He believes a test is as much a check of the instructor as it is the student, and so continuous feedback should be built into a training program with the test as a key component.

5. The test can be a learning tool. Sometimes the trainee can be remediated immediately by the test, as in on-the-job training where corrective feedback is given, performance is adjusted, and learning takes place. In individualized instruction, a few review questions can be given throughout a lesson to help the student assess learning and become more engaged with the program.

Testing as Evaluation

During the 1940s, when military skills training developers created the five-step Instructional Systems Design (ISD) model to systematically produce training products, emphasis was placed on post-instruction evaluation. Today, evaluation is conducted throughout the entire training program—continuous feedback rather than afterthought—and so creates more opportunities to improve training processes.

Evaluation is an essential step in the ISD process. ISD is an iterative cycle with numerous opportunities to gather data and important feedback about the students, instructors, and the instruction itself. Evaluation has formative (pretraining/development work) and summative (post-training assessment) components, and can affect internal and external program indicators. Therefore, it's important to implement evaluation steps during each phase of ISD. Assessments of learning are paramount.

It can be easy to forget to include evaluation in training activities. Placing emphasis on formative evaluation steps, such as testing, makes posttraining evaluation less likely to be overlooked. Donald Kirkpatrick, to many the father of training evaluation, found that the steps build on one another. Evaluating well in the classroom helps support additional evaluating beyond the classroom, such as transfer of training or reviews of business results obtained from educational work research. An approach to ISD as a complete and interconnected cycle can lead to more success—especially with evaluation of students as they complete courses at different stages of development. The training program's end-of-course evaluation test bridges the gap from the island of the classroom to the mainland of the job site.

Future efforts need to look carefully at all of the evaluation steps associated with the ISD model and be particularly concerned with design phase evaluation as it relates to the consideration and preparation of student tests, and with the actual classroom testing in the implementation phase of the program. This is the point, though certainly not an end point, where the instructor or course manager determines whether the student has it, is ready, passes, or crosses the bridge.

Rob Schriver is program manager with Martin Marietta Energy Systems' Center for Continuing Education (701 Scarboro Road, Oak Ridge, TN 37831-8240; 423/576-775; fax: 423/574-2392).

Evaluating Training Results

By Paul R. Erickson

Here's a training-evaluation method that assesses whether participants remember what they were supposed to have learned—and know how to apply it on the job.

It happens all too often. At the end of your training course, participants pass a simple test by reiterating all the buzz words they've memorized. Then they go back to their work stations and immediately begin to forget everything they were supposed to have learned.

Oh sure, they filled out your end-of-course questionnaire: "What did you like about this workshop?" "What would you change about the course?" Such questionnaires may fill your administrative requirements for course evaluation, but they may not tell you what you need to know:

■ whether attendees have retained the information presented in the course;
■ if course objectives have been met;
■ whether trainees' level of knowledge has increased;
■ whether there is enough emphasis on the required material or concepts;
■ the appropriateness of this particular course for teaching certain material, or whether some of the material would be better taught as part of a different course;
■ whether trainees' retention of knowledge taught in the course varies depending on the ability of the instructor.

Many training departments use pre- and post-tests to make up for the inadequacy of end-of-class questionnaires. Such tests may help determine whether training has changed the knowledge level of participants, but they usually provide only short-term assessments—they don't really tell you if trainees are applying on the job the information they've learned in the course.

Erickson is a captain in the U.S. Navy, at the Naval War College, 4 Anthony Place, Newport, RI 02840. The evaluation procedure discussed here was conceived while he was on temporary assignment at the Navy Personnel Research and Development Center, San Diego, California.

What can you do to ensure that training "works"? How can you tell whether participants not only remember what you trained them in, but also are using it on the job—even several months after the course is over? How can you make sure that training isn't a

---

End-of-course questionnaires may fill your administrative requirements for course evaluation, but they may not tell you what you need to know

---

waste of your time and your trainees' time—and the company's money?

What we did

Clearly, many trainers need an evaluation method that shows the long-term relevance of training. Such a method was used to assess three Staffing and Placement classes taught by Navy Civilian Personnel Command (NCPC) instructors from the Southwest and Northwest offices.

Sixty course graduates made up the sample group. All were Personnel Management Specialist Interns, and some had college degrees; their length of time with the federal service ranged from 6 months to 16 years.

The reasoning behind the evaluation method was this: if the course goals could be distilled into competencies, then a series of case studies could be written that would address those competencies. Each participant would be asked, three to six months after the training, to describe how he or she would handle the case studies if they arose as part of the job. Participants

would be scored on their use of the learned competencies to solve the work-related problems.

In designing the evaluation, we made two assumptions:
■ that we could more accurately assess participants' knowledge levels by asking them to use the concepts (or, at least, explain how they would use them in "real life" situations), than we could by getting them to repeat buzz words;
■ that experienced subject-matter experts would be able to distinguish correct responses from memorized statements—and so distinguish actual knowledge from internalized cliches.

We identified nine objectives, or competencies, for the course that was evaluated in the pilot program (see the box, "Competencies or Objectives for Staffing and Placement Course"). The instrument we used was developed to assess trainees' working knowledge of those competencies.

Administration of the questionnaire was divided into two phases.

Phase I: the Interview

After the course goals were defined and worded as competencies, ten case studies were developed, based on those competencies and taken from real-life work situations. Then, the trainee and the subject-matter expert (often the trainee's supervisor) sat down to talk. As the trainee explained how he or she would handle each situation, the evaluator listened for the trainee's understanding of the concepts that were taught in the course.

The evaluator referred frequently to a check sheet that listed the competencies tested in each case study, and checked them off as they were demonstrated. (See "Sample Case Study and Check Sheet for Staffing and Placement Course.") We designed the case studies so that each competency could be used at least three times. If the participant correctly used a competency at least twice during the interview, our expert credited him or her with a working knowledge of that competency.

By the end of the interview, the evaluator would have completed 10 check sheets (one for each case study). Scoring was done from the check sheets; typically, some competencies would have been demonstrated and some would not have been.

Phase II: Checking Up

Phase II was a control on Phase I. In it, an impartial, independent evaluator re-scored some of the interviews, from tape recordings. The independent evaluator's scores were compared with the original scores to check on the reliability of the model and to control interviewer bias.

In our pilot evaluation on the Staffing and Placement course, the scoring by the original evaluators and the blind observers agreed on 48 of the 54 possible scoring combinations (or 89 percent of the time). Because of that consistency, we determined that the test method could be considered reliable, or at least not biased.

Competencies or Objectives for Staffing and Placement Course

1. Employee consults FPM index in order to ensure that all pertinent regulations are considered when dealing with personnel-related problems and issues.

2. Employee understands the various appointment authorities and their applicability to the need of the position to be filled.

3. Employee understands that job applicants must meet basic OPM-designated eligibility requirements as defined by the X-118 and X-118C handbooks, including time-in-grade requirements and uniform requirements (e.g., citizenship, physical requirements).

4. Employee understands the major sources of recruitment.

5. Employee understands the mandatory use of competitive procedures for affecting promotion actions as well as those actions that allow exception to competitive procedures.

6. Employee understands and can apply pay-setting rules and regulations.

7. Employee understands the basic mechanics of reduction in force (RIF), including RIF terminology.

8. Employee understands the concept of priority consideration for placement when taking action on position vacancies.

9. Employee understands the application of the Department of Defense priority placement program in filling vacancies and placing employees adversely affected by management-instigated actions.

Sample Case Study and Check Sheet for Staffing and Placement Course

Case study 9

You have been informed by management that the janitorial function at your workplace is going to be contracted out to private industry. As a result, the 20 employees presently working as janitors must be released from their positions. Ten of those janitors are on temporary appointments, and the other 10 are on permanent competitive appointments.

How would you proceed in "removing" these employees from their positions? Is this a RIF (reduction in force) situation?

Check sheet for case study 9

Detail of Response
This is a RIF situation.

Look up reduction-in-force definition in FPM.
☐ Competency 1: use of FPM index

Check your retention registers to ensure that they are current and accurate.
☐ Competency 7: RIF

Terminate the appointments of the 10 temporary janitors. (If interviewee does not specify the steps in the RIF process, ask "How would you proceed in conducting the RIF?")
☐ Competency 2: appointment authorities

Perform qualifications analyses on the other 10 janitors, who are on competitive appointments.
☐ Competency 3: OPM eligibility requirements

Coordinate with management as to whether vacancies will be utilized in the placement efforts for these 10 janitors.
☐ Competency 7: RIF

Determine the "bump and retreat" rights for the 10 janitors.
☐ Competency 7: RIF

Determine their "best offers." (If interviewee stops here, ask "What about counseling?")
☐ Competencies 6 and 7: pay setting and RIF

Counsel the 20 janitors. (The 10 temporaries may be entitled to state unemployment compensation. Severance pay may be required for the other 10. Inform those placed in other positions of their rights to priority consideration for re-promotion as well as entitlement to registration in the Department of Defense program for the stability of civilian employment.)
☐ Competency 8: priority placement (no vacancy)

Prepare written RIF notifications. Inform the employees of their "appeal rights" to Merit Systems Protection Board (or use of grievance procedures if covered by union contract).
☐ Competency 7: RIF

Perform the proper registrations of the RIFed employees (Department of Defense stopper list, re-employment priority list or RPL, activity's priority consideration file).
☐ Competency 8: priority placement (no vacancy)

Were the responses given in the proper sequence?

What it means

Of the 60 questionnaires we distributed to subject-matter experts for the courses, 21 were completed, returned, and evaluated. We found that trainees had a working knowledge of five out of the nine competencies identified for the course. One competency was eliminated because of poor presentation, leaving three competencies that trainees did not know how to use. Those three competencies can now be re-evaluated, using the following questions as a guide:

■ Should increased emphasis be placed on the course material in the three areas that trainees don't appear to know?

■ Should the material relating to those competencies be removed from the course? (Perhaps it was not retained because it is not important to the course.)

■ Should the material be shifted to another training course because it is not begin understood? (For example, maybe it is being presented at the wrong level.)

■ Are variations in competency retention related to the ability of the particular instructor to present the information?

Of course, this evaluation model cannot singlehandedly determine the reasons for your trainees' lack of knowledge. Few evaluation methods can. But it can point out the competencies that are—for whatever reason—missing from the trainees' operational repertoire.

The distinction between operational knowledge and theoretical knowledge is an important one. A trainee can repeat theoretical knowledge, giving the appearance of learning, but be unable to use the knowledge. It's like being able to explain how a pump works, but not being able to pump water. Pumping water is operational.

After spending a lot of money and time to train employees for a specific set of goals—such as operating a pump or using an administrative program—it is logical to want to determine to what extent trainees have met the goals. With that knowledge, you can intelligently make the decision to continue teaching the program as it is, to overhaul it, or to scrap it entirely. Without that knowledge, you may be just wasting time.

By David S. Bushnell

Input, Process, Output:
a Model for Evaluating Training

IBM's corporate education strategy for the year 2000 uses a new approach for evaluating training effectiveness.

Since World War II, corporations have spent billions of dollars on worker training. As with any corporate investment, training directors are held accountable for the return on that investment; top management is looking for evidence that the dollars spent pay off. Budget justifications are in terms of potential savings generated through productivity gains or improved quality.

Companies are looking for cost-effective training strategies and seriously considering make-or-buy options. The portability and transferability of training materials are issues that multinational corporations wrestle with as global education networks take form and satellite communications proliferate.

Not only is top management becoming more demanding, but trainees are asking for and getting training materials geared to their requirements and delivered on demand. As computer-based training (CBT) and other instructional technologies become readily available (and cost-effective), the challenge for trainers is to deliver course materials in ways that ensure

Bushnell is director of the Center for the Productive Use of Technology at the Human Resources Research Organization (HumRRO), 1100 South Washington Street, Alexandria, VA 22314.

quality products at reasonable prices, tailored to end-user requirements.

Paralleling those trends is the need to link training to future corporate needs and to tie it more closely to other human resource management programs. That is the thrust of IBM's recent effort to project its internal education requirements through the year 2000 (see cover story). Recognizing its responsibility to the corporation to deliver well-conceived and cost-effective training programs worldwide, whenever and wherever they are needed, IBM's education group has opted to overhaul its internal education system dramatically by the end of this decade to meet its anticipated knowledge-worker requirements.

Other advanced multinational companies—Motorola, Xerox, and Federal Express, for example—have also adopted an integrated-systems approach to training improvement. One of their concerns is to anticipate and offer appropriate retraining opportunities before employee skills become obsolete. Another is to maintain the flexibility of their workforces as products and services change to match market opportunities. Still another is to provide appropriate skills training for new hires by means of performance support-systems delivered at workstations. These companies are dramatically lowering the cost of their training programs and, at the same time, increasing their training flexibility and responsiveness by adopting what might be called an input-process-output (IPO) approach to training evaluation.

Why evaluate training?

To put it simply, training directors need to balance the cost and results of training. In the past, much of the cost occurred at the delivery stage. Today, design and development costs are rising rapidly as technology takes more of the responsibility for training delivery.

IBM has found that an IPO approach to training evaluation enables decision makers to select, from several options, the package that will optimize the overall effectiveness of a training program. Those who use the IPO model can readily determine whether training programs are achieving the right purposes. It also enables them to detect the types of changes they should make to improve course design, content, and delivery. Perhaps most important, it tells them whether students actually acquire the needed knowledge and skills.

The IPO evaluation model

If we describe a training system as having an input, a process, and an output, then it encompasses several points (E1 through E7 in the accompanying figure) at which evaluations ought to occur.

At the input stage, the elements (system performance indicators, or SPIs) that could be evaluated in terms of their potential contribution to the overall effectiveness of a training program fall into such categories as trainee qualifications, instructor experience, the availability of already tested instructional materials, the types of equipment and training facilities available, and the training budget.

At the process stage, the evaluator needs to specify instructional objectives, develop design criteria, select instructional strategies, and assemble training materials. At this stage, the training actually takes place and adds value to the human resources.

Output elements include such items as student reactions to training, knowledge and skills gained as a result of the training, and improved performance back on the job.

It is helpful to make a distinction between output and outcomes. Output deals with the short-term benefits or effects of training; outcomes refer to longer term results associated with improvement in the corporation's bottom line—its profitability, competitiveness, and even survival. Outcomes do not always flow directly from the outputs of the training, but in the long term, they do dictate training resources availability.

In the figure, note the feedback loops built in at critical junctions in the evaluation process; they make the training systems somewhat self-correcting. For example, appropriate measures taken at the end of the development phase (E4) should help to ensure corrective adjustment at the design stage before an instructor ever steps up to the lectern.

Steps in the evaluation process

The evaluation process involves four steps.

Identifying evaluation goals. This is a critical stage because it determines the overall structure of the evaluation effort and establishes the parameters that influence later stages of the evaluation. Some evaluation goals are qualitatively different from others. For example, some goals may relate simply to measuring student reactions subjectively, while others may be concerned with measuring changes in trainee performance back on the job.

Developing an evaluation design and strategy. The next set of activities centers on selecting appropriate measures, developing a data-collection strategy, matching data types with experimental designs, allocating the data-collection resources, and identifying appropriate data sources. The choices made at this stage are critical because they determine the likely cost, time, and resources—decisions about which SPIs to measure determine the true value of the evaluation process.

Selecting and constructing measurement tools. At this stage, you want to select or construct the measurement tools that best fit the data requirements. Establishing a match between the data and the tool requires the evaluator to judge in advance the tool's reliability and validity.

Reliability answers the question, "Does the tool provide a consistent and accurate measure of the behavior being assessed?" Validity is a much more complex concept and therefore is much more difficult to establish. A measurement tool is valid if it meets several criteria. The criteria include face validity, content validity, and construct validity. Some of the tools currently in use:

- questionnaires;
- performance assessments;
- tests;
- observation checklists;
- problem simulations;
- structured interviews;
- performance records.

The type of measurement tool you select will vary according to the level of evaluation you need to carry out. If you are concerned with the measurement of individual cognitive abilities, then data collection and analysis techniques need to go beyond the traditional. For example, you may require detailed measures of recall, error, and reaction time; you can build these into the learning module as you study it. At the other extreme, assessing the value of a nuclear-power-plant-operating-room simulator requires the aggregation of data over several months and the handling of emergency conditions.

Analyzing data. This stage of the evaluation process involves the ability to tie the results of the data-gathering effort to the original goals of the evaluation. The following questions come to mind:

- Is the information collected really "need-to-know" information?
- Is the evaluation strategy gathering the right amount of information to answer the key questions raised?
- Is the measurement procedure disruptive to the education activities?
- Are the analytical procedures appropriate for answering the questions raised?

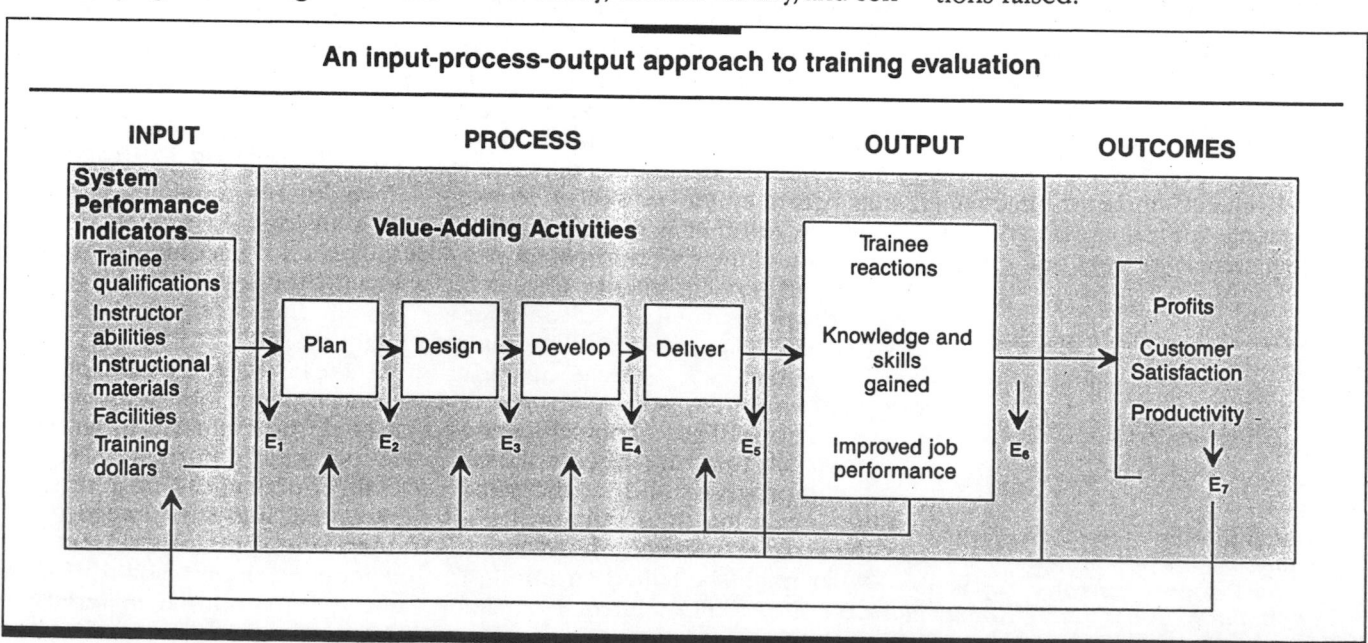

An input-process-output approach to training evaluation

After the data are analyzed, you need to make conclusions and recommendations and present the findings.

A key issue at this stage may involve the potential cost of additional data analysis, especially if the results fail to answer the questions originally posed or if they do not satisfy management's need for information. As with the original evaluation plan, the costs must be weighed against the potential benefits of the additional effort.

IBM's built-in global education network evaluation mechanism

The success of IBM's proposed global education network largely depends on the built-in mechanisms for tracking trainee progress, module completions, and mastery of various learning assignments. The network not only must gather and analyze these essential data, but also must feed the results back into the system to ensure that it can adapt to changing requirements. Thus, the system helps the learner assess his or her own progress in meeting educational goals and guides prospective learners as they attempt to sort out which learning modules they should take on.

IBM's evaluation system will continually and automatically update the employee's profile of skills, aptitudes, and learning preferences. It will assist line managers as they undertake to monitor the progress of their employees. It will provide continuous feedback to instructional material developers on necessary improvements in instructional modules. And it will enable top management to determine the return on its investment in the various components of the education and training system.

The ultimate payoff or added value of an employee's learning experience is how well he or she performs on the job. The quality, timeliness, and effectiveness of the learning experiences are only part of the reason an employee is able to perform effectively; other causes of high performance also must be considered through the use of sophisticated measurement tools and techniques. The goal, however, is to quantify in dollars just what impact a particular course of study had on an employee's ability to perform his or her job.

IBM's master plan for education in the year 2000 is already underway. The company is moving ahead with establishing a common set of guidelines and exploring alternative delivery systems. The outcomes of those programs are being evaluated at several levels. Line management and staff responsible for IBM's education and training programs are beginning to accept standard business measures based on performance improvement, not simply on the number of training-class hours. They are stating learning objectives in ways that make it possible to determine whether they are achieving them at a prescribed level of quality.

IBM is developing a well-articulated, overall corporate strategy for education that preserves the flexibility and responsiveness of a decentralized delivery system. To accomplish all that, line management is actively involved at all stages of the planning process to make sure that they serve IBM's future business requirements worldwide.

HRD's Failure To Sell Itself

As companies slash costs, HRD's head is frequently on the block. Why? Because of our failure to illustrate the value of what we do. HRD professionals need to know how to measure results.

Watch out, HRD! The jaundiced eye of budget cutting is looking in your direction. Are you vulnerable?

We hear a lot about the failure of HRD to demonstrate the value of what it does. Is that criticism just? Yes. Are the budget cuts fair? Probably not. But if budget cuts are based on a lack of information about what HRD can accomplish, HRD professionals have no one to blame but themselves.

Show and tell

Training magazine reports that U.S. organizations spend nearly $50 billion a year on training and other HRD products and services. That expense mandates that HRD link its efforts to bottom-line results. But, in almost 15 years' experience in the HRD field, I've observed that barely one in ten HRD professionals ever produces a bottom-line measure of training's impact—either in dollars or improved skills.

The large amount of effort and money spent on training demands that HRD tie training to measurable results. We must provide feedback other than "smiley-face" program evaluations or lip service on "how great the program was." Reliance on such flimsy justification places HRD programs under threat of the knife.

There's no doubt that today there's a need for continuous employee training, retraining, and basic education. Despite that, HRD professionals face devastating setbacks in these tough economic times.

Lookatch, *an educational psychologist, is an instructional designer at the Agency for Instructional Technology, Box A, Bloomington, IN 47402.*

By Richard P. Lookatch

Ironically, the first casualties are usually technological training tools that dramatically enhance learning and reduce educational costs. Most of us have heard such comments as, "Why do we need interactive video or CBT? We have a VCR!" What follows is usually the elimination of off-the-shelf courseware and support materials. When that happens, HRD professionals may be faced with having to reinvent the wheel.

The cost-cutting actions that penalize HRD seldom save money in the long run. But HRD professionals must come up with the supporting data to prove it.

The employer pays

Employers are the ultimate victims. When HRD programs fail, it's the employer who pays the ultimate price.

Consider this scenario: Management at a chain of electronics retail stores elects not to send a member of its training staff across the country to attend a seminar on high-density television. The HRD manager tries to convince top management that it's important for the sales staff to be able to address customers' questions on this highly publicized technology. The appeal is denied. The trainer doesn't attend the seminar. And the company saves $2,500.

During the next week, five percent of the chain's 600 customers ask salespeople questions about high-density television. Twenty of those 30 customers are disappointed with the responses they get and move on to the chain's competitors. Ultimately, ten of those 20 buy high-density TVs from other retailers. Those customers

spend a total of $12,500, or a net profit of about $2,500. The cost of the training seminar could have been recovered in one week!

Instead, the chain continues to suffer weekly losses of $12,500 in gross sales until management sees the light. As for the sales staff, the loss of sales prospects causes increasing dissatisfaction and, possibly, diminished self-confidence. The latter may be too elusive to measure, but a $12,500 weekly loss in sales is clear and convincing. It wouldn't have been difficult to estimate and predict that loss. Such evidence could have been used to support spending $2,500 on the training seminar.

Take a lesson

HRD professionals have something to learn from academics: basic research skills. For example, when medical researchers attempt to measure the impact of a particular treatment, they use pre-test measurement (performance prior to treatment) and post-test measurement (performance after treatment).

Similarly, if you want to measure the effectiveness of a particular language instruction series, prior to the program the learner should be tested for fluency in that language.

Without a pre-test, there's no way of knowing what gains, if any, are the result of training. Pre-test results also facilitate and document justification for training. A post-test alone may indicate fluency upon completion of the language program, but the lack of a baseline measure renders a trainee's post-test score meaningless.

Gains from training must be measured in relation to performance and skill levels prior to training, and with respect to existing conditions after training. Without those comparisons, we lack objective justification for HRD expenditures.

Performance measurement

In the case of the electronics retail chain, a simple survey of the sales staff could have offered bottom-line justification and demonstrated the need for training. The measure could have been something as basic as a tally sheet on which salespeople tracked the number of customer inquiries and the number of customers who inquired but didn't make purchases. A comparison of that ratio to the overall prospect-to-sales ratio might have provoked a more receptive response from management toward training.

Training objectives are usually performance-based. The measurements for training impact should be performance-based as well, such as in-

studies, which concluded that when subjects are aware that they're participating in a study or experiment, that knowledge alters their performance levels.

Discrete and nondiscrete

There are two recommended approaches to establishing a true baseline. One option is to measure performance discretely at the pre- and post-test stages. The post-measure is not taken immediately upon completion of training. It's delayed for a short period to allow time for new skills to solidify with on-the-job applications. Post-measures taken immediately after training tend to suggest that training has less of an impact than is actually the case.

An additional rise in performance tends to occur when the measurement program is revealed.

direct measure of training's effectiveness. It's unaffected by the circumstances that HRD can't control. For example, if management stresses clerical productivity over customer service, even well-trained personnel may not provide the desired level of service.

Skill measurement also facilitates performance troubleshooting. It provides a means to either pinpoint or exclude training as a problem. And it directs attention to other potential deficits—such as conflicting supervisory or on-the job priorities, or a lack of performance contingencies.

Finally, if the measurement is well validated, it may provide useful information for such activities as employee career development and succession management.

Skills testing facilitates the documentation of HRD's impact on workforce development. Skill measurement may be more difficult to convert to bottom-line justification, but it paints the big picture. Management will get a clearer idea of HRD's overall contributions, such as improving employees' literacy or problem solving skills. Unfortunately, although such benefits clearly indicate HRD's impact, they don't always protect HRD programs from cutbacks.

Unlike performance measurement, skill level measurement is typically free of outside contamination, such as that shown by the Hawthorne effect. Baseline assessment can be made at the onset of training. As in performance measurement, post-testing should be delayed following training to allow for new skills to solidify on the job.

Skill measurements provide invaluable information on training efforts, instructional design, organizational strengths and weaknesses, and individual developmental needs. But such data can be misused. Here are some guidelines.

Performance-based measures easily convert to bottom-line dollars—and to recognition of HRD's value

creased sales volume or lowered defect rates. Performance-based measures easily convert to bottom-line dollars—and to recognition of HRD's value to the organization.

As important as a performance measure is, one that lacks a baseline measure can't be very useful, especially for justifying training.

An HRD professional faced with the task of decreasing defect rates in a particular department—through training and follow-up efforts—should begin by measuring defect rates prior to training in order to establish a baseline measure. He or she then has a basis of comparison from which to calculate bottom-line results after the training is completed.

If the people who are being measured are aware of the measurement program, performance measurement itself has an impact on their behavior. The typical result is an initial spike in performance.

That phenomenon was demonstrated in the 1927 to 1932 Hawthorne

Another option is to implement nondiscrete performance measurement at least 30 to 60 days before training. This approach allows time for the novelty of measurement to dissipate. After this waiting period, you can take a baseline measure and proceed with training.

Performance measures take many forms—such as defect tracking, sales surveys, and shopper surveys. The discrete and nondiscrete measurements apply to most forms. When the guidelines are followed, both kinds of measurement provide invaluable bottom-line evidence of HRD efforts.

Skill measurement

Another type of useful measurement is skill measurement. Unlike performance measurement, skill measurement indicates the level of ability or knowledge that exists independently of on-the-job circumstances and contingencies.

Skill measurement serves several purposes. First, it provides a pure and

Select measures that directly relate to training content and course objectives.

Usually, publishers of off-the-shelf courseware offer competency tests or assessments. They can provide skill measures that are specifically associated with the course. Or, in the case of in-house courses, you can develop an assessment instrument with items based on course content. Development of assessment tools should be conducted by people trained and experienced in that area.

Well-prepared test items, paired with objectively scored role play, and other tests of skill application, provide excellent measures of training's impact on skill level and knowledge.

Use skill measurements on an individual basis for an assessment of the overall training effort and educational remediation.

Individual scores are useful guides for continuing individual development efforts. They're not recommended for hiring, firing, advancement, or other employment-related functions—unless they've been professionally validated for those purposes. Violating this guideline can lead to accusations of discrimination and legal actions against the organization.

Look for trends when assessing the test results.

Examine a variety of groupings, such as results by location or classification. This approach may raise red flags on any weaknesses in the organization. Monitor test items that have a high rate of errors. They may be poorly written or indicate a weak module in the course.

Monitor skill level over time.

Follow-up skill testing or assess-ment provides a measure of long-term learning retention and illustrates the strength of the training effort. Follow-up testing may also indicate the need for (and may justify) follow-up training. Training isn't a one-shot event. Effective training must be a continous process.

Beyond measurement: a case study

Even robust pre- and post-training measurement programs can be enhanced. A case in point involves a large Atlanta-based bank with several locations. It recently evaluated a new, technology-based, interactive video sales training program for its tellers and new accounts personnel.

Receiving positive initial reactions to the system, the HRD staff decided to conduct a pilot test at one location and hold other locations constant with pre-existing training efforts.

Pilot- or beta-testing at the single location provided a low-cost, low-risk opportunity to get the system in the hands of its target audience. If the system bombed, little investment was lost and the HRD department's credibility suffered little damage. This approach also provided for a control group (a representative group of subjects from which treatment or training is withheld).

A control group is compared with the group that receives the treatment or training. Its main purpose is to provide the purest possible evidence that improvement in performance is the result of training rather than other factors, such as an incentive program. In the case of the Atlanta bank, nonparticipating tellers and new accounts personnel were subject to every workplace and environmental condition that could possibly affect performance except the training program.

In the bank's evaluation, pre- and post-performance measures from an established sales-tracking program and the control group provided financial information that formed a basis of comparison. Financial data on the location whose employees participated in training could be compared with data at locations where employees didn't participate. The control factors also provided a means of comparison between the individual performances of those who received the training and those who didn't.

When the results are finally translated into the greater number of dollars earned at the location that underwent training, management may recognize that training can pay off—and that there are potential losses at locations whose employees haven't been trained. The ultimate result should be budget money allocated to training at all locations (see the figure).

HRD departments compete for budget dollars in the same way that

	Location with training	Control location 1	Control location 2	Control location 3	Control location 4
Impact of sales training on bank employee cross-selling performance					
Pre-measure sales ratio	1.2	1.3	1.1	1.2	1.2
Estimated annual margin on cross-sold products (before training)	434,000	517,000	368,000	434,000	434,000
Post-measure sales ratio	2.3	1.2	1.1	1.2	1.2
Estimated annual margin on cross-sold products (after training)	760,000	434,000	368,000	434,000	434,000
Margin gains through training	326,000	0	0	0	0

organizations compete for marketplace dollars. Nice HRD phrases that praise how "great" a program is will be training's epitaph.

It may sound simplistic, but HRD's survival and growth largely depend on its ability to express its contributions in terms of dollar and cents.

Tooting Your Own Horn

You're now armed with cost-saving, profit-building evidence. It's time to toot your own horn. All the bottom-line data in the world are worthless unless you can use them to market and promote HRD's services to the organization.

Here are a few tips.

■ Hold special meetings with line managers and upper-level managers. Show them the results of pre- and post-measures. Express those results in terms of contributions to the bottom line. Point out the overall benefits of a better trained workforce.

■ Publish your results in memos, special announcements, and company newsletters. A sample headline: "First Zero-Defect Course Grads Reduce Defect Rate by 83 Percent!"

■ Use your results to justify further training expenditures. Make certain that you can specify the bottom-line contributions that you expect further training to produce.

■ Formally communicate your results to trainees. By doing so, you reinforce their progress and contributions.

■ Develop a bottom-line file. Maintain an ongoing collection of materials that justify the existence of HRD. Include everything that's relevant—from informal memos to formal program pay-off assessments.

■ Prepare and distribute an annual report. List contributions that the HRD department makes during the year. Distribute the report to upper-level managers. Use the right language. If possible, express the benefits in terms of dollars saved or earned. For example, an increase in the cross-selling ratio is a "feature." The "benefit" is $326,000 in profits.

Five Uneasy Pieces in the Training Evaluation Puzzle

The clarion call for results-oriented training evaluation may be misguided; not all training is results-oriented. Before you can decide how to evaluate training, you need to make five important decisions.

For the last 30 years, a steady stream of articles has lamented the shortage of rigorous training evaluation efforts. Some recent popular publications have even made it appear that evaluation is relatively easy and inexpensive to do, and that recent advances in evaluation technology have significantly eased the process for training and development practitioners.

But such advances in evaluation technology are not readily apparent in the literature of training evaluation. Written guidelines for evaluation have, in fact, remained quite consistent over time and across authors.

The standard "textbook" coverage of training evaluation starts with the topic of criterion development. That encompasses criterion relevance (including deficiency and contamination), levels (such as reaction, learning, behaviors, and results), time frame, and reliability. Typically, a discussion

By Glenn M. McEvoy and Paul F. Buller

of internal and external threats to validity follows, as well as an examination of the pros and cons of experimental and quasi-experimental designs.

Anyone who conducts a training evaluation effort should be familiar with these issues, but our experience suggests that other, less well-examined issues are equally important.

Furthermore, evaluating training is definitely not easy, at least not in our experience. It is easy to write about evaluation and to exhort others to undertake the task, but it is not a simple matter to conduct a training evaluation. Our training work has led us to explore some of the problems and examine some of the key decisions that must be made in training evaluation.

OMT as a case example

Outdoor management training, or OMT, is a good example. OMT is an

McEvoy *is a professor of management and human resources at Utah State University, Logan, UT 84322-3510.* **Buller** *is a professor in the School of Business Administration, Gonzaga University, Spokane, WA 99258-0001.*

outgrowth of Outward Bound training, which was founded in Britain by Kurt Hahn and Lawrence Holt as survival training for naval recruits in World War II. Outward Bound spread to the United States in 1962 and gained popularity as a means of developing character, self-confidence, and leadership among young people.

Management trainers have adapted the Outward Bound experience because of its emphasis on such areas as risk taking, challenge, teamwork, problem solving, self-confidence, and trust.

OMT programs typically consist of a series of perceived high-risk activities or "initiatives," most requiring teamwork and problem solving skills. Interspersed with the activities are debriefing sessions in which partici-

- work versus perk
- substantive versus symbolic
- external versus internal
- behaviors versus outcomes
- self-ratings versus ratings from others.

Work versus perk

 The first issue is the extent to which the training program is truly an attempt to improve work performance, rather than a perquisite for job performance that has already been judged successful. Off-site programs, in particular, run the risk that organizations will select participants on the basis of prior achievement rather than the expectation of improved future performance.

managers—who have not participated in our program—have referred to the program as "a walk in the woods." And, it does have a reputation for serving excellent food!

But our program is conducted at a remote mountain site, a former Forest Service camp with "rustic" accommodations (it's cold, with mice living in the bunkhouse). There is no hot tub, alcohol is prohibited, and the nearest golf course is 25 miles away.

Is the program a perk? Participants would answer with a definite "no." A typical reaction is this: "I'm glad I came, but I wouldn't want to do it again." Because our OMT program was not a perk, we decided that our evaluation study had to go beyond the mere reaction level.

Substance versus symbol

 Of course, many training programs are work-related rather than being designed as rewards. Even so, they may not be intended to improve "performance" in the traditional sense of the word.

Considerable attention has been paid of late to the concept of corporate or organizational culture: shared values, rites, rituals, heroes, myths, symbols, and ceremonies. All organizational practices, including training, serve both *to do* things (the practical side) and *to say* things (the symbolic side). Training can be considered part of the basic underlying culture of the company, or the "expressive side" of an organization.

Rites are the prescribed or formal procedures associated with certain ceremonies. Janice Beyer and Harrison Trice place organizational rites into three categories ("How an Organization's Rites Reveal Its Culture," *Organizational Dynamics,* 1987, no. 4). Each type of organizational rite can also be expressed in terms of training, as shown in parentheses:
- rites of passage (initial training and orientation)
- rites of enhancement (seminars to enhance power and social identities)
- rites of renewal (small-scale organizational development activities such as MBO, team building, and quality circles).

The cultural or symbolic consequences of some training programs may be more important than technical

---

"I send my people to training as a reward—a chance to get away for a few days of R&R on the company's money"

---

pants analyze their feelings and experiences and share their learning with colleagues. Most programs also provide short lectures on management skills and action planning to help participants transfer their new-found knowledge and skills back to the job.

Over the past two years, we have been involved in the implementation and evaluation of a university-based OMT program for a client in the aerospace business. The conclusions we have drawn about the difficulty of training evaluation come primarily from that experience.

The five pieces of the puzzle

We have identified five levels of issues in training evaluation. They exist in a hierarchy in a sense; you must resolve issues at the top before you can move on to issues at lower levels. Here are the five levels in order:

Such objectives are seldom stated for a training program. Rare indeed is the executive as honest and direct as the one who said, "To tell you the truth, I send my people to those things as a reward—a chance to get away for a few days of R&R on the company's money. If they happen to get anything out of the seminar that proves useful on the job, so much the better."

If the true purpose of a training program is to reward good performers or to renew sagging spirits at company expense, an extensive performance-based training evaluation is misguided. A simple reactions measure, or "smile sheet," may be all that is really necessary. Such evaluations are extremely common in practice but are uniformly vilified by authors. But they are entirely appropriate when the training is actually a perk.

OMT programs are necessarily off-site, and thus can easily be used as perks rather than serious training. Indeed, some of our client company's or substantive results. In one training evaluation study, for example, the desired improvements in managerial skills were minimal compared to the reduction in anxiety about organizational downsizing—an outcome that was completely unintended. Managers insensitive to the expressive or cultural side of these rites may be tempted to discontinue them on purely technical grounds, and thus lose the substantial intangible benefits they accrue.

Off-site training programs such as OMT are particularly likely to carry ritualistic or symbolic overtones. For example, being selected to attend such training can be seen as a symbol of the organization's confidence. These programs can also build an esprit de corps among participants. The aerospace employees who graduated from our OMT program formed an informal network, meeting occasionally, wearing distinctive pins to work, and providing each other with quick access and support on the job.

In fact, joining that network appeared to be one of the most beneficial outcomes of the program. Some typical participant remarks:

■ "Those of us who have shared the common experience of OMT look for each other and roll out the red carpet."

■ "Probably on at least a weekly basis, someone who attended OMT with me will call requesting help dealing with his or her area of responsibility. The feeling of comradeship that was developed at OMT prevails and together we solve the problem."

■ "On several occasions I have requested support from other departments through normal channels. After some hassle and delays, I have contacted someone in the OMT network and suggested that I'm on the rock and looking for someone 'on belay.' This has worked well."

■ "I have been called on often by members of my OMT class to answer questions or help work out problems relating to safety."

External versus internal

 A related issue is the extent to which desired training program outcomes are internal or external to the program itself. Once again, authors have railed against the use of internal evaluation measures, because they assume that all training must be designed to effect some external change. This is not always the case.

For example, some training may be designed primarily to instill company pride, or perhaps to inculcate company history or philosophy. In such cases, it would be unreasonable to expect measurable changes in behaviors or results external to the training setting, and one would correctly rely on internal evaluation of changes in attitudes or knowledge.

OMT again provides an example of this subtle expectation for training outcomes. Organizations that send employees to such programs may be attempting to emphasize some unique aspects of corporate philosophy, such as the importance of individual stretching and risk taking, or the balance between the needs for individual excellence and group teamwork.

Again, such changes may not manifest themselves obviously in behavior or results back on the job. In such cases, internal measures (such as diary entries over the course of the program that describe a participant's attitude toward taking risks) may be appropriate.

Our client was in fact attempting to overcome a conservatism and risk aversion that was deeply imbedded in the organization's culture and was hindering efforts to improve competitiveness. A common experience of OMT participants was that they benefited from "going beyond the comfort zone" when the climate was one of trust and support. One participant said, "As a result of the OMT experience, I am pushing myself out of my comfort zone regularly now."

Behaviors versus outcomes

 Once it has been determined that the purpose of training is truly work-related and that the desired outcomes are more substantive than symbolic and more external than internal, the question becomes how to measure the changes back on the job.

Basically, the issue is whether to use a more objective quantitative criterion or a more subjective measure. It is not true, as is often assumed, that "results" measures are superior to "behavior" measures. Results measures are only possible when the skills being taught can be directly translated to on-the-job effects (as can sales techniques) or when considerable time and attention

The trainees formed a network, meeting occasionally, wearing distinctive pins, and providing each other with quick access and support on the job

are devoted during the training to establishing ROI (return on investment) targets.

Like other clients, the aerospace company was interested in results, particularly in their dollar value. The managers selected to attend OMT were responsible for such major areas of the firm as safety, quality, project management, finance, and training.

Certain "hard" measures could be gathered in those areas: forms processed, lost-time accidents, inventory utilization, budget variances, unit costs, turnover and absenteeism, equipment downtime, scrap and rework, and so forth. But the goals of the OMT program were to improve participant skills in these areas:

■ increasing self-awareness and insight

■ developing supportive communication with others

■ diagnosing and solving performance problems of subordinates

■ delegating and individual decision making
■ building and maintaining effective work teams.

Improvement in participant skills are first-order outcomes; results are second-order outcomes. Many other factors besides managers' skill levels can cause results. Furthermore, in two years only 37 managers out of a total workforce of more than 6,000 have been trained in the OMT program. How great an impact on measurable organizational results can the company expect from such a small group? New knowledge and skills in teamwork and team building, in particular, may only translate to observable improvements after a critical mass of managers has completed the OMT program.

Our client was concerned about the time it would take to collect such measures. Given such considerations, our client decided on behavioral measures (ratings), instead of results memasure.

Since the client expressed an interest in dollar-valued outcomes, we offered to convert improvements in job-performance ratings to dollars. To do that, we used the general utility formula explained in W.F. Cascio's book, *Costing Human Resources: The Financial Impact of Behavior in Organizations.*

Training evaluators have been heralding the advent of utility formulas as a major breakthrough in the justification of human resource development programs. We should have been delighted to calculate the dollar-valued outcomes of our program.

We weren't.

The utility formula requires that assumptions be made about the standard deviation of job performance in dollars and the length of time the training effect can reasonably be expected to last. We used conservative assumptions; still, the calculated net benefit of the OMT program was a whopping $510,731.

We faced a dilemma: Should we share this number with the client? It was so large that it was likely to be dismissed as unbelievable or as program hype. How could we explain the way this dollar amount would accrue to the client organization?

An accounting-oriented client would be likely to say, "OK, when will I see an additional half million dollars in cash flow?" The client was also likely to ask how we determined the net utility figure. Of course, all we did was plug numbers into the general utility formula. This formula is not at all intuitive; it is derived from a linear regression of job performance (measured in dollars) on predictor scores. But the derivation is difficult to explain, leaving us to tell decision-makers to just "trust the formula." That certainly wouldn't help our credibility.

Another problem is that numbers tend to drive out other considerations. If the client buys the notion that OMT has a net benefit of more than $500,000, that tends to become the major rationale for the program. Little attention will be paid to more subtle effects, such as changes in corporate culture or improvements in teamwork.

One participant did volunteer a specific example of a results-oriented outcome for the OMT program:

"The first week or two after returning to the job, my OMT partner and I as a team were able to influence past policy and implement changes in the quality discrepancy reporting requirements that saved a total of $800,000 to $1 million."

That anecdote floated throughout the company and was often cited as justification for the program.

Self-ratings versus other ratings

 Given the limitations of ratings from any single source, we suggested to our OMT client a criterion involving combined ratings from multiple sources: self, supervisor, subordinates, and peers. We were acutely aware, in particular, of the limitations of self-ratings when used alone. Compared to other ratings, self-ratings tend to exhibit less variability, more leniency, and less agreement with other ratings sources.

These weaknesses were less of a concern to the company than we expected. The client was more concerned about the time, cost, and organizational intrusiveness of collecting ratings from sources other than program participants; the client ultimately prevailed.

Of course, a client unwilling to invest the time and effort to collect ratings from other sources may really be saying that the perquisite, symbolic, or internal outcomes of the program are the most important. Or the client may be expressing the widely held opinion that performance ratings are not valid anyway.

So we were left with self-ratings. One of the purported advantages of such ratings is that they suffer less halo error than ratings from other sources. We expected this to be advantageous in an initial study of the effectiveness of an OMT program, because the lack of halo would give us good feedback on which skill areas the OMT greatly affected and which were affected only slightly or not at all.

Unfortunately, that advantage of self-ratings never materialized. The self-rating instrument we used had five questions in each of the five skill areas cited earlier as program objectives (self-awareness, communication, performance problems, delegation and decision making, and teamwork). The responses on all 25 questions were highly correlated. Therefore, we had to collapse all questions into a single scale, and it was impossible to tell if the OMT had different effects in the five skill areas.

Hidden issues

Overall, our experience suggests that training evaluation research is not as straightforward as "textbook" discussions would suggest. Many subtle issues are not typically addressed; even the organizational sponsors of a training program may be unaware of some of them.

For example, the "work versus perk" aspect of training may never be stated by an organization, but is part of the decision process for at least some program sponsors and participants. Also, many managers may be oblivious to the significance of the symbolism involved in attending and completing certain training programs. Several of our program participants said that being sent to the OMT was one of the few signs they had ever received that the organization cared about them.

It is possible that the emphasis on increased accountability in the training function may drive out programs that cannot demonstrate quantitative, measurable improvements in performance or results. Such an outcome would be unfortunate indeed.

EVALUATION THAT GOES THE DISTANCE

BY PAUL R. BERNTHAL

Four-level training evaluations often stop short of reaching meaningful, long-term results. Here's a road map for adding on measures that can go the distance.

In 1959, Donald Kirkpatrick published a paper that classified training outcomes into four levels: reaction, learning, behavior, and results. Kirkpatrick's classic model has weathered well. But it has also limited our thinking regarding evaluation and possibly hindered our ability to conduct meaningful evaluations.

Too often, trainers jump feet first into using the model without taking the time to assess their needs and resources, or to determine how they'll apply the results. When they regard the four-level approach as a universal framework for all evaluations, they tend not to examine whether the approach itself is shaping their questions and their results.

Other options

The simplicity and common sense of Kirkpatrick's model implies that conducting an evaluation is a standardized, prepackaged process. But other options are not spelled out in the model.

First, it's important to reexamine some faulty assumptions about four-level evaluations and evaluations in general.

Assumption: Evaluations are definitive. Most training evaluations are based on the philosophy that a single study can answer all questions about the effect of a training effort. Most evaluators aren't prepared for ambiguous findings. But in any evaluation, the degree of certainty regarding the results depends on such variables as the rigor (reliability) of the design, the measures, and the sampling strategy.

Most importantly, credibility depends on whether an evaluation's findings can be replicated. If only one evaluation is conducted, it probably can't stand on its own merit. Other studies are likely to point out flaws or alternative explanations for the findings.

Assumption: Evaluation equals effectiveness. Evaluation focuses on the learning aspect of training. It answers the question, "Have the requisite skills and knowledge appeared as a result of training?"—a level-2 evaluation in Kirkpatrick's model. An evaluation can become problematic when it also tries to measure effectiveness. Effectiveness focuses on whether the training has produced the intended outcomes (levels 3 and 4). To answer the effectiveness question, the evaluator must measure several organizational, individual, and training-related variables.

Evaluation and effectiveness are linked. But they shouldn't necessarily be arranged on a continuum, as they are in Kirkpatrick's model.

Assumption: Trainers are accountable for effectiveness. Many trainers who conduct evaluations don't have the skills, time, or resources to do an in-depth study. They may not be knowledgeable about the training topic. In such cases, does it make sense for trainers to be held responsible for the success of all training programs, especially those instituted by senior managers? Still, trainers often have the most to lose when results aren't positive.

TRAINERS HAVE THE MOST TO LOSE WHEN RESULTS AREN'T POSITIVE

Assumption: Level-4 evaluation is superior. Some studies show that Kirkpatrick's levels 1 through 4 may be correlated. But they measure different things.

So, why is level 4 often described as a higher level of evaluation? In fact, many level-4 evaluations are conducted only because they're viewed as the toughest assessment of training, even when the measures used have no real link to training. One shouldn't choose a level-4 evaluation and then try to tie it back to the training, just because the training involves some level-4 variables.

Suppose that the training is on interpersonal skills. Such variables as operational costs and equipment downtime may have some relation to employees' interpersonal skills. But those variables wouldn't be the best measures to use in the training evaluation.

Typically, trainers believe that level 4 is the pinnacle of training evaluation. But each level can provide equally valuable information, depending on the type of trainees being evaluated. Level 1 or 2 outcomes can provide some of the most useful information because those outcomes are often the easiest to measure and change.

Assumption: You just have to measure it. Many measures used to assess training are inappropriate and not sensitive enough to detect changes in trainees' behaviors. It's also difficult to know what questions to ask, how to phrase them, and to whom to direct them. And all measurement methods aren't equally reliable or valid.

Building onto the model

Instead of choosing a particular level and jumping into the evaluation process, you may want to make these add-ons to Kirkpatrick's model:

Consider the context. Understand how the training fits within the organization's operations and culture. Many things in addition to training can affect the work environment. In evaluating level-3 or level-4 changes, it's important also to measure contextual variables. For example, the lack of management support can undermine even the most effectively designed and delivered training program.

Establish a link. Draw a cause-and-effect path between training interventions and such outcomes as job behaviors and productivity. The path can show where and in what ways the training results in measurable changes.

Make appropriate choices. It's important to choose an appropriate evaluation design and appropriate measurements. Ask what you really need to know. Let the answer determine your approach. Choose one level at a time to evaluate specific results.

An appropriate initial evaluation might be a post-training assessment on changes in trainees' knowledge, using a paper-and-pencil test. Another approach would be to use a multi-rater tool to compare trainees to a control group.

Inventory your resources. No matter what level approach you use, make sure that you're realistic about what you can accomplish. Consider the costs, the amount of time you can spend, trainees' downtime (to fill out questionnaires or other evaluation instruments), and the expectations of your customers (trainees, senior managers, and others).

Set goals and long-range plans. Don't just do an evaluation; establish an evaluation program. Many organizations and their training programs grow and change in ways that make most evaluations obsolete in a few years. Establish a program that includes multiple evaluations at various levels on the effect of each training effort.

For example, in *In Action: Measuring Return on Investment* (American Society for Training and Development, 1994), editor Jack Phillips recommends evaluating different percentages of programs at the four levels. He says you could evaluate 100 percent of all programs at level 1, 70 percent at level 2, 50 percent at level 3, and 10 percent at level 4.

The training-impact tree

The first task in setting up a long-range evaluation program is to create a training-impact tree. It will help identify the variables that could affect a training intervention and help establish links between the training and organizational values and practices. (See the box "A Training-Impact Tree.")

But before you create the training-impact tree, establish your team. Assemble a group of people with diverse perspectives on the effects of training on the organization. Team members should be knowledgeable about the organization's values and practices, the scope and appropriateness of the training, and the factors that could affect training transfer. The team should spend at least one entire day developing the tree.

Step 1: Identify the organization's values and practices. Many organizations publish a list of values. But they don't always practice them. The team should make a list of the organization's main values and associated practices. An example of an organizational value is teamwork. It is manifested in such practices as formally recognizing team efforts, linking individual goals to group goals, and creating an environment of open communication throughout the organization.

Step 2: Identify skills, knowledge, and attitudes. Once the organization's values are linked to practices, it's easier to identify the type of training that will enable employees to perform effectively under current conditions. To that end, the team should tie each practice to a list of skills, points of knowledge, and feelings that people can be trained in and about.

University of Colorado professor Kurt Kraiger has written that a training evaluation should focus on three areas of learning: skills (technical and motor), "cognitions" (knowledge and thoughts), and feelings (attitudes and emotions). In other words, training affects what you do, how you think, and how you feel.

For example, if the team's goal is to achieve a free exchange of information within the organization, team members should identify the skills, cognitions, and feelings associated with open communication. A skill would be knowing how and when to share ideas without being asked. A cognition would be understanding how the individual members of a group can affect the group as a whole. A feeling would be team members' concern about the success of the team.

Step 3: Define the scope and purpose of the evaluation. In addition to helping identify how training fits within the organization, the training-impact tree also can help generate a list of questions to include on a training evaluation.

An evaluation should measure more than reaction, learning, behavior, and results. Those levels focus mostly on outcomes; they don't take into account the process leading to the results. For example, one can evaluate behavior changes without recognizing that they might depend on people's motivation, the degree of managers' support after training is completed, or the extent to which the training was appropriate for meeting needs.

Here are several areas beyond the four-level scope that can be evaluated:
▶ the quality, delivery, or retention of the training
▶ how well the training cut deficiencies in a particular work group
▶ the usefulness of parallel training for managers and their staffs
▶ variables in the work environment that discourage or facilitate the effect of training.

Most evaluations can benefit from measuring organizational context. The training-impact tree can show such context by listing the barriers to training and the factors that facilitate

---

A Training-Impact Tree

Here's an example of a training-impact tree. It can help identify the variables in an organization that might affect training outcomes.

Barriers to Training

▶ Organizational values aren't clearly communicated.
▶ Senior managers send mixed messages.
▶ Achieving goals requires competitive behaviors from employees.
▶ Managers aren't held accountable for their teams' success.
▶ Team members work on different floors or in different buildings.

Organizational Values

Example: teamwork

Supporting practices include the following:
▶ recognizing team efforts
▶ participating in cross-functional teams
▶ linking individual goals to group goals
▶ sharing responsibility and accountability for team outputs
▶ exchanging information openly and frequently
▶ establishing partnerships with other departments and teams.

Factors in Facilitating Training

▶ The organization has a good relationship with its unions.
▶ The organization has a stable profit margin.
▶ The organization's values are logical.
▶ The job descriptions are clear.
▶ The selection system includes effective teamwork as a criterion.
▶ Managers serve as positive role models.
▶ The performance-management system rewards team accomplishments.

Training

Cognitions (knowledge and thoughts)
▶ the role of team leaders
▶ the effect of individual members on a group as a whole
▶ the meaning of empowerment
▶ the way teams and groups work together
▶ the importance of sharing activities and tasks among group members
▶ the importance of diversity.

Skills
▶ expresses thoughts clearly
▶ listens and responds with empathy
▶ seeks input from others
▶ offers help when others need it
▶ makes sure all team members are heard and understood
▶ shares ideas without being asked
▶ regularly recognizes individual efforts in support of team goals.

Feelings
▶ the motivation to grow and develop
▶ a sense of team spirit
▶ self-efficacy
▶ a concern for group cohesion.

training next to their associated values and practices. If the team isn't sure what contextual variables to consider, it can seek direction from trainees or from focus groups made up of senior managers, frontline leaders, customers, or people from all three groups.

In the article, "Individual and Situational Influences on the Development of Self-Efficacy: Implications for Training Effectiveness (*Personnel Psychology,* spring 1993), authors John Mathieu, Jennifer Martineau, and Scott Tannenbaum say that training doesn't occur in isolation from employees' job responsibilities and personal lives. Just providing time for employees to receive training doesn't ensure training effectiveness.

Organizations that use such standard reaction measures as smile-sheet responses to revise a program may be making fruitless "improvements." Trainees' negative reactions to the training may have nothing to do with the training itself. They may resent the time it takes or the fact that it is mandated. They may be distracted by personal problems. Or, they may not be interested in the topic.

Step 4: Identify data sources. The quality of the evaluation data depends heavily on the source. For example, self-assessment is rarely the best way to determine whether a person's behavior actually changed as a result of training. The criteria for choosing the best data sources include a source's objectivity, accessibility, and reliability. Data sources should be unbiased, should provide understandable information, should be easy to access, and should produce information that's immune to irrelevant influences.

It's best to collect data from a variety of sources. People are more likely to believe similar findings from different sources. Even disagreements among sources can provide valuable information by offering different perspectives.

Step 5: Choose the best method for collecting data. One of the most difficult aspects of developing an evaluation is selecting and implementing an appropriate design. There are books that can help. An evaluation design can use almost any traditional research method. But the choices may be limited due to practical considerations.

Ask yourself these questions:

▶ How frequently do you want to collect data? It's best to survey trainees immediately after training is completed. But the findings will be more conclusive if you also conduct a pretraining assessment and a long-term follow-up. It's fairly easy to detect changes measured over time.

▶ Would data from a control group strengthen the findings? If trainees show improvement over time, it might be due to factors other than the training. A control group (people who did not receive the training) can provide a basis for comparison.

▶ How many people should I collect data from? There are no rules. The criteria for determining the appropriate sample size include randomness, representativeness, the size of the trainee group, and the desired statistical results.

Ideally, the sample should be randomly drawn from the population being evaluated. But training rarely occurs randomly across an organization. Focusing on specific departments is all right, as long as they are similar to other departments in the organization.

If the population is fairly homogenous, fewer people will be needed in the sample. But the sample should always be representative of the target population. Every classification or demographic in the population will require more people in the sample.

Most statistical tests require at least 30 participants in each group being evaluated. For example, a training evaluation with a control group should involve 60 participants—30 in the control group and 30 trainees. Over time, there's usually a high rate of dropouts among participants. So for the benefit of follow-up assessments, it's best to pad the sample from the start.

Step 6: Select the best measurement approach. Typically, the easier the measurement, the less objective it is. To increase objectivity, use different measurement methods in the same evaluation or conduct several evaluations, each using different approaches—including self-assessment, multirater assessments (supervisors, peers, and subordinates rate trainees' performance), focus groups, and behavioral simulations (trainees use role play to practice new skills).

The assessment-center approach can yield valuable data, though it can also be difficult and time-consuming to implement. Alone, each of these evaluation methods presents unique advantages and problems that might affect your conclusions about an evaluation. Combined, they represent a diverse and powerful approach for painting a complete picture.

Step 7: Gather and inventory your resources. Identify the people who will help conduct the evaluation. What are their skills? Which parts of the evaluation should they be responsible for? Do they have enough power and

The Trade-Off

The evaluation approaches in the figure represent typical implementations. The ease and rigor of an evaluation can vary greatly, according to how extensive the evaluation is.

	Difficult	
Subjective (less rigorous)	Focus Groups / Self-Assessment	Behavioral Simulations / Multirater — **Objective** (more rigorous)
	Easy	

influence in the organization to implement the evaluation results?

The hardest part of establishing a long-range evaluation program can be gaining buy-in from people across the organization. Expect to do some internal selling of the evaluation. Try to create partnerships with your internal clients. You might want to develop a matrix listing the people who will serve as the doers, approvers, and reviewers. The matrix can show the points in the evaluation at which each person will be involved.

If you don't have the in-house resources to conduct the type of evaluation you need, you can hire an external consultant. Though many aspects of an evaluation can be conducted without outside help, more complex ones tend to require experienced evaluators. Be sure that the consultant understands the evaluation. Also make sure that he or she thinks in broader terms than just a four-level approach.

Many consultants charge $1,500 to $2,500 a day to conduct an evaluation. You can cut costs by administering the evaluation yourself. But you should expect to pay the consultant for several days of analysis and report generation. Typically, expect the consultant to spend four to five days planning, designing, and creating materials.

Training evaluations aren't one-time events. It's important to develop a schedule for periodic assessments. As you collect data over time, look for trends. Some evaluations may be "quick and dirty"; others may be more extensive. Remember: No one evaluation stands alone; almost all evaluations yield valuable information.

Once you've conducted several evaluations using various methods, you can fine-tune your approach and do fewer evaluations. The results can lead the way toward the desired final destination: a strong training program. Let the program evaluations show where you've been and where you need to go.

Paul Bernthal *is a research consultant with Development Dimensions International, 1225 Washington Pike, Bridgeville, PA 15017-2838. Phone 412/257-0600; fax 412/257-0614.*

Involving Managers in Training Evaluation

A simple, six-step training evaluation can help you generate greater commitment and support for employee training programs.

By Linn Coffman

Do your training programs suffer from predictable-results syndrome? Do you find yourself perplexed over which training programs will best suit the needs of your diverse employees? Do you have a good grasp of the value of your training? Most training managers don't, and the predictable-results training blues seem to be catching. Consider the following based-on-life situation, and discover a prescriptive method for overcoming training blues.

No one-size-fits-all training course

After sending three of his line supervisors to an interpersonal-skills training course, Ian Keen, a senior manager at a medium-size tool-and-die plant, felt frustrated. The results were just as he would have predicted.

Gail Wilson, his top supervisor, made good use of the learning experience. But Gail was the type who could be stranded on a deserted island and still learn something. It didn't make much difference what she learned in the training; she'd apply it to the job somehow.

Then there was Walt McFarland—

Coffman *is a senior consultant with Development Dimensions International, 1225 Washington Pike, Box 13379, Pittsburgh, PA 15243-0379.*

an average supervisor whose initial response to training was generally positive. When Walt attended training, he would say, "It's useful," or "I got a lot out of it." But back on the job, nothing happened that showed he was doing anything differently. Walt's track record never changed—no matter how good or bad the training was.

Ian's third supervisor, Anita Rodriguez, was an enigma. Sometimes she took initiative and used what she learned in training; other times she seemed to do worse after training. If Ian showed an interest in what Anita was doing, she'd improve, but only for a while.

Ian knew of no sure way to tell how training would affect his supervisors. With or without training, Gail would figure out the task and get it done. Walt

always did the usual, which was the same as what he'd done the day before. And Anita might do the job just fine, or she might ask Ian to share in the work. Ian never could tell.

Are you sure that the training your employees receive is valuable or at least helpful? Do you have sufficient data or expertise to evaluate the training effectively? If Ian's problem sounds familiar, you could probably benefit from a simple, effective method that can help you gauge the real value of training. The secret? In-

Are you sure that the training your employees receive is valuable or at least helpful?

volving employees in the process of selecting and evaluating training programs.

Training Impact Assessment

The Training Impact Assessment (TIA) is a process that requires managers to look collectively at what happens to their employees as a result of training. The thinking is that if most of the "Anitas" and even a few of the "Walts" use the training effectively, then the training has value. If the "Anitas" haven't changed or developed, then the training is probably ineffective. TIA helps eliminate managers' uncertainties about the effectiveness of training programs.

The TIA approach also results in managers supporting and committing to effective training. By communicating their support to higher management, they can influence executives who may not recognize effective training programs.

TIA has been used to evaluate programs for both technical and nontechnical activities in Australia, Canada, South Africa, and the United States. In at least one case, the technique was used to evaluate a proposed marketing plan—with excellent results. In another case, a director of employee development who used TIA to assess a management-development program was personally rewarded with a bonus check for using the high-impact evaluation process.

Steps to success

The TIA method follows six steps:

1 Invite key clients to participate in the assessment sessions. In practice, the key client is often the trainee's boss. The boss is responsible for evaluating or gathering relevant data on the impact of the training on the job and on the unit or person. In some cases, other managers may participate as key clients, provided that they are willing to undertake the data-gathering assignments.

2 At the first session, ask the key clients to gather data on the effectiveness of employee training. Be sure to clarify what clients are being asked to do and to emphasize the importance of gathering ample data before the next session in two weeks.

3 At the second session, ask subgroups to share positive results of training. Divide the clients into subgroups of four to twelve people, six being an ideal group size. Having three to five subgroups increases the validity of common conclusions. (More than five subgroups may become cumbersome, though one client reported success with seven.) Have each subgroup discuss and list the advantages and positive aspects of the training program.

4 Ask subgroups to list negative or unachieved results. Keep the list of weaknesses separate from the strengths to avoid confusion and debate and to help clients focus on specific areas of needed improvement. Post all lists in the general-session room for all to see.

5 Have the entire group reconvene to share overall results. All clients should be present during the general session so that everyone has access to all possible data on which to evaluate the training program. For the first time, many managers will be able to see for themselves the value of evaluation and will feel their initial uncertainty about the usefulness of assessing training vanishing.

Senior executives may also want to attend this session to witness the excitement and intensity. Their presence can help them sense the impact—both positive and negative—of the training, as well as get a clear idea of the intensity of the assessment process.

One executive said later that she had never realized just how excited about and committed to employee training her managers were until she saw it for herself. The paper reports she received just didn't capture the enthusiasm conveyed in the group session.

For people who cannot attend the assessment session, try videotaping the entire process and making available an edited version for later viewing. Videotape is a good medium for communicating the impact and energy of the TIA assessment process.

6 Consolidate lists, agree on actions, and set a follow-up date. Make one comprehensive list of strengths and one of weaknesses, including input from all clients. Tell the clients that within two weeks, they will receive a follow-up written report, summarizing the results of the meeting and the agreed-on plans for action.

But does it work?

Since its introduction in 1979, the TIA process has been used many times. It has been proved to be as valid today as it was 10 years ago. The process is only as strong, however, as the organizational commitment behind it.

Some practitioners report they are nervous about the first session. They fear their clients won't gather enough data to bring to the second session. There are several ways to ensure that enough data are gathered. In evaluating its management-development program, one organization provided key clients with a patterned questionnaire. It also arranged for 39 training participants to be available for interviews immediately after the first session. Each client agreed to interview one other participant before the second session.

In Chicago, a hospital client mixed program trainees with supervisors during the second session, ensuring that the supervisors had firsthand data about the actual program results. The client reported favorable results and recommended TIA as a way to gather plenty of accessible data. Other clients have discussed data-gathering methods during the first session.

Another approach to guarantee data on the training is to have the facilitator contract with key clients in front of the entire group as to exactly how and by when they will gather the data. The public agreement tends to motivate the clients to follow through on their commitments.

The validity of the evaluation is another area of concern. As mentioned, having three or more subgroups increases the validity of common conclusions. Another benefit of having several subgroups is that personal biases among the group usually become obvious when the lists are posted in the general-session room.

As the subgroups work on their lists and attempt to gather, analyze, and reach objective conclusions about their data, they realize that their results are going to be posted and compared with results from peers in other subgroups. In effect, two dynamics are at work: the desire to do well in front of peers and the motivation to compete with other subgroups.

Evaluation factors

In evaluating the training program, pay particular attention to conclusions that appear on most of the subgroups' lists, especially the negative lists. Such conclusions as "insufficient interviewer training" or comments about a particular training module that did not work are often the primary criteria for determining the importance and validity of certain factors to the clients.

To consolidate the lists, first ask the clients if any conclusions—positive or negative—are common to all the subgroups' lists. Write down those common conclusions. Then ask if there are conclusions common to all the lists except one. Consider those conclusions as somewhat important. Conclusions that appear on only an occasional list are not worth pursuing.

Time is another important consideration when evaluating training programs. Distribute a report to all participants within two weeks after the second assessment session. Two weeks gives you time to analyze the results and to determine what actions need to be taken. If you wait longer, the clients may sense a lack of interest in their evaluation conclusions; it may become more difficult to capitalize on their enthusiasm and to gain their commitment.

Finally, you should recognize the potency of this process. For the first time, key clients (usually managers or supervisors) may be involved in researching the value of an activity that they aren't really sure about. But their involvement in the process increases their certainty of the value of evaluation. Once they have seen it and done it, they are more certain that it works. Your immediate action on their evaluations will reinforce that certainty.

In one management-development program assessment, a common conclusion among the subgroups was that the supervisors needed training in interviewing and selection. The facilitator suggested that the next group of managers to go through the evaluation program receive interpersonal skills training. That quick response to the clients' assessment showed real commitment to improving the process and appreciation for their input.

No more guesswork

So what happened to Ian Keen? Using the Training Impact Assessment process, Ian learned how to evaluate the training programs that were best suited to his particular line supervisors' needs. He subsequently scheduled Gail for management training, Walt for assertiveness training, and Anita for decision-making training—without worrying about whether the programs would be valuable or helpful. He no longer had to guess which types of training would be best for his people or how they would respond. He had answered his own question before even signing them up.

Using the TIA process in your own organization can help you evaluate and select specific, targeted training programs for your people. You'll find you no longer suffer from that deadly malady—predictable-results syndrome. By involving managers and executives and showing them how evaluation can work, you'll generate greater commitment and support for employee-training programs—from the people who will ultimately benefit most.

By Anne F. Marrelli

TEN EVALUATION INSTRUMENTS FOR TECHNICAL TRAINING

Consider these advantages and disadvantages before selecting evaluation instruments.

This article explains how to select and use 10 types of evaluation instruments to assess technical training programs and materials. For easy reference, the information is presented in a series of charts that describe the instruments and their possible applications, advantages and disadvantages, and guidelines for use.

First, though, you need to know some ground rules for selecting evaluation instruments. For instance, to collect accurate information with evaluation instruments, you need a basic knowledge of statistics and research methods. Such knowledge helps you plan the evaluation design, sampling method, and data analysis procedures.

Next, you should know how to use various instruments. Expanding your range of instrument proficiency enables you to select the most appropriate instrument for each evaluation. After all, trainers who can use only a few tools tend to use them in all situations—whether appropriate or not.

You should also use more than one instrument in any evaluation. Multiple instruments offer several benefits:

▼ Each instrument has strengths and weaknesses. You can compensate for the weaknesses of one instrument with the complementary strengths of another.

▼ The more indexes that can point to similar benefits or problems, the more accurately you can assess your training program. Your results will also be more credible.

In This Story

▼ evaluation
▼ games and simulations
▼ tests

▼ Different instruments highlight different results of the same training program. You might miss some of these results by using only one instrument.

QUESTION QUINTET

Before selecting an evaluation instrument, you should also consider these five questions:

▼ *Will the Instrument Answer Your Questions?* The first and most important step is to define the objectives of your evaluation clearly and specifically. What information do you need? The instruments you select must be appropriate for the questions you ask.

Consider the following four categories of evaluation information:

—Learner acceptance and reaction. How do the learners perceive and feel about the program?

—Achievement of learning objectives. At the end of instruction, did the learners achieve the defined performance objectives?

—Change in behavior. Did the program help change the learners' work performance?

—Benefits to the organization. Has the program resulted in measurable benefits to the organization—such as decreases in absenteeism or turnover, cost savings, increased production, or lower error rates?

▼ *Does the Instrument Suit the Evaluation Design?* Which instrument will allow you to make the necessary comparisons to answer your evaluation questions? Will you compare a group that received training to a group that did not receive training? Will you compare a group to itself before and after training? Or if you're not interested in prior per-

A face-to-face interview involves an individual responding orally to oral questions asked by another person. In the structured version, the interviewer asks each interviewee the same list of predetermined questions. In the unstructured version, the interviewer typically begins with standardized questions but bases subsequent questions on the interviewee's responses to the previous questions.

Examples

▼ Structured interview question: *What is the most important thing you learned in the supervisory skills training class?*

▼ Unstructured interview question: *Please give me an example of how you handle the type of situation you described?*

Applications

▼ obtaining a comprehensive understanding of learners' reactions in pilot testing of course materials and content

▼ obtaining information from senior managers and executives—who are often unwilling to complete written questionnaires—regarding training programs in which they or their subordinates participated

▼ collecting more in-depth information from a subset of a questionnaire's original group of respondents

▼ collecting information from respondents either by building rapport with them or by lessening their anxiety

▼ obtaining information from trainees who have trouble completing written questionnaires or tests.

Advantages

▼ Skillful interviewers can elicit information that respondents would not provide in a questionnaire.

▼ Interviews allow much flexibility.

The interviewer can both pursue additional topics that arise during the interview and tailor the questions to the interviewee.

▼ The interviewer can observe the interviewee's gestures, tone of voice, and posture that reveal feelings and attitudes not available from a questionnaire.

▼ The interviewer can establish rapport with respondents—or determine when rapport has not been established, thus judging the reliability of the responses.

Disadvantages

▼ Interviews are time-consuming and expensive.

▼ Qualitative data can be difficult to analyze.

▼ Productive interviews require either skilled interviewers or extensive training for novices.

▼ Many biases are possible, from the interviewer or interviewee, situational variables, or the interaction of all of these. In unstructured interviews, it can be difficult to record the interviewee's responses accurately.

▼ Some people find interviews threatening and will reveal less information than in a questionnaire.

Guidelines

▼ Remember that structured interviews are more efficient and accurate for collecting factual information. Unstructured interviews provide the most in-depth information for complex or elusive issues.

▼ Consider basing questions on a rating scale with only a few responses. Limiting responses can make analysis easier, but you could lose much of the interview's flexibility.

▼ Prepare concise and clear questions that encourage interviewees to express their views.

▼ Be sure interviewers are well trained in asking questions, recording responses, building rapport, and easing interviewees' anxiety.

▼ Prepare an interviewer's guide with both structured questions and acceptable questions for probing further. Include a form on which the interviewer can record responses.

▼ Consider recording the interview if the interviewee does not feel threatened.

▼ Test the interview questions, and the response-recording and analysis techniques, by means of several practice runs.

▼ Tell interviewees the purpose of the interview, how the information will be used, and if their responses will be confidential.

---

formance, you might measure just the amount of learning that took place.

▼ *Is the Instrument Valid?* Will the instrument accurately measure the perceptions, behavior, or benefits you are trying to measure? The closer the measurement simulates the objective—such as a targeted behavior—the more valid the instrument is.

▼ *Is the Instrument Reliable?* Will the instrument provide consistent information? Would you get the same results if you took a second measure-

ment? The instrument must provide a stable measurement over time.

▼ *Is the Instrument Practical?* Carefully consider the instrument's suitability to your program and learners. Make sure the reading and vocabulary levels of printed questionnaires or tests are appropriate for your learners. Allocate sufficient time and resources to develop or purchase and administer the instrument. And remember to choose an instrument that is acceptable to program participants and to

readers of your evaluation report.

Keep these selection issues in mind as you examine the 10 evaluation instruments that follow (see Figures 1 to 10).

Anne F. Marrelli is a senior human resources management specialist in the Chief Administrative Office of the County of Los Angeles (500 W. Temple St., Room 555, Los Angeles, CA 90012; 213/974-2674).

A questionnaire is a printed or computerized form using one or more types of these questions: multiple-choice, ranking scale, rating scale, or open-ended. Respondents either record their answers on the printed form or use a computer keyboard, touch screen, or other input device.

Examples

▼ Multiple-choice. Respondents select a response from a list of two or more choices or check all applicable items in a list:

Which kind of educational presentation helps you learn best?

a. Lecture

b. Videotapes

c. Computer-based training

d. Interactive video

e. Don't know

▼ Ranking scale. Respondents order a list of items according to priority:

Five possible reasons for attending our training program today are listed below. Please rank them in order of importance to you by placing a 1 by the most important, a 2 by the second most important, and so on.

—— My supervisor told me to attend.

—— I need to learn the information for my job now.

—— I want the continuing education credit.

—— I want to further my professional development.

—— I want a break from the work routine.

▼ Rating scale. Respondents select the description that best matches their perceptions:

How would you rate the quality of training materials?

—— Very good

—— Good

—— Fair

—— Poor

—— Very poor

—— Don't know

▼ Open-ended. Respondents construct their own answers to questions:

What is your most common obstacle to applying what you learn in training to your job?

Applications

▼ assessing learners' reactions to courses, materials, and instructors either in pilot testing or after an actual training session

▼ measuring training-related changes in attitudes

▼ obtaining supervisors' assessment of training-related changes in subordinates' work performance

▼ obtaining trainees' reports of how training has affected their work performance

▼ obtaining instructors' estimates of the quality and amount of trainees' learning.

Advantages

▼ Questionnaires are convenient and inexpensive to administer to large groups of people, especially if the people are geographically dispersed.

▼ Questionnaires provide anonymity and encourage honest responses.

▼ Responses to multiple-choice and rating scale questionnaires are easy to summarize and analyze by using percentages or proportions.

▼ Questionnaires are flexible enough to use for either assessing attitudes or collecting facts.

▼ Users may easily standardize questions for all respondents or tailor questions for subgroups.

Disadvantages

▼ The quality of information depends on respondents' honesty, awareness, memory, and perception.

▼ Users have no assurance that respondents understand the questions or know enough to respond appropriately.

▼ Low return rates are common unless respondents are strongly motivated.

▼ Responses to open-ended questions are difficult to summarize and analyze.

Guidelines

▼ Precisely define the information you need before writing questions.

▼ Use short questions because they are easier to understand.

▼ Avoid negative terms such as "not" because they are often overlooked.

▼ Avoid jargon or technical terms.

▼ Avoid questions that lead people to a particular answer.

▼ Avoid potentially threatening questions.

▼ Include in the instructions an explanation of the survey's purpose and how the information will be used.

▼ Assure respondents of anonymity.

▼ Combine open-ended questions with multiple-choice and ranking questions to let respondents explain their answers. Such combinations also help you interpret the data more accurately.

▼ Use a pilot group to test the clarity of questions and instructions.

▼ Hold a group orientation session prior to distributing the questionnaire. Use the session to encourage respondents to complete the questionnaire and give it their best effort.

A facilitator meets with a group of five to 12 people to collect in-depth qualitative information about the group's perceptions, attitudes, and experiences. The group discusses the issues on a planned agenda following specific procedures. Before the group meets, methods for recording, reviewing, and synthesizing discussion comments should be established.

Examples

Discussion group questions:
How did the role-playing exercise on dealing with hostile employees help you?
How effective do you think the diversity training program was in defusing multicultural conflicts among your staff?

Applications

▼ assessing learners' reactions to course materials, exercises, or instructional techniques in pilot trials
▼ evaluating participants' initial perceptions immediately after training
▼ soliciting previous participants' suggestions for improving a training program
▼ discovering in follow-up evaluations how trainees feel the training program affected their work performance
▼ obtaining input separately from diverse groups on their reactions to pilot training sessions to ensure that the training is appropriate for all members in each group.

Advantages

▼ Group discussions are inexpensive and require minimal preparation time.
▼ In the group process, individuals build on each other's ideas and comments to provide an in-depth view not attainable from questioning people individually.
▼ Unexpected comments and new perspectives can be easily explored.
▼ The rapport the facilitator builds with the group and the rapport the group members build with each other can encourage participants to express their feelings fully.

Disadvantages

▼ Group discussions produce a large amount of qualitative data that is often difficult to analyze.
▼ The information collected is subjective.
▼ More outspoken individuals tend to dominate the discussions. Viewpoints of less-assertive people are often difficult to assess.

▼ Both the quality of the discussion and the usefulness of the information depend on the skill of the facilitator.

Guidelines

▼ Carefully plan the discussion questions and discussion procedures, as well as data collection and analysis procedures.
▼ Choose a facilitator who knows how to work with a group to encourage full participation and interaction among members.
▼ A person acting as an observer or recorder should record all comments made by the group, and note any significant gestures or behaviors.
▼ Select group members who represent the target population.
▼ Select group members who are at comparable salaries and power levels in the organization.
▼ Limit group size to between five and 12 people. A group this size allows everyone to participate, yet provides enough diversity of opinion for a well-rounded perspective.

Critical incident reports are oral or written anecdotes of key events that demonstrate the training program's value. These events can be included in a questionnaire or interview, or requested separately.

Example

Here is a typical critical incident report question:
Could you give me an example of how your customer service training helped you cope with a very difficult customer?

Applications

▼ discovering how training affected people's day-to-day work behavior
▼ starting group discussions, perhaps by having each person relate one critical incident
▼ complementing the objectivity of quantitative measures with personal experiences

▼ using a small number of respondents to develop the questions for a questionnaire.

Advantages

▼ The reports provide an in-depth look at the differences that training creates in work behavior, either positive or negative.
▼ Information in the reports can give managers or others without educational expertise a sense of the training's value.
▼ Little preparation time is required to develop or administer reports.

Disadvantages

▼ Reports are purely subjective, open to the personal biases and goals of the respondents.
▼ The data collected are difficult to summarize and analyze.

Guidelines

▼ Ask general, open-ended questions to encourage respondents to select the incidents they see as most important.
▼ Avoid using critical incident reports as the primary evaluation tool—but they are a good supplement to more objective instruments.

WORK DIARIES

FIGURE 5

Participants fill in a daily work log or diary listing their activities and the times started and stopped for each. They make entries in the diaries for several consecutive days, weeks, or months, depending on the variation in job activity.

Example
Here is a sample excerpt from a work diary completed by a supervisor:

Monday, June 14, 1993

8:00 - 8:15	*Opened and read mail*
8:15 - 8:18	*Subordinate walked in to ask for time off*
8:18 - 8:25	*Returned two telephone calls*
8:25 - 8:40	*Looked for subordinate's personnel file*
8:40 - 8:41	*Secretary reminded me of 9 a.m. meeting.*

Applications
▼ measuring changes in work patterns before and after training, such as the number of minutes required to complete tasks or the incidence of desired behaviors
▼ providing an overall picture of how training has altered the daily work of individual staff members.

Advantages
▼ Diaries provide valuable insights into how training affects people's work on a daily and individual basis.
▼ Most people can easily understand and fill out diaries.

Disadvantages
▼ Reporting errors are possible.
▼ Making diary entries can be tedious.
▼ The information collected is difficult to summarize.

Guidelines
▼ Remember that the diary must be maintained longer when work activities vary.
▼ Keep the log's design as simple as possible.
▼ Remember that a work diary is rarely 100 percent accurate. Use the information as an approximation of actual work activities.

PERFORMANCE RECORDS

FIGURE 6

Performance records evaluate a training program's effect on work behavior. They include existing records routinely maintained as part of work procedures or records expressly developed for evaluation purposes. Various data—covering aspects such as quality of work, amounts produced, costs incurred, revenue generated, or time required to complete tasks—are measured before and after training.

Examples
Performance records come in various forms:
▼ personnel records—absenteeism, turnover, terminations, grievances
▼ productivity records—output, defects, complaints, backlog, equipment use
▼ cost and revenue records—amount of sales, cost reductions, overhead costs.

Applications
▼ measuring changes in work performance before and after training to estimate the training's return on investment
▼ measuring changes in work performance before and after training to evaluate the transference of new skills to the job for individual trainees.

Advantages
▼ Performance records offer a direct measure of work performance.
▼ The information is easy to collect and analyze.
▼ The results are easy to understand.
▼ Performance records are objective and reliable.
▼ Data collection is part of the regular work routine and does not disrupt the work environment.

Disadvantages
▼ Because they provide limited information about the total impact of a training program, performance records work best in combination with other evaluation methods.
▼ Developing and maintaining new records can be expensive.
▼ Inaccuracies in recording and reporting data are common.
▼ The records may reflect many variables other than the training.

Guidelines
▼ Review existing performance records to identify any that will supply the information to meet your evaluation objectives. In some cases you may be able to extract relevant data from records that contain more information than you need.
▼ Work with the people who will keep any new records to be sure they understand how and when to record information.
▼ Count the incidence of relevant events before and after the training program, making sure that the length of time for the count is the same both before and after training.
▼ Use a long enough counting period to avoid seasonal, cyclic, or chance fluctuations.
▼ Use any of the following to analyze data: numerical counts, percentages or proportions, dichotomous categories, time studies, or other forms of quantitative comparisons.

FIGURE 7

SIMULATIONS/ROLE-PLAYS

Participants enact real-life situations to demonstrate their mastery of the skills and knowledge presented in training. The situation should be similar to challenges the participants will face on-the-job, with participants assigned specific roles to play. The evaluation may focus on group interactions (such as team problem solving) or on individual behavior (such as a role-play in which the individual being evaluated plays a key role while actors play the other roles needed to evoke a targeted behavior).

Example
At the conclusion of a train-the-trainer program designed to help new instructors develop classroom presentation skills, each participant prepares and presents a short lesson to the other participants.

Applications
▼ evaluating trainees' mastery of interpersonal skills such as management and communication skills
▼ evaluating a complex set of skills presented in a lengthy training program.

Advantages
▼ Simulations and role-plays offer an excellent method of evaluating complex interpersonal skills because the actual behavior is demonstrated.

▼ Most trainees see simulations and role-plays as valid indicators of their mastery of skills.
▼ The evaluator controls the types of behavior likely to occur, thus evaluating specific skills.

Disadvantages
▼ The simulated situation may not accurately predict how a person would behave in a real-life situation.
▼ Simulations and role-plays are time-consuming to develop and administer, and they require trained observers to score.
▼ Some participants find simulations and role playing threatening.

Guidelines
▼ Design the simulation or role-play to mirror actual simulations as closely as possible.

▼ Provide clear instructions for administrators, participants, and evaluators.
▼ Choose evaluators who are carefully trained in observation techniques, including analyzing and recording behavior.
▼ Give evaluators a standard form—such as a behavior checklist, rating scale, or behavior coding form—on which to record their observations.
▼ Minimize anxiety for trainees who do not enjoy being in front of a group by having them perform their role-play with only the most essential participants present—unless the behavior being simulated involves group interaction.
▼ Give all participants a private oral and written review of their performance.

FIGURE 8

OBSERVATIONS

The work behavior of trainees is observed before, during, or after training. A trained observer watches and records the behavior. Sometimes, the behavior is videotaped or audiotaped to study later.

Example
A training program is designed to help supervisors conduct better performance appraisal interviews. An observer sits in on several interviews with a supervisor—both before and after the training program—to measure any changes in how he or she conducts the interviews.

Applications
▼ measuring changes in behavior from before to after training
▼ evaluating trainees' mastery of the skills taught in training
▼ evaluating proficiency in physical skills.

Advantages
▼ Observation is an excellent method of evaluating behavior change because actual behavior is measured.
▼ The trainee's interactions with others can be evaluated.
▼ Verbal and nonverbal behavior can be evaluated.
▼ Observation offers direct measures of real behavior.

Disadvantages
▼ The presence of an observer may modify the participants' typical behavior.
▼ Some participants are uncomfortable being observed.

▼ A poorly trained observer may collect unreliable data because of biases or poor recording techniques.
▼ Observations are expensive and time-consuming.

Guidelines
▼ Choose observers who are carefully trained in observation techniques, including analyzing and recording behavior.
▼ Give observers a standard form—such as a behavior checklist, rating scale, or behavior coding form—on which to record their observations.
▼ Minimize the presence of the observer—but explain the presence of the observer to participants.

A written test is a printed or computer-based series of questions designed to evaluate the respondent's knowledge or skills. Such tests include one or more of these types of questions: true/false, multiple choice, matching, short answer, or essay. Respondents record their answers on the printed form or in the computer.

Computer-based tests sometimes use interactive video technology to display still pictures or live action video sequences that simulate on-the-job situations to which the respondent reacts by answering questions.

Examples

Written tests come in various forms:

▼ True/False. Respondents indicate "true" or "false" for each statement:

A test is valid if it measures what you are trying to measure.

▼ Multiple-choice. Respondents select the answer of their choice:

The developers of a new selection test for clerks compare the test scores of clerks with their job performance ratings one year after being hired. Which type of validity are the test developers measuring?

A. Face validity
B. Content validity
C. Criterion validity
D. Construct validity

▼ Matching. Respondents match corresponding items in two lists:

Match the type of validity with the example of validation strategy by writing the letter of your choice in the blank following the example.

A. Face validity
B. Content validity
C. Criterion validity
D. Construct validity

1. Trainee scores on a new test of reasoning ability are correlated with their scores on a standardized reasoning test. ___

2. The developers of an employment selection test for clerks compare the selection test scores of clerks with their job performance ratings one year after being hired. ___

▼ Short answer. Respondents write an answer of a few words or sentences:

Name the four types of validation strategies.

▼ Essay. Respondents write one or more paragraphs in answer to the question:

Compare and contrast the four types of validity.

Applications

▼ measuring trainees' acquisition of knowledge

▼ evaluating written communication skills such as letter, memo, and report writing; outlining; documentation; and project and task planning.

Advantages

▼ Most written tests are easy and inexpensive to administer and score for large groups.

▼ Written tests are a quick method of assessing knowledge.

▼ Experienced test developers can design written tests to simulate real-world conditions.

▼ Most written tests provide objective assessment.

Disadvantages

▼ Written tests indirectly measure skills, so they may not accurately indicate the application of knowledge and skills to actual job performance.

▼ Written tests are often hastily developed by someone lacking the required knowledge and skills.

▼ Essay tests can be difficult to score.

▼ Most written tests are not suitable for someone with poor reading or writing ability. Interactive video tests are an exception. They permit the test-taker to respond to oral questions by touching the selected answer on the screen.

▼ Written tests make many people anxious.

Guidelines

▼ Carefully define the knowledge and skills you wish to evaluate and compare these with the content of the test, especially when purchasing a commercial test.

▼ Use professional test developers when developing your own test.

▼ Use criterion-referenced tests (in which trainee performance is compared to predetermined standards or objectives) rather than norm-referenced tests (in which trainee performance is compared to the performance of other trainees).

▼ Run a pilot trial of the test content and instructions with a small group.

Performance tests simulate work experience by letting trainees demonstrate their actual skills and knowledge on assigned tasks. Performance tests can evaluate both verbal and nonverbal skills.

Examples

Performance tests come in various forms:

▼ An in-basket test assesses verbal skills. For example, a trainee who has completed a supervision class may be presented with the simulated contents of a supervisor's daily in-basket. The contents might include requests for time off, employee grievances, customer complaints, memos from the boss, and the like. The trainee reads and prioritizes each item, prepares responses in the form of memos and letters, and notes other actions to take.

▼ Nonverbal skills are assessed in a skills test in which computer repair technicians are given a defective computer and asked to identify and repair the problems.

Applications

▼ evaluating trainees' ability to apply learned knowledge and skills

▼ assessing physical skill performance

▼ evaluating trainees' ability to perform complex combinations of skills.

Advantages

▼ The testing situation closely resembles actual work demands.

▼ Measuring proficiency in physical skills is easy.

▼ Trainees and evaluation report readers accept performance tests as valid measures because they directly measure realistic behavior.

Disadvantages

▼ Performance tests are often expensive to develop, administer, and score—as well as time-consuming.

▼ Financial or time constraints can make performance tests impractical to administer.

Guidelines

▼ Make the test closely resemble the actual work situation for greater accuracy in evaluating performance.

▼ Design a series of performance tests to evaluate a constellation of skills for when you need a comprehensive training program.

▼ Design performance standards for expected products and processes, acceptable solutions to the problems posed, and a method of recording and scoring trainee performance.

▼ Choose well-trained observers and give them a checklist that covers the objectives, standards, and scoring procedures.

▼ Provide clear instructions to the trainee. Include objectives, performance standards, and scoring methods.

Training Investments:

The Global Perspective

By Edward E. Gordon

U.S. manufacturers have a lot to learn from their counterparts in Europe and Japan, where investments in worker training are paving the way to profitability. Moving to a return-on-investment emphasis will help steer U.S. businesses on the same path.

In This Story

▼ competitiveness
▼ technical training trends
▼ return on investment

Corporate America has been on a diet. Businesses are becoming lean and reengineering, outsourcing, and downsizing have contributed to record profits. A recent *Wall Street Journal* article suggests, however, that the shrinking corporation has gone too far, in many cases producing "corporate anorexia." There is a great danger here that companies can gradually lose their ability to grow. "You can't shrink to greatness," says Jim Stanford, president of Petro-Canada, based in Calgary, Alberta.

Training has been particularly hard-hit during this present business cycle. Even though sales and profits have increased, corporate training and development often face a struggle merely to survive. The immediate return-on-investment practices of American corporations are threatening contemporary workplace innovation. American managers do not equate investment in human capital (through education and training), which has the potential for producing dazzling innovations, with capital investment in equipment or buildings.

Many U.S. business leaders now believe that it is too costly to train or reeducate the American worker. The chief concern that seems to drive most American boardrooms to this conclusion is the quarterly financial return. In the short term, it seems far cheaper for many companies either to relocate overseas where they can find well-educated, technically ready employees, or to sell off their technology for a quick profit.

Unless this pattern changes, the United States will develop into a two-tier society

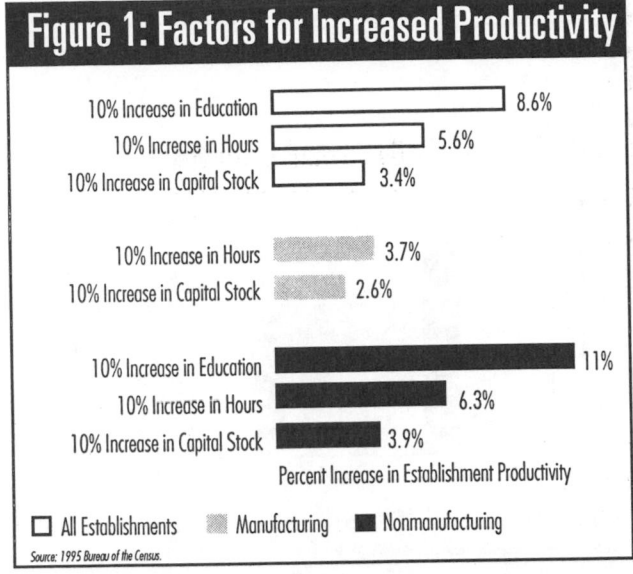

Figure 1: Factors for Increased Productivity

- 10% Increase in Education — 8.6%
- 10% Increase in Hours — 5.6%
- 10% Increase in Capital Stock — 3.4%

- 10% Increase in Hours — 3.7%
- 10% Increase in Capital Stock — 2.6%

- 10% Increase in Education — 11%
- 10% Increase in Hours — 6.3%
- 10% Increase in Capital Stock — 3.9%

Percent Increase in Establishment Productivity

☐ All Establishments ▨ Manufacturing ■ Nonmanufacturing

Source: 1995 Bureau of the Census.

of high-wage/high-educated employees versus low-wage/low-educated workers. Left unchecked, this manufacturing trend will gradually undermine the American middle class and its consumer-driven economic system. Signs of this change have already begun: witness the ongoing wage erosion of middle-class workers, who comprise the bulk of U.S. consumers. With middle-class buying power waning, who will be left to purchase these products being made overseas? Will U.S. business continue to thrive regardless of these long-term economic shifts? Or will American manufacturing be bought off by foreign investors, leaving the United States to become another "post-industrial" nation like Great Britain, where Jaguar is owned by Ford and Rover by BMW?

Is the average German, Japanese, or Korean manufacturing worker innately more intelligent or more technically astute than the average American worker? The many case studies of financially successful skills training and educational programs within American companies suggest otherwise.

Simply put, American manufacturers need to learn from their chief foreign competitors that their employees are key resources capable of development. By adopting a workforce education policy, any organization can harness a proven economic-value-added concept: Human knowledge equals profit.

Competitive Practices

German and Japanese track records for training expenditures during a genuine market downturn are the exact opposite of that of the United States. Our chief foreign competitors increase training investments for employees as they reorganize/retool for the new services/products demanded by the subsequent business cycle. Employees have more time available to be trained. The business culture sees leveraging an organization's "intellectual capital" as a principal management strategy.

Unlike the United States, neither Germany nor Japan possess abundant natural resources. Businesses see people development as their most important national asset. They do not believe a company's competitiveness will be guaranteed by slicing, dicing, and chopping. Instead they seek to better manage by leveraging collective organizational knowledge. Entire business sectors cooperate to educate the next generation of workers, or reeducate the current workforce. Even BMW and Mercedes have learned how to collaborate in this manner without sharing their proprietary secrets.

U.S. Manufacturing Practices

In sharp contrast with the German and Japanese examples, a newly released U.S. Department of Labor survey of nearly 12,000 U.S. businesses found that, as of 1993, fewer than half were offering their workers the formal job-skills training essential for improved productivity. Worse yet, less than three percent were offering training in basic reading, writing, math, or English as a second language.

Equally depressing news comes from

Our chief foreign competitors increase training investments for employees as they reorganize

the National Education Goals Panel, which estimates that only 52 percent of the U.S. adult population scored at level 3 or higher (high school/college) on the prose portion of the National Adult Literacy Survey (NALS) conducted in 1992. Nearly half of all American adults are not able to perform the range of complex literacy tasks (i.e., reading and comprehending at grade 13) considered important for successful U.S. competition in a global economy.

We do not have to look far to see how these results are playing out in daily American life. According to recently released U.S. Census Bureau data, the median income of the average American family has declined each year since 1991. This trend persists despite recent increases in employment, productivity, and total national income. If current thinking regarding worker training does not change, how will the middle class be able to drive the American economic machine? It seems that hard times are ahead for many American businesses.

Why do so few manufacturers attempt to reeducate their own workers in light of these potentially grave economic consequences? There seem to be three major roadblocks:

▼ Providers of skills training and the results of such training lack credibility.

▼ Management is skeptical about whether the concept of economic-value-added or return-on-investment methodologies can accurately measure the monetary results of education programs.

▼ Manufacturers' accounting practices continue to show training as a cost (charge against earnings) rather than a capital investment (plant/equipment) that can be depreciated over time.

Credibility Factor

In the minds of senior management, worker retraining programs have produced poor results. Many manufacturers don't think they work. Basic skills programs in particular are viewed as social service programs. This management viewpoint continues to accelerate U.S. businesses' flight to overseas locations. A 70-mile corridor between Glasgow and Edinburgh, Scotland is now home to more than 300 American manufacturing plants. U.S. high-tech manufacturers are increasingly conducting operations in Singapore, Korea, China, Ireland, and Holland due to the availability of better educated workers. Of course, labor costs are also a factor. However, average U.S. labor costs are now below those of many European nations.

Most skills training programs offered by American manufacturers exemplify a lack of confidence that the average American employee can be reeducated for the high-tech workplace. The majority of these educational programs are not conducted by the same professional sources that have long provided American businesses with their training and development programs.

Almost $50 billion is spent annually on management training programs, which are provided by in-house corporate trainers, outside consultants, such learning technologies as multimedia, and academic and governmental resources. Yet, the majority of basic skills programs are established through some form of governmental grant or subsidy, and are often sustained only as long as a cheap training source is available. This has created a Catch 22 of sorts, since low-cost training usually produces poor results.

This was clearly not the intent of many well-intentioned governmental grant programs. Such programs were established to encourage businesses to see the worth of employee skill improvement programs, so that they would then adopt them as part of their overall training and development budgets.

Today's reality is far different. Many companies only seem interested in beginning or continuing skills training if it remains heavily subsidized by a governmental agency. Furthermore, companies fail to see these programs as permanent features of doing business in a high-tech

environment, unlike their view of management development.

Retraining American workers will remain difficult until American manufacturers use the same variety of quality providers for technical and skills training they now use for management training. This will happen only when we can realistically measure the economic results of quality worker education programs.

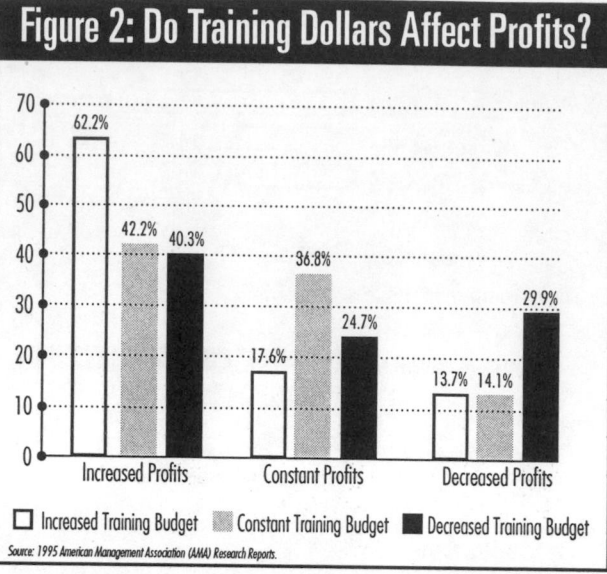

Figure 2: Do Training Dollars Affect Profits?

Increased Profits: Increased Training Budget 62.2%, Constant Training Budget 42.2%, Decreased Training Budget 40.3%
Constant Profits: Increased Training Budget 17.6%, Constant Training Budget 36.8%, Decreased Training Budget 24.7%
Decreased Profits: Increased Training Budget 13.7%, Constant Training Budget 14.1%, Decreased Training Budget 29.9%

☐ Increased Training Budget   ▨ Constant Training Budget   ■ Decreased Training Budget

Source: 1995 American Management Association (AMA) Research Reports.

Changing the Economic Rules

There are many positive signals supporting a basic change in the economic rules regarding investment in training. A major 1995 U.S. Census Bureau survey demonstrated that a direct connection can be made between employee educational gains and increased company profit. This first detailed American business survey of its kind polled managers at approximately 3,000 establishments with 20 or more employees. The National Center on the Educational Quality of the Workforce at the University of Pennsylvania designed the survey to document the degree to which education is linked to productivity. Specifically, the survey asked about workers' educational levels and various measures of output.

The results showed that a 10 percent increase in the average education of all workers within an organization (equivalent to slightly more than one additional year of schooling) is associated with an 8.6 percent increase in output for all industries. This effort rises to 11 percent for the nonmanufacturing sector—nearly three times the boost for a similar increase in capital stock.

By way of comparison, the study also found that a 10 percent increase in the value of such capital stock as tools, buildings, and machinery produced only a 3.4 percent increase in productivity. (See Figure 1.)

A recent American Management Asso-

ciation (AMA) survey also reported a correlation between increased training budgets and improved productivity. (See Figure 2.) Among the findings:

▼ Of the companies that showed increased profits after downsizing or restructuring, 62.6 percent had increased their training budget.

▼ Of the companies that saw their profits decline after downsizing or restructuring, 29.9 percent had cut their training budgets.

This strongly indicates that the amount an organization invests or does not invest in training has a positive or adverse affect on profits. Linked to the Census Bureau survey, this is good news. There seems to be a clear relationship between increasing employee knowledge and increasing profit.

The current trend of measuring costs throughout a business means that senior executives are attempting to determine the economic value added by every department—including training. However, senior managers tend to see training and education as an 'ivory tower.' Finance managers also argue that it is not realistic to measure long-term, quantitative effects of training and educational programs.

I beg to differ. Here are several systems that trainers can use to calculate realistic economic value added (EVA) for their company's employee education and training programs.

Utility analysis (added value). How much is any training program worth? Utility analysis is a hard-data method for determining return on investment (ROI) for training by calculating the value of an intervention (i.e., a training program) minus its cost. Economic gain equals the training effect times the monetary value of that effect. This produces the following utility analysis equation:

Where $F = N[(E \times M)-C]$

F = financial utility
N = number of people trained
E = effect of the training on the business
M = monetary value of the training effect
C = cost of the training per person.

Both Wayne F. Cascio (*Training and Development in Organizations,* Jossey-Bass, 1989) and Michael Godkewitsch (*Training,* May 1987) have written clear and detailed descriptions of how to use the utility analysis method over a broad range of training content and methods.

Time-value of money. The National Planning Association has addressed training ROI in Frederick W. Crawford's and Simon Webley's *Continuing Education and Training of the Workforce* (British-North American Commitee, Issue Paper No. 1, 1992). The added-value approach described is determined by calculating the opportunity cost of training. The authors offer a set of economic formulas that compare a company's investment in a specific training program to other potential forms of investment. They give a step-by-step method for a company's financial officer to determine if the training productivity increase will give a greater ROI than investing the company's money in capital improvements, commercial paper, etc.

Performance value. A third method is offered by Richard A. Swanson and Deane B. Gradous in *Forecasting Financial Benefits of Human Resource Development* (Jossey-Bass, 1988). They argue that it is better to forecast the potential results than to later evaluate the effect of training. Performance value helps to choose among training program options before investing in any program, rather than waiting to evaluate after the training has been completed.

This forecasting approach forces a business to determine the fiscal value of the operational problems to be addressed by the training program. How will quality, time, costs, and output issues translate into specific quantifiable training results? The authors offer a worksheet to calculate the performance value of a training program.

How Practical?

To measure ROI effectively, trainers first must receive approval from senior management and their financial department for a specific financial analysis model and an acceptable rate of return. Remember that achieving a good, feasible rate of return (profit) in senior management's eyes is what the business is all about. Bringing the training department on board as another revenue center rather than just another cost means you are finally providing the monetary proof that investing in people will be profitable. This holds equally true for either basic skill programs or more traditional management development programs.

Total quality management (TQM), ISO 9000, and other quality efforts are areas in which ROI for training is gaining widespread interest. Poor quality in products or services may often cost tens of percents of economic value added. High potential training payoffs will be reached only if the specific education programs precisely uncover local TQM operational problems. European Union (EU) members and the Japanese have demonstrated clearly how high productivity gains can become by using training to correct substantial business problems.

Changing Accounting Procedures

The final step to applying EVA successfully in a manufacturing environment is to change how training dollars appear in the company ledger. Training needs to be viewed by management as a long-term, ongoing, capital investment similar to plant and equipment, rather than a short-term operating cost charged against quarterly earnings. In other words, training must be moved from the cost side of the ledger sheet to the capital investment side. To some extent, manufacturers have been doing this for many years.

For a new equipment purchase, accountants generally establish a depreciation schedule that covers its useful life. Interviews with manufacturing managers have uncovered the common industrial practice of including related employee technical training costs in the purchase

price of the equipment. This is an acceptable accounting practice. However, other associated training costs also need to be added. For example, if a new equipment purchase requires employees to use statistical process control (SPC), such training also should be considered part of the depreciated capital investment. The same is true if your employees now require algebra, geometry, or trigonometry to operate new equipment. In many cases, even higher levels of reading comprehension or written or oral communication skills are essential in a high-performance work organization. The associated computer-driven or robotic equipment will increase a manufacturer's productivity and profit only if employees use it to the full extent of its design capabilities. Too often, limited employee know-how can undermine a major capital equipment/plant investment.

Future EVA Practice

At a recent meeting of the Conference Board, Phyllis Eisen, a senior policy director for the National Association of Manufacturers (NAM), announced a major project for members to report training EVA/ROI results back to NAM. This effort is designed to help determine simple, usable EVA procedures that will encourage manufacturers to use education in order to increase productivity and profits.

To make EVA happen, manufacturers need the close collaboration of their managers, trainers, engineers, and accountants. Their joint goal: to quantify increased employee knowledge in relation to increased company profit. Hopefully, this EVA effort will help begin a new era of human capital investment, ensuring that U.S. employees remain highly competitive and that ongoing worker education makes a significant contribution to every organization's bottom line.

Edward E. Gordon is president of Imperial Corporate Training & Development (10341 S. Lawler Avenue, Oak Lawn, IL 60453-4714; 312/881-3700; fax: 708/424-9956). He is the author of *FutureWork:The Revolution Reshaping American Business,* and *Closing the Literacy Gap in American Business.*

The Team Approach to Formative Evaluation

This strategy helps keep instructional design on time and within budget.

By James Heideman

Evaluation is both a systematic process and a systemic part of all training development. In the instructional systems design (ISD) model for developing training, two types of evaluation occur during training design: end-of-course evaluation and formative evaluation (see accompanying figure).

End-of-course evaluation, sometimes referred to as summative evaluation, occurs in step 5 after training has been completed. This evaluation helps assess the worth of the training intervention.

In conducting end-of-course evaluations, organizations typically consider three factors: quality of the training materials, timeliness of delivery, and adherence to budget. But sometimes these factors become trivialized. As a result, organizations that need training are often offered programs that promise to be "good, fast, and cheap," but really only deliver two of the three. Sadly, this suggests that the development of training programs is really a zero-sum game in which we must select the factors we can do without.

If we accept this notion, we're presented with a clear dilemma. Training must have inherent value. It must benefit the organization in some measurable way, such as reducing or eliminating job errors, increasing productivity, introducing greater efficiencies in job performance, etc. Organizations also demand that training must stick to an established time frame for development and delivery, while staying within budget. So which essential factor will be sacrificed? Do time and budget concerns outweigh the necessity for an ideal level of training quality? Or can we have it all?

In This Story
▼ evaluation
▼ instructional systems design
▼ teamwork
▼ project management

I believe we can have it all. Instructional systems design offers an effective strategy for delivering quality training on time and within budget because it provides a built-in means for monitoring the instructional design process. This monitoring system—called formative evaluation—is so critical to designing instruction that it constitutes a separate "step" in the ISD model.

In general, formative evaluation is a systematic process in which assessments are made while development takes place. This differs from the end-of-course evaluation in that it focuses on continuous improvements to the instructional materials and instructional design process itself. To further clarify, consider the following examples of formative evaluation activities.

During needs assessment, learner needs are defined so that instructional materials bridge the gap between existing and desired levels of learner performance. When you conduct a formative evaluation of your needs assessment, you step back from gathering information to ask the following sorts of questions:

▼ Have the best information sources been contacted?

▼ Have data been collected from enough sources?

▼ Are the data consistent from one source to another?

▼ Are the needs defined in sufficient detail for the learners?

By conducting a formative evaluation while the needs assessment is underway, you improve the instructional design by ensuring that subsequent steps are based on accurate and valid data.

During the design phase, formative evaluation is applied to the written training objectives around which you develop the course. Don't write training objectives only to set them aside as you rush on to development. Look at the objectives, return to your

Instructional Systems Design (ISD) Model

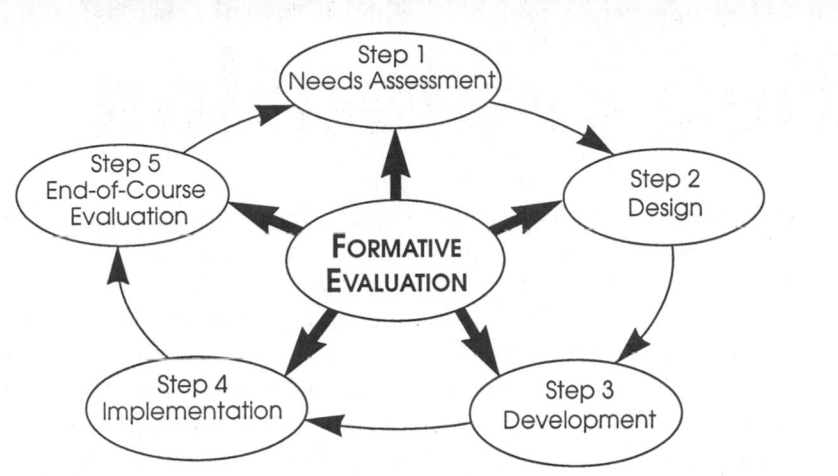

needs assessment data, and make comparisons. You might ask these questions:

▼ Are the objectives written in sufficient detail for the specified learning to occur?

▼ Are the objectives consistent with the needs identified?

▼ Have any needs been overlooked?

Formative evaluation requires that each step of instructional design be both consistent with its predecessor and form the basis for the next step.

These examples provide an idea of how formative evaluation binds the steps of the instructional design process together to produce efficient and effective training. For formative evaluation to deliver its full potential, however, it must be more than just a set of activities plotted on a production timeline. The perspective from which formative evaluation is viewed must focus on matching appropriate responses to concerns that emerge during instructional design.

The Strategy

While the instructional designer works mostly alone, effective instructional materials are not developed in a vacuum. As you plan your instructional design, you will gain the greatest benefits from formative evaluation by building an evaluation team. Assemble an ad-hoc team that includes instructors, subject matter experts, other instructional design professionals, and perhaps even training managers from the organization. The team's purpose is to provide valid input throughout the project.

But remember that team members have their own responsibilities and jobs. Do not try to monopolize their time. Instead, develop an accurate plan that details your expectations for each individual. A team consisting of three or four individuals with diverse skills should be enough to give you the information you need, while keeping input at a manageable level.

There are a few key points to remember about your team:

▼ Empower team members by making them your primary source of information for decision making. This does not limit you to their input alone, nor do you abdicate any responsibility. But if you have selected team members wisely, they will provide good judgment and quick answers to your formative evaluation questions.

▼ Build a team with diverse skills and talents. Selecting a mix of individuals with expertise in media production, instructional design, presentation skills, and knowledge of the subject matter will reduce the influence of the group's most vocal members. While selecting individuals with differing points of view may lead to disagreements, this is not necessarily so. As the principal figure in the instructional design process, you should handle conflicts to avoid unnecessary confrontations or alienating any team member. You may find that opinions cannot be reconciled, so allow all members to express their objections and suggestions. Through such discussions you may also find that team members are not as far apart as they first seemed. And such discussions may uncover ways to increase efficiency.

As the instructional designer, you are responsible for reaching closure, making the decision, and moving on. If you work to foster open, honest, and trustworthy relationships from the beginning, you will build a valid consensus, rather than reaching "group-think" decisions that might push the instructional design in the wrong direction.

The Concerns

There may be no right or wrong way to deliver instruction, only some ways that work better than others. Formative evaluation should be viewed in the same way. And remember that some formative evaluation activities may not fit neatly on your development timeline. For example, presenting training objectives to your evaluation team, reviewing the instructional media that will support the instruction, and conducting pilot tests of instructional materials are all parts of formative evaluation. These activities will assist in the delivery of quality training and will keep the project on time and within budget. But while these activities describe formative evaluation activities, you must also focus on the qualitative aspects of formative evaluation, manifested in three concerns: flexibility, validity and reliability, and specificity.

Flexibility

Remaining flexible while conducting a formative evaluation means recognizing that questions will arise during the instructional design process. The frequency of such questions will depend on numerous factors, including the following:

▼ how familiar you are with the content area of the training

▼ how well you know the target population for the instruction

▼ your confidence in the instructors' abilities to deliver the training

▼ the clarity of direction you've received from the significant decision makers who are supporting or underwriting the project.

Demonstrating a concern for flexibility also means not defaulting to previously used formative evaluation techniques. Every project is unique. It may be tempting to revert to familiar patterns of activities that worked comfortably in the past, and then quickly get back to the "real" business of instructional design. But you may miss some important cues or insights that need investigation.

Finally, be flexible as you respond to questions raised by team members. You might worry that using formative evaluation to uncover time-consuming problems will delay the scheduled delivery of materials to the client. But uncovering such problems early in the instructional design process actually helps increase production efficiencies. That's because it is more efficient to deal with small issues during the early stages of the design than to rework large parts following the pilot test.

Validity and Reliability

As you plan your formative evaluation, consider these two research terms: validity and reliability. Results are considered valid when they measure what you intended them to measure. Results are not valid when you use invalid sources of information. For example, to determine the tasks required for effective job performance by technicians who are employed by an automobile manufacturer, you should survey the manufacturer's training instructors, not high-school or college automotive instructors.

And don't think that your evaluation team is the sole source of valid information. Instead, interview trainees' supervisors, contact present or prior job incumbents, and if possible, review the organization's records pertaining to past job performance. But if you use past records, remember that written data can become quickly outdated, especially in emerging technologies. There can also be gaps or missing data concerning important elements in the training or work environment. If not addressed, these inaccuracies can adversely affect the acceptance and viability of your training materials.

Validity also involves asking the right questions and getting the answers you seek. Focus on the training outcomes you want to achieve, not on trivialities. For example, when you distribute preliminary training materials to team members, explain that you do not want comments on misspellings, grammar, and punctuation. Although these are important concerns, they are not valid if you want to determine the appropriateness of instructional tasks, sequencing of instructional events, and depth of content coverage.

Reliability is attained through sample size. It refers to the degree of consistency your data provide. To improve reliability, gather data from multiple sources and continually check the information. Seek comments from many people, even from people outside your team. This will yield a more complete picture of how well instructional materials will meet the trainees' needs. Continue to ask questions even if the first source gives you the answer you expect. Using limited sources of feedback or hearing only the good comments can easily take you in the wrong direction. Validity depends on reliability.

Specificity

Specificity involves evaluating little pieces of your instructional design. When you evaluate worksheets, test early versions, even if just partially outlined. Picking out errors and inconsistencies early can save you from repeating the same errors in subsequent worksheets.

This is particularly critical in situations where training involves a unique presentation technology or is structured around a new product or innovative process. While you may be tempted to keep the training program secret to enhance its dramatic effect when released, don't do it. Circulate the objectives as soon as they are written.

Specificity requires objectives that are accurate and concise statements of what trainees should be able to do. They must match the needs of your target population. For example, a course on locating the electrical circuit test points and connecting the meter leads will benefit no one if trainees do not have the necessary skills for meter scale selection and interpretation.

As you develop training content, verify the logic and sequence of the instruction with your team members. For specificity, have your evaluation team verify that all necessary information, instructions, and artwork are included. If the training is modular, you must also check the sequence of the information.

Though you may feel confident to do this yourself, get a reality check. Sometimes other team members can spot minor details that have slipped past. Don't wait for the course's pilot test to find these mistakes. Use the pilot test to "fine-tune" materials. You enhance your ability as an instructional designer and developer by conducting a pilot test that is free from major surprises.

Summary

Sometimes, formative evaluation is like the medicine we choke down when ill: The bad taste is quickly forgotten when physical well-being returns. Used often and in every phase of the design of training, formative evaluation can keep the development process healthy.

Can we have it all? Indeed we can. We have the tools, starting with sound instructional design principles and hard work. Will there be unexpected problems? Of course. Building an evaluation team and responding actively to the concerns for flexibility, validity, and specificity in the application of formative evaluation are actions you can take to address these concerns and deliver quality training on time and within budget.

James Heideman is a technical training project specialist for Nissan Motor Corporation. He may be reached at 10821 Holly Dr., Garden Grove, CA 92640; 714/530-5635.

How To Ensure Transfer of Training

BY PAUL L. GARAVAGLIA

HERE ARE SOME METHODS FOR MAKING SURE THAT THE MONEY YOU SPEND ON TRAINING ACTUALLY IMPROVES TRAINEES' JOB PERFORMANCE.

More than ever, training evaluation must demonstrate improved performance and financial results. As HRD professionals, we need to show organizations that they're getting good returns on their investments in training. To do that, we need to find out whether the skills and knowledge taught in training get transferred to the job.

We've been asking that question for more than 30 years. Still, studies show that only about 15 percent of companies measure training transfer, defined as the effective and continued application to trainees' jobs of the knowledge and skills gained in training.

One reason measuring training transfer has become so important is the rise in training costs. In 1992, organizations in the United States invested more than $45 billion in employee training, according to an October 1992 *Training* magazine report. Yet, few firms can show that their training expenditures result in observable behavior changes on the job. As HRD practitioners, we can't allow the 1990s to be remembered as the decade in which so many were trained at such a great cost with so few results.

The job of HRD practitioners is to assess the value of what participants gain from training and the extent to which training increases job productivity. In reality, training evaluation often assesses only whether immediate instructional objectives have been met—specifically, how many items trainees answer correctly on posttests.

That isn't enough. The people evaluating training must examine job-skills transfer and effectiveness.

Most traditional evaluation models are divided into four levels: reaction, learning, behavior, and results. The value of the information gathered at each level increases as the evaluation moves from measuring reaction to measuring results. Generally, results or follow-up evaluation is considered the most intensive level. But many organizations do only reaction-level evaluations.

In measuring training transfer through evaluation, consider the following aspects:
- why training transfer should be measured
- when transfer should be measured
- who should measure transfer
- how to measure transfer
- which instructional- and performance-technology techniques can be used to increase transfer.

Why measure transfer?

In order to demonstrate the value of training to an organization, it's important that the HRD department plan, budget for, and implement transfer measures. When a company is cutting costs, senior managers can start to view training as a frivolous expense. To avoid losing training dollars, the HRD department must demonstrate through evaluation the ways in which training improves productivity. Simply put, transfer measurements can provide data to justify training costs.

Another reason to measure transfer is to verify the effectiveness of training curriculums. Feedback from such evaluations can help trainers and instructional designers update and redesign training programs.

In *Evaluation: A Tool for Improving HRD Quality*, Nancy Dixon writes that measurements taken after training must answer three questions: How should the training program be changed? What kinds of assistance do trainees need after they return to their jobs? What prevents trainees from implementing what they learned in training?

Transfer measurements can account for and document the nature and extent of on-the-job transfer and also lead to measuring organization-wide results. Gaining a degree of competence matters only if the competence translates to improved job performance. Transfer measurements compare responses at the completion of training with responses in a later follow-up, thus gauging the longevity of new skills.

When should transfer be measured?

Trainers disagree on when to measure the transfer of training, including the time frame for the initial measurement, as well as intervals for ongoing assessment.

Some evaluators collect the initial data immediately after training ends; others wait one, three, or six months. Those who prefer to take the first measurement at three months tend to think it's important to give trainees enough time to apply the new skills at work, but not enough time to forget where they learned them. Generally, it's appropriate to measure the initial transfer of training three to 12 months after training, with six months being the most common time frame.

In order to measure the longevity of behavior changes, most training evaluators recommend follow-up transfer measures at six-month or yearly intervals.

Who does the measuring?

Many training staffs lack the skills, knowledge, and unbiased viewpoints about evaluation that are necessary to

perform one. Needless to say, a poorly developed or improperly administered evaluation instrument will provide inaccurate measures. Some training staffs neglect evaluation because they don't have the time or because they lack the support of senior managers. So it may be necessary to contract with professional evaluators to determine whether on-the-job behavior has changed.

The instructional designer who designed a training course should not be involved in evaluating it. A colleague or consultant who wasn't part of the original development process is likely to be more objective. Also, the trainers who conducted a course should not evaluate it. They tend to think that if trainees mastered a skill during training, then trainees are prepared to implement it on the job.

To save money, consider using in-house training staff to measure existing programs and external professional evaluators to measure new training programs, where internal staff are more likely to be biased. Trainees can create self-reports, and line managers and supervisors can measure the degree of transfer. After all, supervisors are the end-user clients.

How can transfer be measured?

You can choose from many methods for measuring transfer of training. Each needs to focus on the training outcome, which usually must be changed behavior.

One method for measuring changed behavior is to obtain reports from supervisors. Supervisors are in excellent positions to provide data about trainees' strengths and weaknesses. Also, supervisors can report changes in the duties and tasks trainees perform on the job. Requesting information from supervisors gets them involved in the process, and the data they provide tend to be relevant.

Supervisors can provide information that can help trainers revise training programs to meet managers' needs. But sometimes supervisors don't have the time or motivation to complete written reports. Also, the reports can be more subjective than objective. And it can be costly to con-

duct and control a reporting system.

An alternative method of evaluating behavior changes on the job is to conduct surveys and questionnaires. You can use different questionnaires for trainees and supervisors, or the same questionnaire. If you need further evaluation, try surveying trainees' peers.

Surveys and questionnaires provide different perspectives from all of the trainees and supervisors, without costing a lot in time or money. But sometimes questionnaires come back incomplete or inaccurate. Or they don't come back at all, particularly when they are mailed to potential respondents rather than distributed in person.

Another method for measuring transfer is to develop action or implementation plans. Before completing training, participants create action plans and send copies to their supervisors.

By creating the plans, trainees make contracts with themselves to implement the skills learned in training, thus increasing the likelihood that transfer will occur. By sending copies to their supervisors, trainees get them interested in their plans and progress. And after trainees have achieved their initial objectives, they can make additional plans of action.

Action plans are similar to performance contracts. They show what is to be done, by whom, and at what time in order to achieve goals. But it can be difficult to implement a plan and to get commitment throughout the entire process. And response rates tend to be low.

Another low-cost way to get evaluation information is to interview trainees and supervisors in order to validate certain evaluation findings. The interviews can be conducted with individuals or groups, face-to-face or on the telephone. Interviews are easily adapted to different jobs and departments, but it's not always easy to find time to schedule interviews.

Observation involves viewing trainees in their work areas in practice situations, actual on-the-job situations, or job simulations. Such "qualification sessions" enable evaluators to observe actual behavior in work situations.

Observation is very effective for repetitive tasks. And it avoids the bias that participants, supervisors, and managers can bring to reporting systems. But observers can be biased, too. Observations require constructed instruments in order to avoid observer bias.

Observations can be costly, time-consuming, and obtrusive, due to the fact that an on-site observer must be present. It's preferable that trainees don't know they're being evaluated; observers can use such unobtrusive measures as measuring the use of reference materials by looking for frayed edges, bent pages, and so forth.

Self-reports and performance appraisals are used less often than other approaches for measuring training transfer. Nevertheless, trainees can be accurate, sensitive judges of their own achievement levels in their self-reports. Performance appraisals can provide base lines with which to compare changes in on-the-job performance, as long as the appraisals are specific, accurate, and up-to-date.

No matter which method is used to measure training transfer, it's best also to measure the responses of a control group that didn't receive the training. That increases the validity of the evaluation, but it isn't always feasible.

How can transfer be increased?

In *Transfer of Training,* John Newstrom writes that 40 percent of skills learned in training are transferred immediately, 25 percent remain after six months, and only 15 percent remain a year later. Instructional designers can increase the likelihood of training transfer by building certain techniques into the training's design.

Using many different examples increases training transfer by presenting various contexts in which trainees can expect to use the skills and knowledge learned in training. In addition, varying the training setting shows trainees that a classroom isn't the only place in which knowledge and skills can be learned.

Analogies also help increase training transfer, by showing how important principles can apply in various situations. Training transfer tends to be better when learners understand the general principles behind the skills they are learning. Understanding

those principles can turn neophytes into experts, giving trainees a broader, deeper knowledge and more tools for problem solving.

For example, in *Training: Program Development and Evaluation,* Irwin Goldstein cites an experiment that shows how the transfer of principles improves the transfer of learning and skills. Two groups practiced shooting at an underwater target until each group was able to hit the target consistently. Then the depth of the target was changed, and one group was taught the principles of light refraction through water. In the next target-shooting session, the group that learned about refraction performed significantly better than the group that did not.

One innovative technique for teaching principles is computer simulation. Computer simulations help trainees learn principles that apply to the real world, not just to the simulations.

When trainees don't have relevant information about new skills and knowledge, or when they have information they don't realize is relevant, it's advisable to use "advance organizers"—relevant concepts that are taught to trainees before they learn the actual training material. For example, if the training content is about supply and demand, a relevant advance organizer might be a discussion of the high cost of advertising for the Super Bowl.

Though advance organizers have been known to produce mixed results, results tend to be positive when the organizers are properly designed and the effects are measured by a transfer test. Advance organizers provide a stable framework of knowledge to which new learning can be anchored.

Drill and practice techniques can help trainees reach a level of automatic implementation on the job. Intensive drilling is especially effective with tasks and environments that remain the same from situation to situation. When different elements of the task do vary, it's important that trainees practice in different situations. Training transfer can increase when trainees practice an original task, labeling and identifying its important features.

Using mnemonics (memory aids)

can increase transfer of training. Mnemonics create mental images that make instruction meaningful. Examples of mnemonics include connecting new information to something that is already known and forming bizarre, unusual, or exaggerated mental associations.

Visual displays of information boost training transfer. Trainees tend to group together in their memories elements that are similar or are presented in close proximity or in quick succession. Objects that are viewed together become linked in a learner's mind; a learner who remembers one of them is likely to remember the others, as well. Consequently, grouping tasks into one process helps trainees transfer their knowledge of that process to the job.

Transfer is more likely to occur when identical elements appear in two different situations. For instructional designers, that means that tasks taught in training should closely match the tasks people do on their jobs. The training's content and activities should reflect the real world.

Another technique that helps increase training transfer is letting participants explore the training content before training actually begins—through pretraining reading assignments, for example.

In situations in which instructor-led training isn't possible, a self-training system can be used to increase transfer. The training materials should include job aids to be used as on-the-job cues, activities to be completed on the job, and follow-up activities to be conducted after training ends.

Transfer increases with types of training in which learners produce real outcomes. This kind of learner-centered or problem-driven training can take place in a classroom—or the classroom can come to the learners in the form of on-the-job training.

At the end of a training session, trainees should write down the ways they've benefited from the training and the changes they anticipate making when they return to their jobs. They should put the information in self-addressed, stamped envelopes so that trainers can mail the envelopes to them one or two months later to provide positive reinforcement.

Performance technology and instructional design

Performance technology is the selection, analysis, design, development, and evaluation of programs, with the goal of influencing job behavior and performance in the most cost-effective way. To ensure training transfer, performance technologists look beyond instructional design.

Performance technologists depend on the participation of supervisors and senior managers to ensure that skills and knowledge learned in training transfer to employees' jobs. Traditionally, HRD practitioners have concentrated on improving training design. Now, they need to focus more on seeing that managers support training on the job.

Positive actions are more effective than negative actions for maintaining desired job behaviors. A strong correlation exists between training transfer and the quality and amount of managerial support. When supervisors use a positive approach, employees' behavior changes are still evident six to 12 months after training. When supervisors use a negative approach, behavior changes practically disappear in six months to a year.

Managers can show their participation and support of the training in several ways:

▶ making sure supervisors know the training requirements
▶ getting supervisors' input on the training content
▶ showing supervisors how to reinforce desired behaviors on the job
▶ telling supervisors the benefits and expected outcomes of the training
▶ enlisting supervisors' help in collecting evaluation data.

Supervisors should be given reports on the progress of the training and trainees. In addition, supervisors should change job-performance expectations to adjust for new skills and knowledge taught in training. They should plan practice activities to facilitate trainees' transition from training back to their jobs. And supervisors should assign new tasks that involve the training content.

Performance technologists focus on organizational systems that support new behaviors and skills. Often, the organizational system is more important than the training itself. Generally, 20 percent of critical job skills are learned from formal training and education; 80 percent are learned on the job or within organizational systems.

A training program may contain well-stated learning outcomes, appropriate media, excellent materials, and effective instruction. But if the training addresses the wrong performance area, the training won't be reinforced by consequences and feedback, it won't be supported by a well-designed work process, and it won't be linked to the organization's strategic direction. That kind of training isn't worth the investment.

Consequences and feedback, or rewards and incentives, should encourage trainees to apply on the job the behaviors they have learned in training. It may seem too obvious to say, but the performance of behaviors learned in training should not be discouraged or punished. When nonperformance is rewarded more than performance, or performance doesn't matter to trainees, it's essential for trainers or managers to create positive consequences.

Well-designed work processes can streamline the transfer of behaviors learned in training, especially behaviors that define roles and responsibilities, empower employees, and link with organizational values and culture. Well-designed work processes provide road maps for trainees to follow in transferring to their jobs the behaviors learned in training.

Having the proper equipment, tools, and materials in the work environment is crucial to the transfer of new skills and knowledge. Otherwise, it's unproductive for trainees to try to transfer behaviors learned in training to their jobs. In fact, it might even be unproductive for them to attend the training in the first place. Removing such obstacles will increase training transfer.

When learning doesn't transfer to the job, the two most likely reasons are that the work environment doesn't support the learned behavior and that trainees think the training was irrelevant. Organizations with environments that nurture training effectiveness circumvent the possibilities by providing safety nets for their employees.

One measurement of training transfer is actually part of the instructional-design process. The first phase is analysis, in which the company develops job-performance measures, to be compared later to actual job performance.

In the design phase, developers consider training objectives and design evaluation instruments. Then they develop training materials and evaluation instruments; in the testing phase, they run pilots or field tests of those materials and instruments. The final phase includes implementing the training and evaluation process and measuring training effectiveness.

Evaluators should take into account the variables that may prevent trainees from implementing what they learn in training. Knowing the obstacles, evaluators can make more informed choices about when and how to measure transfer.

Evaluating training transfer can be challenging. But training can have a significant effect on an organization, so measuring its results is worth the effort. It's up to HRD practitioners to increase training transfer through enhanced instructional materials and organizational systems. It's also up to us to ensure that evaluations justify training costs. We can become more knowledgeable about evaluation methods, or we can use the services of evaluation experts. It doesn't matter how we do it, as long as we do it.

Paul L. Garavaglia *is a performance analyst and instructional designer in the performance technology group at Electronic Data Systems, 26555 Evergreen, Suite 907, Box 5121, Southfield, MI 48086-5121.*

An Integrated Evaluation Model for HRD

If you evaluate your HRD programs only after the fact, you may miss the chance to nip problems in the bud. This integrated approach ties evaluation into every part of the HRD process.

By ROBERT O. BRINKERHOFF

Imagine this scenario: An HRD director is sitting in her office when the telephone rings. It is the chairman of the board, who says, "Say, some of the other directors and I were just sitting around, and one of them asked if all the fancy HRD programs we have around here actually do something and whether it makes any difference. Seemed like an interesting question, so we decided to ask you to sort of drop by, say sometime in the next 10 minutes, and just bring along any data you have that shows what we get out of our HRD effort. We're up in . . . Hello? Hello? . . . Are you there? . . . Hello?"

This fictitious scene highlights what I think is a common and unfortunate condition in HRD today: the HRD manager up the creek without an all-in-one model for measuring program quality.

That's a tall order, but such a model *does* exist. It's an evaluation system that integrates with needs analysis, design, and delivery. Armed with information produced by this model, our heroine could have raced gleefully up to the chairman's office with an armload of convincing data. More to the point, had she been using this model, the chairman never would have called in the first place, for the HRD director would have asked and answered many times over the chairman's very good question.

Brinkerhoff is a professor in the College of Education of Western Michigan University at Kalamazoo.

Yesterday's evaluation system

There is no doubt that the HRD profession is enjoying the best of times—never before have organizations invested so heavily in our services. And in such good times, it can be difficult to think of evaluation and accountability. Most of us are too busy serving the next client's demands to bother to see whether the last client benefitted. But as the HRD profession does more, promises more, and inevitably requires more resources to support itself, we need—now more than ever—comprehensive and effective evaluation approaches.

In 1967 Donald L. Kirkpatrick proposed a four-step model that has, in many respects, provided a sound and simply understood conceptual base for evaluating HRD programs. Appearing as part of the *Training and Development Handbook*, edited by R.L. Craig and L.R. Bittel, Kirkpatrick's model very clearly articulates four levels of outcome for any training session. Each level, he wrote, demands separate evaluation: Did trainees react favorably to the training—did they like it? Did trainees learn? Did trainees use what they learned? Did using the learning make a difference? This definition of four levels of outcome is extremely useful, for it pushes the focus of evaluation beyond mere favorable reactions and learning to where it rightfully belongs: on payoff to the organization.

Yet I find important shortcomings with the Kirkpatrick model. First, its definition of HRD covers training only, programs in which trainees learn discrete skills that transfer readily to the workplace, produc-ing immediate results. Today's HRD efforts are much broader than that.

Consider, for example, applying the four-step model to a program teaching people how to save lives by using CPR techniques. Barring a workplace heart attack, we would find no on-the-job application of the skill learned. Does this failure to detect broad third-level effects negate the value of the program? Probably not, so we need a different way to evaluate this program properly. Some HRD programs that do not produce behavioral results may nonetheless have value. Others may deserve to be thrown out. We need an evaluation process to help us decide not only which programs to keep and which to discard, but also how to revise those we keep to make them more cost effective. The Kirkpatrick model's narrow focus prevents us from doing this.

The Kirkpatrick model is entirely outcome oriented, reflecting a legitimate bottom-line bias. Yet there are many reasons to be concerned with evaluating HRD programs as they happen, well before they have had a chance to produce results. In fact, to look for effects only after the program is to perpetuate trial-and-error learning. If evaluation during the early developmental stages can show that a program is ill conceived or poorly executed, then there may be good reason to revise or even abort it. Evaluation made part of the program development process can *help* programs succeed, as well as measure whether or not they do.

We need to think about evaluation just as we think about any other portion of the

HRD process. No professional would conduct a workshop without telling participants where and when it is. Yet, how many of us plan and conduct HRD programs without communicating criteria for their success to upper management? Or without reporting the programs' impact to our bosses or bosses' bosses?

The logic of HRD

An evaluation system such as the six-stage evaluation model presented below must be part of the entire HRD process; it must tie into program planning, development, and operation. Evaluation begins at the outset, when HRD is first considered, and continues throughout the remainder of the process. To understand how the six-stage model works and how it meshes with the rest of the HRD process, we need to examine the logic of HRD.

I believe that all HRD programs should be designed to produce beneficial results on an organizational level. A sales training program, for example, should not simply train salespeople. It should increase sales volume, open new markets, or have some other positive effect on the company's goals. All programs should share the same logic: trainees go through training in order to learn something that will eventually benefit the organization.

The six stages

This basic logic suggests six stages of HRD program development and operation:
■ **Stage 1.** A need, problem, or opportunity worth addressing exists that could be influenced favorably by someone learning something.
■ **Stage 2.** An HRD program capable of teaching the needed something is designed or located.
■ **Stage 3.** The organization successfully implements the designed program.
■ **Stage 4.** The participants exit the program after successfully acquiring the intended skills, knowledge, or attitudes.
■ **Stage 5.** The participants retain and use their new learning.
■ **Stage 6.** The organization benefits when participants retain and use their learning.

Analyzing an HRD program in terms of these six stages can show whether and how programs benefit an organization. This analysis also helps trace any failures to one or more of the six stages.

Program failures could derive from flawed logic. Trainees might learn and learn well, but conditions on the job could

Figure 1—Six-stage model for evaluating HRD

Evaluation Stage	Key Evaluation Questions	Some Useful Procedures
I. **Goal Setting** *(What's the need?)*	—How great is the need, problem, or opportunity? —Is it amenable to HRD solutions? —Would the HRD difference be worth making? —Would HRD work and be likely to pay off? —Are criteria available to judge whether it paid off or not? —Is HRD better than alternative approaches?	Organizational audits; performance analyses; records analysis; observation; surveys; study of research; document reviews; context studies
II. **Program Design** *(What will work?)*	—What kind of HRD might work best? —Is design A better than design B? —What's wrong with design C? —Is the selected design good enough to go with?	Literature review; expert reviews; panels checklists; site visits; pilot tests; participant review.
III. **Program Implementation** *(Is it working?)*	—Is it installed as it is supposed to be? —Is it working on schedule? —What problems are cropping up? —What really took place? —Did they like it? —What did it cost?	Observation; checklists; trainer and trainee feedback; records analysis.
IV. **Immediate Outcomes** *(Did they learn it?)*	—Did they learn it? —How well did they learn it? —What did they learn?	Knowledge and performance tests; observation; simulations; self-reports; work sample (product) analyses.
V. **Intermediate or Usage Outcomes** *(Are they keeping and/or using it?)*	—How are they using it? —What part(s) of it are they using?	Self, peer, and supervisor reports; case studies; surveys; site visits; observation; work-sample analysis.
VI. **Impacts and Worth** *(Did it make a worthwhile difference?)*	—What difference does using it make? —Has the need been met? —Was it worth it?	Organizational audits; performance analyses; records analysis; observation; surveys; document reviews; panel reviews and hearings; cost/benefit comparison.

still conspire against skills transfer, preventing trainees from using their learning. Perhaps trainees learn and use their learning; even so, poor initial needs analysis at Stage 1 could have misidentified the problem, making the chosen intervention irrelevant.

HRD failures might also result from flawed operations. If, for instance, an unskilled trainer ran the program participants might not learn and the effort would fail. Sloppy follow-up could discourage retention and use of the learning, squelching potential benefits.

The logic of the new model

The six-stage evaluation model follows traditional HRD logic. It emphasizes evaluation—defined as the collecting of information to facilitate decision making. Evaluation, for example, can help HRD staff decide at Stage 1 whether the problem or opportunity is worth addressing in the first place, and whether the assump-

decisions about whether to cease, continue, curtail, or expand HRD programs. Figure 1 summarizes the model's six stages and lists several data-collection procedures useful during each of the stages.

Articulation

Using the six-stage model requires articulating the assumptions about why and how each HRD activity is supposed to work. Without such articulation, comprehensive evaluation is impossible. And with such articulation, HRD practitioners and consumers can define and clarify expectations.

The broad view

The success of an HRD effort in producing results of value to the organization depends on the quality of the decision making at each of the six stages listed

the crucial developmental stages of HRD programs, from assumptions about needs and potential payoff; through design; to implementing, debugging, and controlling program operations. These stages cast the die that determine payoff. Outcome evaluation alone can improve neither programs nor results. To this end, the six-stage model emphasizes a formative evaluation role and encourages the recycling of evaluative information from and to each of the six stages. In this way, all programs are *made* to work as best they can, and good programs are made even better.

Finally, of course, careful evaluation at all six stages enables HRD practitioners to educate HRD customers about their trade and practice, and enables them to justify their existence. Expectations at each stage are clarified, and accountability and effectiveness can be assessed and reported.

An HRD practitioner who uses the six-stage model can convincingly tell management, "The training we do is important and needed! The training designs we use are the best available! The training we run operates smoothly and people like it! People learn what we teach! What we teach lasts, and gets used! And our training pays off!"

Wouldn't it be nice if all HRD practitioners could make these claims? Shouldn't we all be able to?

Adapted by permission from Achieving Results from Training *by Robert O. Brinkerhoff. Copyright 1987 by Jossey-Bass, Inc., 433 California St., San Francisco, CA 94104; 415/433-1767.*

> There are many reasons to be concerned with evaluating HRD programs as they happen, well before they have had a chance to produce results

tions about causes of the problem are accurate and sound. Thus, evaluation at Stage 1 is needs assessment. Remember, deciding whether or not a training need exists is essentially a value-laden, judgmental process.

At Stage 2, evaluation can help a trainer decide whether a training design is sound enough or which of several competing alternatives will work best. Once a program is underway, Stage 3 evaluation procedures can help a trainer decide whether the program operates as intended and whether it really produces the desired results.

Deciding whether and how much trainees have learned is clearly an evaluative activity at Stage 4. At Stage 5, evaluation can assess retention, endurance, and learning transfer. Here it can also begin to track the complex process by which HRD accomplishes its objectives.

Finally, Stage 6 evaluation provides the input for making value decisions. Was the HRD program worthwhile? Did it accomplish its organizational goals? What other value did it produce? Answers to these questions are crucial to making good

above. Mistaken needs assumptions, for example, can make entire programs worthless. Many critics of the human relations training popular in the late sixties and early seventies blame the failure of these programs to improve productivity on failed needs analyses. Deciding to implement an HRD program with a critical design flaw or requiring trainees to remain in a program long after they have mastered content can seriously jeopardize the opportunity for the program's payoff.

The six-stage model forces a view of HRD in an organizational context and requires HRD professionals to articulate the logic of any program, from its roots in desired organizational benefit to its final payoff. The six-stage model precludes defining a program as successful because it is popular, or because it is easy to teach, or because it uses state-of-the-art technology. Careful attention to stages 4, 5, and 6 of this model keeps the HRD practitioner focused on the proper elements of HRD results and recognizes organizational impact as the final arbiter of worth.

The six-stage model also emphasizes

Evaluation Techniques that Work

Using Kirkpatrick's evaluation matrix as a base, the author discusses how trainers can evaluate their programs from beginning to end.

By HERMAN BIRNBRAUER

Every HRD professional pays lip service to the idea that evaluation is important to successful training, but few conduct complete and thorough evaluations. Evaluation can be time consuming and the attendant methodologies sometimes overwhelm those who lack a statistics background. And let's face it, evaluation often seems an anticlimax to the excitement and creativity that goes into developing and delivering a new course or program.

Nonetheless evaluation is mandatory, for it's the only way of determining whether or not our efforts have paid off. The technique that follows is a practical one that works. It uses a matrix and a model developed by Donald L. Kirkpatrick. Although this model is 28 years old and the processes involved are far from revolutionary, it remains valid because of its comprehensiveness, simplicity, and applicability to a variety of training situations.

The heart of this technique consists of the four evaluation levels arrayed vertically on the left-hand side of the matrix shown in Figure 1. The evaluation matrix is merely a tool, a way of displaying Kirkpatrick's four-point framework in an easily understood and workable form. Kirkpatrick's model considers the following elements of evaluation:

■ *Reaction*. Were the trainees satisfied with the program?

■ *Learning*. What facts, techniques, skills, or attitudes did the trainees understand and absorb?

■ *Behavior*. Did the program change the trainees' behavior in a way that improves on-the-job performance?

■ *Results*. Did the program produce the results desired?

Moving down the column, the matrix presents these levels in order from simple and inexpensive to complex and costly.

Each level has advantages and disadvantages, so it is important to plan the evaluation process as you plan the training. Depending on the purpose, difficulty, or importance of the skills and knowledges to be taught, you may choose to use all four levels or you may elect only one or two. But regardless of which ones or how many eventually apply, you need to consider all of them at the outset. The questions that run horizontally at the top of the matrix can guide you in looking systematically at the implications of Kirkpatrick's model.

Trainee reaction

Kirkpatrick's first level—*trainee reactions to the course*—measures how trainees feel about the program. Most often trainees record this information on postsession forms or questionnaires, noting their impressions of the instructors, curriculum, homework and assignments, materials, classroom or facilities, and value and depth

Birnbrauer is president of the Institute for Business & Industry, Inc., in Bensalem, Pennsylvania.

of the course content. Some trainers refer to these questionnaires as "happiness sheets."

While gauging trainees' reactions can be valuable to course presenters and designers, this level of evaluation may lack the precision necessary for meaningful revision. There is little correlation between how trainees feel about a program and what they have learned—or more importantly, what they will do on the job because of it. The data collected by questionnaires can't be validated.

But despite the drawbacks, trainee reactions are the most widely used of any of the four evaluation levels, probably because they are relatively inexpensive, quick, and easy to administer. They also can indicate the need for minor mid-course corrections during lengthy programs. Early evaluation of reactions can tell instructors whether or not their presentation techniques and pace are appropriate.

Trainee learning

Kirkpatrick couched his second evaluation level—*trainee learning*—in terms of the principles, facts, and techniques the trainees understand and absorb as a result of the training. As a practical matter, though, it may be more useful to consider knowledge, skills, and attitudes as shown on the matrix.

What do these terms mean and how can they be measured? All the stakeholders in the training effort should agree on definitions. If they don't, it will be difficult to make sense of the evaluation results.

Knowledge refers to what trainees know as a result of attending a program. This sort of information might include facts, principles, or rules relevant to the training topic. You can measure what trainees know by testing them before and after the program. If the tests validly and reliably cover critical information taught, you can determine a program's effectiveness by comparing the pre- and posttest results.

Use the *skills* category to evaluate skills trainees acquire from the program. This demands a little more effort than knowledge evaluation, but the resulting data provide a better indication of how well the training accomplished its objectives. The best skills tests require trainees to perform tasks using their newly acquired skills in an actual or simulated on-the-job environment. Trainee proficiency indicates the program's effectiveness.

The *attitudes* category presents a stickier problem because attitudes are difficult to define. Attributes like cooperation, quality-mindedness, and innovation are subjective; they can't be observed, measured, or verified in any conventional way.

Robert Carkhuff, in the 1984 text *Instructional Systems Design*, says that a three-step technique called *favorability scaling* can minimize this problem. First label the subjective concept, using words like "cooperation" or "innovation." Then define it in meaningful terms. Finally, establish a scale for levels of trainee adherence to the concept; you might select a most-favorable level, an acceptable level, and a least-favorable level. Even with favorability scaling, it is impossible to measure subjective attitudes precisely. But if everyone involved in the evaluation process understands the scale, this technique can provide useful information.

Kirkpatrick suggested procedural guidelines for measuring the amount of knowledge-, skills-, and attitude–learning that takes place. He suggested evaluators measure each trainee's learning by using a before-and-after approach to establish a relationship between the learning and the program. As much as possible, you should measure with an objective basis, statistically analyzing the results to correlate learning with the training.

These guidelines emphasize that

Figure 1—Evaluation Matrix

Levels (Kirkpatrick)	What might be measured	What are the sources of data?	How should data be collected?	What are potential problems?
1. Trainee reactions to the course				
2. Trainee learning • Knowledge • Skills • Attitudes				
3. Trainee behavior on the job				
4. Organizational results				

evaluation in terms of learning is more difficult than evaluation in terms of reactions. This second-level approach requires more thorough planning, analysis, and interpretation.

Trainee behavior on the job

Level three—*trainee behavior on the job*—is what trainees actually do as a result of the training program. Do they use their newly acquired knowledge, skills, and attitudes in the desired manner at the appropriate time? To find out, you must measure trainees' behavior when they return to their workplaces after training.

While this type of evaluation—measuring a fairly exact transfer of training to performance—sounds ideal, behavior evaluation is more difficult than reaction or learning evaluation. Behavioral evaluation requires that program developers devote more skillful, up-front analysis of the performance objective. Evaluators must know the task in detail in order to determine if trainees know how to perform appropriately. And since the evaluation must happen at the workplace, evaluators can't control when, or even whether, the trainees will have to use their new skills. After all, it is hard to predict when learning will be necessary and when it will be used. In addition, some organizations do not require trainees to apply newly acquired learning.

These factors seem to conspire against successful behavior evaluation, but it can be done. Kirkpatrick suggested several techniques:

■ Conduct a systematic appraisal of on-the-job performance before as well as after training.
■ Involve the trainees, their supervisors and subordinates, and their peers in both pre- and posttraining performance appraisal.
■ Make a statistical analysis to compare "before" and "after" performance and to correlate changes to the training program.
■ Conduct the posttraining appraisal three months or more after the training so trainees have an opportunity to put into practice what they have learned. Subsequent appraisals may add to the study's validity.
■ Evaluate a control group who did not receive the training. This experimental approach may not be realistic in all organizations.

These techniques illustrate how complex behavioral evaluation can be, but if properly done they can give a true indication of a training program's effectiveness.

Organizational results

The fourth level—*organizational results*—helps identify how training changes organizational functions. It looks at things such as reductions in costs, turnover, absenteeism, and grievances. It can also help organizations gauge quality and productivity improvements. Evaluating training's effect on the organization is often part of a cost-benefit analysis. The resulting data can demonstrate a good return on the training investment if evaluators can relate improvements directly to the training.

Unfortunately, direct correlations may be very difficult because so many factors can complicate the process. Measuring organizational results requires considerable front-end analysis of desired objectives. Identifying specific objectives such as reducing grievances or costs makes this task easier. Even so, it may be difficult to prove a training correlation.

Summary

Keep in mind that training can have a significant impact on an organization even if it can't be evaluated as an organizational result. Measuring a training program's effects at the reaction, learning, or behavioral levels can be just as revealing and just as important. You don't have to apply all four of Kirkpatrick's evaluation levels to determine how well, or how poorly, training worked.

But it is important to consider all four evaluation levels during the planning process. Be sure to think about them when you establish course objectives because that's what your evaluation will measure. Think about them when you design the course materials, and think about them when you begin training or give the program to the instructor.

Above all, don't wait until the course is over. Kirkpatrick's model and the evaluation matrix can help guide the entire evaluation process, but only if you use them early and thoroughly.

References

Kirkpatrick, D. L. "Techniques for Evaluating Training Programs—Part 2: Learning." *Journal of the American Society of Training Directors* (December 1959): 21-26.

Kirkpatrick, D. L. *Evaluating Training Programs.* Madison, Wisconsin: American Society for Training and Development, 1975.

The best skills tests require trainees to perform tasks using their newly acquired skills in an actual or simulated on-the-job environment

American organizations today are experiencing significant turbulence, and HRD functions are feeling it. On the one hand, HRD practitioners are receiving the recognition they've sought for a long time; on the other, this recognition means they must live up to their potential. They have to deliver!

Gone are the bit parts they've been used to playing: They're now expected to perform on center stage. The lights go up and the curtain opens, but too many HRD professionals still have a "chorus-line" mentality. A recently completed doctoral dissertation concluded that top management excludes HRD managers from strategic planning

Your New Role in the Organizational Drama: Measuring Effectiveness

By Neal E. Chalofsky and Carlene Reinhart

because they perceive these individuals as reactive, service-oriented functionaries—a sort of supporting cast.

But HRD professionals needn't continue playing the same role. They *can* help their organizations if they understand the play and the production and how the stage should be set. They *can* be effective.

Structuring the model

We recently completed research in which we attempted to identify and define "effectiveness" of HRD functions in a useful way. Our thoughts

Chalofsky is visiting assistant professor of HRD at The George Washington University in Washington, D.C. Reinhart is manager of computer-based training design with Xerox Corporation's U.S. Marketing Group in Leesburg, Virginia. This article is excerpted from the authors' book, Effective Human Resource Development, *published by Jossey-Bass, Inc.*

prior to starting this project were that, according to many indicators, American business and industry is in a period of chaotic and sometimes violent change. We also believed that these indicators point to human incompetence and inefficiency as perhaps the most critical problems to solve if we are to overcome productivity problems and decrease business failures.

The charter of *most* HRD functions is to help the human resources of organizations become more competent and productive and to help organizations meet their goals. But, we reasoned, if HRD functions were performing as effectively as they could be, there would be fewer organizations in trouble and more organizations successfully meeting their goals. Therefore, by designing a blueprint or model of effectiveness for HRD functions, we believed we would have an opportunity to help them—which in turn will help their organizations.

Our first step was to develop a definition of HRD effectiveness. We speculated that once we identified what effectiveness looks like, we'd be able to develop a model and a process for evaluating and increasing it in HRD functions.

We built a knowledge base from which to create this model. This base took the form of a multiphased research study that included

■ a literature search for articles about successful HRD functions;
■ a delphi panel of HRD experts who identified 10 elements that are critical to HRD effectiveness;
■ an organizational survey that identified four major criteria of HRD effectiveness as defined by successful HRD functions;
■ interviews with managers of effective HRD functions;
■ hours of analyzing and massaging the data into a meaningful and useful model and process.

We concluded that the overriding goal of the effective HRD function was to *build a responsive resource*—a goal that captured for us the essence of all our findings. This goal evoked a vision of a thoroughly competent, professional HRD team that is constantly building and maintaining a track record of high-quality products and services through close relationships with line and staff management and that enjoys the credibility it has worked extremely hard for and deserves. We believe this is what HRD effectiveness is all about.

The foundation blocks for this vision are three criteria that need to be achieved to make the vision a reality:
■ close relationships with line and staff management;
■ a highly professional HRD staff;
■ a track record of high-quality products and services.

Our research determined that the way to achieve these criteria and fulfill the vision is by meeting the standards in Figure 1.

We found that the standards all apply to some extent to each of the three criteria; based on this conclusion we structured the model as a matrix, shown in Figure 2. To give HRD functions a useable tool, we developed questions for each box in the matrix. An HRD manager seeking to improve HRD functional effectiveness can answer the matrix questions as they relate to his or her own organization; compare these answers to the variety of input from our research; identify trends, discrepancies, and similarities; and begin the process of functional development and growth by improvement in specifically targeted areas.

The effectiveness improvement process

The degree to which the HRD function achieves the goal of being a responsive resource greatly determines the degree to which the organization

---

Measuring your effectiveness is a bit like throwing three pebbles into the water and watching how the ripples interact

---

perceives the function to be effective. The quality of HRD staff and its management, its programs, and its organizational relationships will determine the function's level of effectiveness. In other words, what you put into HRD is what you get out of it.

Effective HRD functions are made, not born, and those that are not fully effective can improve. Figure 3 presents a flow chart that depicts how to actually use the standards and criteria of the model to boost effectiveness.

Planning for effectiveness

One of the critical standards of effectiveness our research clearly identified was an organizational needs analysis to determine present and anticipated problems and opportunities. Once needs are uncovered, the HRD effectiveness model can act as a template to develop a mission statement for the HRD unit and to identify its goals and objectives.

For example, if you were concerned that your organization's word processing system was quickly approaching obsolescence, you would feed back that information to management. This situation should also cause you to think about whether management perceives your staff as internal consultants (Standard 7) and as experts (Standard 9). You would want to know this so that when your organization begins to discuss changes before actually moving to a new, more sophisticated system, you could be sure your staff could be involved in the planning stages of the changeover. Your involvement in the changeover would cause you to look at standards 7 and 9 as your functional objectives.

The mission and objectives formulated for the HRD function should be based on identified needs and should drive the development of strategic action plans. You should base your staff's objective of becoming internal consultants with expertise in sociotechnical approaches on real requirements of your organization, and you should take specific actions to achieve this objective. For example, your strategy may be to educate management about the sociotechnical approach and to increase your staff's visibility as internal consultants. The actions you plan may include conducting an executive briefing on sociotechnical systems change and adding it to your management development program.

Figure 1 — Standards for fulfilling the vision of HRD effectiveness

1. The HRD function has the ability to diagnose problems and anticipate needs.
2. The HRD function is supported by a corporate HRD mission statement and organizational culture.
3. The HRD function is committed to strategic planning and supporting organizational change.
4. The roles and responsibilities of the HRD staff are clearly defined.
5. The HRD function is committed to front-end analysis and evaluation.
6. The HRD function is strongly committed to its own staff development.
7. The HRD function is perceived as an internal consultant to management.
8. The HRD function has a strong marketing and public relations capability.
9. The members of the HRD staff are perceived as experts.
10. There is a high level of HRD staff teamwork, creativity, and flexibility.
11. There is a high level of HRD staff ethical conduct.
12. The HRD function is perceived as "part of the business."
13. There is a high level of congruence between the HRD function and organizational goals and objectives.
14. The HRD function is perceived as conducting reality-based programs.
15. There is a high level of networking for all levels of management.

Managing for effectiveness

In order to carry out your strategic action plan successfully, you need to take into account the three HRD effectiveness criteria:

■ *Close relations with line and staff management*—Initiate a new network or improve an existing one by setting up management-level steering committees and task forces. Insure that the HRD function's priorities are based on top management's goals. Build an HRD network among specialists in other organizations in your industry or in your geographic area. Listen to people in your networks, and benefit from their expertise.

■ *Highly professional HRD staff*—Insure that your staff really knows the business of the organization. Develop your staff to the highest possible level of expertise; at the same time develop your own managerial skills to the optimum. Be constantly aware of changes in technology—high technology in general and training technology in particular. Make sure your HRD activities are realistically results oriented.

■ *High-quality track record*—Conduct accurate and thorough front-end

---

Figure 2 — HRD effectiveness model

STANDARDS	CRITERIA		
	Close relationship with line and staff management	Highly professional HRD staff	Track record of high-quality products and services
1. HRD function has ability to diagnose problems and anticipate needs.	• What methods are available for gaining the close relationship that will allow access to organizational data? • How can trust be achieved to allow the HRD function into the larger organization to collect necessary data?	• What kind of expertise is needed, such as ability to diagnose organizational and individual problems and recommend appropriate solutions? • How is needed competence achieved?	• How can HRD function make recommendations and deliver appropriate solutions to diagnosed problems? • What is at stake when needs are diagnosed correctly but are not what the organization wants to hear?
2. HRD function is supported by corporate HRD mission statement and organization culture?	• How can the HRD function ensure that its mission statement and understanding of HRD purpose is clearly understood at all levels and across all subunits of the organization?	• How can HRD professional staff facilitate the development of an HRD mission statement? • Can the HRD staff influence organizational culture?	• What needs to be done to ensure that the HRD function's output is congruent with mission statement? • Are some public relations efforts better than others?
3. HRD function is committed to strategic planning and supporting organizational change.	• How can HRD staff first get involved in strategic planning? • What can they do to *stay involved*?	• How can professional staff be better prepared to support strategic planning and organizational change?	• What can HRD staff produce to demonstrate commitment to strategic planning and organizational change?
4. Roles and responsibilities of the HRD staff are clearly defined.	• How can HRD staff best model the role and responsibility definition process for the larger organization?	• What techniques and resources are available to facilitate this process?	• What are some of the ways to demonstrate payoff of clear role and responsibility definition to the larger organization?
5. HRD function is committed to front-end analysis and evaluation.	• How can line and staff management be involved in front-end analysis? How can they be involved in ongoing evaluations?	• What specific competences and skills are required in today's organizations to conduct front-end analyses and evaluations? • How can the HRD function develop these skills and competencies in staff? • Are valid instructional design principles and processes used and understood by all personnel?	• How can time spent in front-end analysis and evaluation be depicted to the larger organization as producing a valid return on time invested in the effort? • Can the HRD function correlate a financial return on investment to evaluation?

(continued)

analysis. Link the outputs of the HRD function to the bottom line. Publicize your accomplishments. Build in both management support and accountability. Again, insure that your programs are reality based. Compare your activities with HRD functions in similar businesses, and ask for their critiques of consultants and packaged programs.

Lastly, have fun building your track record.

Once you have built in the quality control of managing against the effectiveness criteria, then you can develop, conduct, and monitor the needed programs. Make mid-course corrections as needed.

Let's continue our example of your

need to build expertise in sociotechnical systems and improve your staff's visibility as internal consultants. During this phase your staff may do some reading on sociotechnical approaches to change or go to a workshop on consulting and change. You would also identify key managers who will be the most influential in supporting your ac-

STANDARDS	CRITERIA		
	Close relationship with line and staff management	**Highly professional HRD staff**	**Track record of high-quality products and services**
6. HRD function is strongly committed to HRD staff development.	• What role does management play in HRD staff development? • Is staff development integrated into performance evaluations? If it is, how is this done?	• Do effective HRD functions "make" or "buy" professionalism? • How is staff development managed for maximum efficiency and effectiveness? • How are personal and organizational goals aligned in effective HRD organizations?	• How can the HRD function show a direct connection between professional products developed by staff and ongoing professional development?
7. HRD function is perceived as internal consultant to management.	• What kinds of legitimate "networking" activities can HRD staff perform with line and staff management to institutionalize the function as viable internal consultants?	• What types of professional development activities prepare HRD staff to function as internal consultants to management? • How can the internal consulting capability best be communicated to management?	• How can the HRD function capitalize on its track records to establish itself as an internal consultant?
8. HRD function has a strong marketing and public relations capacity.	• Should the HRD function market to line and staff management?	• Are there ethical limits to marketing professional staff? • How can marketing and PR capabilities be developed?	• Is a track record dependent on PR? Can an HRD function establish a track record without marketing and PR activities? If so, how can this be done?
9. Members of the HRD staff are perceived as experts.	• What roles does the relationship between the HRD function and line and staff management play in the organization's perception of the HRD function as expert?	• Can this be done without a highly professional staff?	• Can a track record be established if the HRD function is *not* perceived as a group of experts by the larger organization?
10. There is a high level of HRD staff teamwork, creativity, and flexibility.	• Can teamwork, creativity, and flexibility occur in HRD functions when these elements are not present in the larger organization? • How can close HRD function relationships with line and staff management help line and staff to be more creative and flexible?	• Is creativity, teamwork, and flexibility a result of professonalism? • Do effective HRD functions spend time developing teamwork, creativity, and flexibility?	• Do factors such as teamwork, creativity, and flexibility contribute to high-quality products and services in effective HRD organizations?

Figure 2 — HRD effectiveness model

tivities and your image as internal consultants. You would conduct the programs and sessions you developed on sociotechnical systems and do some short-term evaluations. You would invite the key managers to these sessions and cultivate their involvement by asking them to review the sessions for you.

Evaluating for effectiveness

Most HRD professionals believe in evaluating training and development activities whether they use simple questionnaires that identify how well participants like a particular workshop or a sophisticated evaluation that measures behavioral change. But evaluating courses doesn't tell you if your HRD *function* is effective or not; course evaluation is just one piece of the picture.

You can use a performance audit process to evaluate and improve HRD functional effectiveness. The steps in this part of the HRD effectiveness improvement process are based on a combination of performance analysis, auditing, and HRD management concepts and techniques. This evaluation process is basically a form of discrepancy analysis and is concerned with the impact of any discrepancy—positive or negative—between the criteria and standards and actual accomplishments. The process assumes that

■ a program evaluation is warranted in order to make sound decisions regarding the need to improve, change, or maintain HRD practices and activities.

■ a problem-solving or goal-setting activity will be necessary to improve the HRD function.

■ the HRD staff is committed to the change process required by the results of the evaluation.

The evaluation process follows these steps:

1. *Identify an evaluator and establish an evaluation team.* Ideally someone outside the HRD function, preferably someone with program evaluation or management auditing experience, should conduct the evalua-

STANDARDS	CRITERIA		
	Close relationship with line and staff management	**Highly professional HRD staff**	**Track record of high-quality products and services**
11. There is a high level of HRD staff ethical conduct.	• Where should HRD staff loyalities lie—to the profession on to the organization?	• Is there a relationship between professionalism and ethical conduct? Is either necessary to the other?	• Is it possible to produce high-quality products and services *without* a high level of HRD staff ethical conduct?
12. The HRD function is perceived as "part of the business."	• What is the balance in effective HRD functions between being perceived by line and staff as professional and being perceived by staff as a "field relevant" or "part of the business?"	• What types of professional development activities for HRD staff are required to insure that the function is perceived as "part of the business?"	• In effective HRD functions, how are quality products and services best produced as "part of the business?"
13. There is a high level of congruence between HRD function and organizational goals and objectives.	• Is the congruence a direct result of the close relationship of line and staff management to the HRD function?	• In effective HRD functions, is the congruence a direct result of HRD professional development?	• What is the relationship between congruence and high-quality products and sevices?
14. The function is perceived as conducting reality-based programs.	• What is the relationship between the function being perceived as conducting reality based programs and the degree of HRD functional relationship with line and staff management?	• Is this dependent on an HRD professional staff, or can reality-based programs be bought "off the shelf?"	• Can a track record be established with anything other than reality-based programs?
15. There is a high level of networking with all levels of management.	• How can a new HRD function develop these "network/relationships?"	• Do professional practicitioners network more easily with other professionals than with management?	• How does "networking with all levels of management" contribute to the production of high-quality products and services?

Figure 2 — HRD effectiveness model

tion; most of us, however, can't afford the luxury of an outside evaluator. The next best alternative is a small team composed of HRD professionals with at least one person from another staff function and one person from a line unit. This mix of personnel will give the analysis added objectivity and give the results added credibility.

2. *Plan the evaluation effort activity.* Set up areas of responsibility for the team and time lines for the tasks. Insure that both needed documents and HRD staff are available.

3. *Identify the HRD function's planned activities and outcomes.* You

What you put into HRD is what you get out of it

should base these on the goals and objectives of the HRD function, which, in turn, you should have based on the mission and goals of the organization. If the goals and objectives of the HRD function are not aligned with the mission and goals of the organization,

then you already have one discrepancy to investigate.

The criteria and standards of the HRD effectiveness model should serve as a framework for establishing a new HRD unit's goals and objectives, but they can certainly be useful in setting performance objectives for an existing HRD function.

The HRD effectiveness model's criteria and standards should stimulate the HRD unit's objectives or planned activities. For instance, if by looking at the model you feel that one of your priorities should be "close working relationships with line and staff manage-

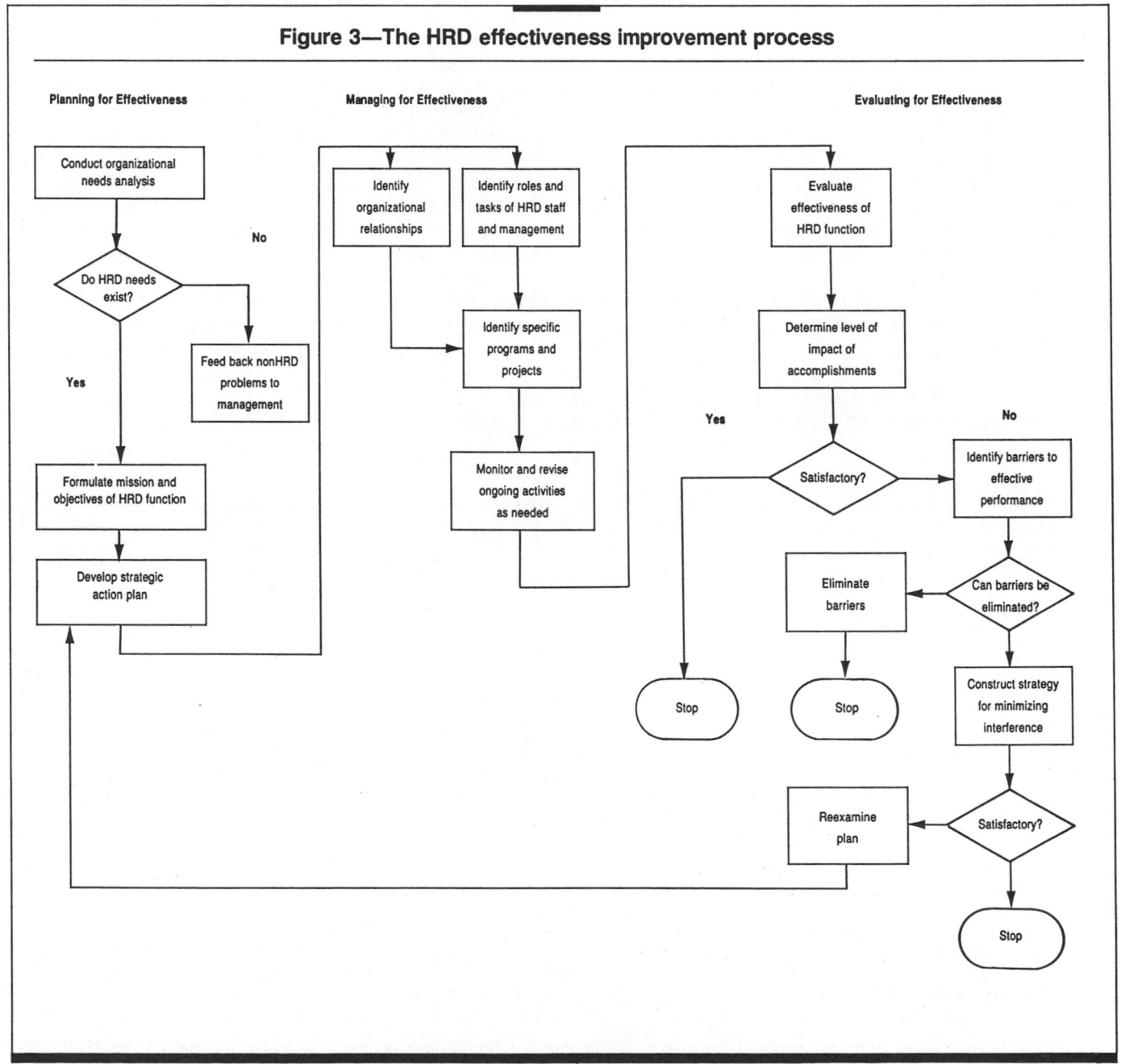

Figure 3—The HRD effectiveness improvement process

> **M**easuring your effectiveness is like throwing pebbles into the water and watching the ripples interact

ment" and you decide that you want to establish a management steering committee, then this activity becomes a focus of the evaluation.

4. *Design an evaluation format.* We suggest that you design your own form and format to actually conduct the evaluation.

5. *Conduct the evaluation and analyze the results.* Let's look at a scenario of how this process would work.

You're concerned about the level of professionalism of some of your HRD staff, so one of your planned activities is to send several of your staff to an upcoming HRD workshop on organization development skills. They go to the workshop and come back with a fairly sound theoretical approach to OD and some new ideas about combining OD and career development efforts.

But they also come back concerned that they really didn't learn "skills" but that they gained more knowledge about techniques that will turn into skills with practice. When you conduct the evaluation, you note the discrepancy between the outcome of learning new OD skills and the actual learning about techniques. The staff's understanding of the theoretical basis for OD and ability to plan a new OD and career development effort may be more significant than obtaining new OD "skills" in the long run. What the staff learned may have positively affected their level of professionalism.

The importance of any program evaluation is to get a holistic view of the effectiveness of the HRD function. You not only want to know if your activities and programs were successful, but also whether there were other ways the activities related to your effectiveness, whether there were unintended spin-offs from your activities, and what the implications are of the findings. In determining the level of effectiveness of your accomplishments, you look at both the specific changes and improvements and at the long-term implications that resulted from the evaluation. HRD is a system like any other organizational system. You therefore need to take a systems approach to identifying the interrelationships of the short- and long-term findings.

In the final analysis

Measuring your effectiveness is a bit like throwing three pebbles into the water and watching how the ripples interact. The idea is to step back and observe the intersecting ripples and get a total picture of the pattern they create. If everything is satisfactory— great! If it isn't, can you identify the barriers, and can you eliminate or minimize them? If you've done everything you can to try to improve the situation, then reexamine your strategic action plan to see if there were any problems you weren't aware of at the time. Remember: We aren't suggesting perfection, but thoroughness.

To finish with our sociotechnical consulting example, this is when you want to step back and determine if you've been able to achieve the expertise and the image you want. The final proof would be whether you were called in to advise and work with the technical experts in designing and implementing a new word processing system.

The purpose of this HRD effectiveness improvement process is to create, maintain, and improve your HRD function's output and consequently its credibility. Your actual achievement in helping your organization's employees reach their highest potential while you also help the organization as a whole adapt to change is what HRD effectiveness is all about.

Management Training Evaluation: An Update

Results of this study indicate that evaluation of management training programs is still not what it could be.

By WILLIAM H. CLEGG

Although management training evaluation is often a difficult process, it must be done. Management training requires an investment of time and money which, like any other type of investment, must be justified on the basis of the return from that investment. The following study indicates that current management training program evaluation methods are not as comprehensive as they could be. This lack leads the author to recommendations for improving evaluation methods and choosing and designing more suitable training programs based on ease of evaluation.

Study background

The study covers management training programs for first-line supervisory positions and above that are sponsored by large industrial corporations. The programs include on- and off-site, in-house programs, programs developed by outside agencies, programs involving either two- or four-year colleges or universities, and "canned" programs developed by outside agencies for attendance by supervisors and managers in any interested organization.

As its base, the study used A.P. Sullivan's 1970 dissertation, "An Analysis of Management Training Program Evaluation Practices in American Industry," which obtained information from 50 random *Fortune*-500 industrial corporations. A seven-page questionnaire was sent to those same companies in an attempt to obtain

Clegg is associate professor of information systems and operations management in the college of business administration at the University of Toledo.

current information on the status of their management training programs. The questionnaire was very similar to the previous one, with a few added questions, opportunities for additional responses to several questions, and some minor changes in the instructions.

Because turnover since 1970 made it improbable that the same individual would be completing the questionnaire, it was directed to to president of each of the organizations with a cover letter explaining the purpose of the study. Forty-three—or 96 percent—of the companies contacted returned the questionnaires. Five companies responded to the cover letters but did not complete the questionnaire, listing insufficient time and corporate policies against participation in studies as their reasons.

Response summaries

The following are the survey questions asked and tables showing percentages or rankings of responses given. Keep in mind that even though the questionnaire was sent to the presidents of the participating corporations, the questions were directed to the chief training officer.

1. *Does your company conduct in-house management training programs?*

This question was designed to determine if any companies that previously utilized in-house courses had discontinued them. Responses showed that 83.7 percent of the companies still conducted in-house programs and 16.3 percent do not. Respondents answering *no* to this question were directed not to answer questions 2, 3, 4, 7, and 10.

2. *Who conducts the evaluation effort on your in-house management training programs?*

Over three-fourths of the evaluations of management training programs were conducted by the training staff. There apparently was no reluctance on the part of large industrial corporations to have the same group which conducted or sponsored management training programs to also evaluate those same programs.

The *Other* category includes top management, participants, and some combination of these groups.

Who Evaluates Management Training	N	%
The training staff	28	78
An ad hoc committee	3	8
A special group of in-house measurement or control experts	1	3
An outside consultant or specialist	1	3
Other	3	8
Totals	36	100

3. *Indicate your degree of responsibility for management training evaluation.*

Over three-fourths of the chief training officers were either fully responsible for management training evaluation or shared that responsibility with others at the same or higher levels.

Degree of Responsibility	N	%
Fully responsible	20	55
Share responsibility with others at same or higher level	10	28
Share responsibility with others at lower level	6	17
No responsibility in this field	0	0
Totals	36	100

4. *Indicate your degree of authority for management training evaluation.*

Nearly three-fourths of the chief training officers either authorized all or some of the final decisions in matters pertaining to the evaluation of management training programs.

Degree of Authority	N	%
Make all final decisions	11	30
Make some final decisions	15	42
Make decision recommendations to higher authority	10	28
No authority in this field	0	0
Totals	36	100

5. *What do you consider to be the most pressing problem, weakness, or shortcoming you are encountering with respect to the evaluation of formal in-house management training courses?*

Nearly one-half of the large industrial corporations listed lack of standards and yardsticks as the most pressing problem, weakness, or shortcoming with respect to evaluation of in-house management training programs.

Answers in the *All other* category includes "accuracy or honest of comments," "not behavior oriented," and "application of ideas taught to on-the-job situations."

Problem, Weakness, or Shortcoming	N	%
Lack of standards and yardsticks	18	42
Follow-up evaluation	4	9
Trouble with management	3	7
All other	5	12
None given	13	30
Totals	43	100

6. *What short- or long-range changes do you plan to make in your in-house management training courses evaluation program?*

Over one-half of the large industrial corporations did not plan to make any changes in the evaluation of management training programs.

Eleven separate items were mentioned in the *Other* category.

Planned Changes	N	%
None	23	53
Behavioral objectives and changes	5	12
Improve evaluation one to two years following training	2	5
Perfect what are now doing	2	5
Other	11	25
Totals	43	100

7. *Rank, in the order in which you actually rely on them in practice, the following course effectiveness indicators. Use a 1 to indicate which aspect you use most, a 2 to indicate the next most used aspect, and so on until you reach the least often used aspect. Use a 0 to indicate those aspects you do not normally use.*

There was no single overwhelming industry-wide criterion in ranking the order of importance in evaluating a given management training program. The most frequently cited criteria were change in performance on the job, reaction of students to training, and changes in knowledge, skills, or attitudes possessed by the students.

Percentages were calculated after rankings were assigned weighted values.

How Evaluation is Based	Weight in points	%
Change in performance on the job	95	15
Reaction of students to training	68	11
Change in knowledge possessed by student	55	9
Change in skills possessed by student	54	8
Change in attitudes possessed by student	53	8
Change in company operating results traceable to training	53	8
How well the course was planned	46	7
How well the course was presented	38	6
Extent of continued demand for the course	35	6
Reaction of management to the training product	32	5
Extent course was recommended by company management	30	5
Degree of student motivation	17	3
Degree of student participation	16	3
How much the course cost	12	2
Reaction of instructors to course effectiveness	8	1
Reaction of other students, peers, and subordinates to the student	7	1
How well the course was attended	6	1
Reaction of the local training group	6	1
Completion of projects	3	0
Results or needs	3	0
Quality of students selected	2	0
Totals	639	100

8. *Rank, in order of severity, the following possible constraints to your management training program evaluation effort. Use a 1 to indicate the greatest constraint, a 2 to indicate the next, and so on.*

The primary constraint to in-house management training program evaluation efforts was lack of time by the individual responsible for evaluation.

Percentages were calculated after rankings were assigned weighted values. The percentage of each constraint was then multiplied by 43 (the number of respondents to the question) to arrive at the average number of respondents for each type of constraint.

Type of Constraint	Average N	% of Total Weighted Points
Lack of time	13	29
Lack of adequate evaluation methodology	9	22
Lack of money	8	18
Lack of necessity to evaluate	7	16
Lack of expertise	6	15
Totals	43	100

9. *How do you justify management training fund requirements?*

Three-fourths of the large industrial corporations justified management training fund requirements by whatever amount is required for the courses planned, be evaluating results of previous training programs, or by using the previous year's expenditure levels.

The *Other* category included such justifications as "needs identified and approved," "needs analysis," and "ask for funds on programs demanded by top management."

Type of Justification	Average N	% of Total
Whatever is required for courses planned	18	43
By evaluating results of past training conducted	9	21
By using previous year's levels	5	11
By insuring a favorable return on investment	3	6
As a fixed percent of some figure such as last year's profit or sales	1	3
Other	7	16
Totals	43	100

10. *Indicate by a check in the appropriate spaces the degree to which you actually utilize any of the following methods to evaluate your typical management training courses. For some of the questions related to timing, these definitions apply:* before *means prior to course start or during opening sessions of course,* end-of-course *means during closing sessions, and* post-course *means some period of time after course is completed.*

See Figure 1 for results of this question. Less than one-half of the evaluation methods that could be used in evaluating management training programs were actually being used.

Figure 1

Type of Objectives	Evaluation Method	Degree Method Used to Evaluate Typical Management Courses			
		Always Used	Used 50–90% of the Time	Used 10–50% of the Time	Never Used
INTERNAL	Informal collection of passing comments	12	8	6	10
	A set of anecdotal records is kept	3	3	9	21
	Evaluation is based on attendance records	1	4	10	21
	Evaluation is based on student participation	6	14	8	8
	Evaluation is based on short quizzes given during course	1	6	12	17
	Hour-by-hour student reaction reports during course	0	1	1	34
	Observe degree to which principles of learning observed in conducting courses	4	4	11	17
IMMEDIATE	End-of-course student course evaluation sheet	20	11	2	3
	End-of-course report by instructor	11	14	6	5
	End-of-course achievement or performance tests or attitude measures:				
	With a control group used	3	5	4	24
	Without a control group used	1	6	7	22
	Before- and end-of-course achievement or performance tests or attitude measures:				
	With a control group used	0	4	6	26
	Without a control group used	2	6	9	19
INTERMEDIATE	Before- and post-course reaction of superiors, subordinates, and peers to changes observed:				
	With a control group used	1	4	5	26
	Without a control group used	4	7	10	15
	Post-course reaction of superiors to before-course expectations stated	4	8	6	18
	Post-course reaction of superiors, subordinates, and peers to changes observed:				
	With a control group used	2	2	4	28
	Without a control group used	6	4	11	15
	Pre- and post-course achievement or performance tests:				
	With random assignment to control groups	0	3	3	30
	Without random assignment to control groups	1	5	7	23
	Post-course on-the-job survey of trainees (questionnaire or interview)	5	10	13	8
	Post-course achievement or performance test or attitude scale to measure retention and on-the-job behavior	2	3	12	19
ULTIMATE	Before- and post-course survey or operations audit of various aspects of company operations that training is expected to influence	1	4	10	21
	Post-course survey or operations audit of various aspects of company operations that training is expected to influence	1	6	8	21
	Evaluation is based on continued demand for courses	14	10	8	4
	Determine extent to which trainees are progressing (promotion and more pay and responsibility)	4	4	11	17
	Spot checks by outside consultants	1	0	6	29
	Return on the dollar investment in the course	3	3	6	24

11. *Which of the following 12 reasons for management training evalua-tion do you think the evaluators in industry regard as the three most im-portant reasons? Which three would you regard as the most important if you were going to make a case for evaluation? Use a 1 to indicate the most important reason, a 2 to indicate the second most important reason, and a 3 to indicate the third most important reason.*

The principal reason for evaluating management training pro-grams was to determine if there was a payoff from such programs. Also of importance were determining where improvement is re-quired and measuring progress toward objectives.

Results of this question are based on a weighted point system. The six most important reasons are listed in the table below. The other six possible reasons provided in the questionnaire were: re-quired by higher authority, evaluation is intrinsically good, to help sell training throughout industry, to give trainers a sense of ac-complishment, to make trainers feel important, and to determine effectiveness of training staff.

Reasons for Evaluating	Top 3 Industry Reasons Rank	Top 3 Personal Reasons Rank
To determine if there is a payoff*	1	3
To justify existence of the training function	2	—
To measure progress toward objectives	3	2
To find out how training can con-tribute more	4	1
To find out where improvement is required	5	4
To establish guidelines for future programs	—	5

Most probable reason

12. *If your answer to question 1 was no, check the reasons you do not now conduct in-house management training courses.*

Of the seven companies which indicated they had discontinued in-house management training programs (in question 1), four in-dicated that such programs were being conducted by outside agen-cies, at colleges and universities, or through self-study. The other three companies gave lack of interest by top management, finan-cial reason, or elimination of the training function as the reasons for discontinuance of programs. Some companies listed more than one reason.

Reasons for Discontinuation	N	%
Conducted at outside agencies (AMA-type)	4	30
Conducted at colleges or universities	3	23
Conducted through self-study	1	8
Sub-total—conducted by other than in-house	8	61
Discontinued due to lack of interest by top management	2	15
Discontinued for financial reasons	1	8
Training function eliminated	1	8
Now developing an in-house program	1	8
Totals	13	101

13. *Which of the following 13 reasons for not evaluating management training programs do you think the non-evaluators in the industry regard as the three most important? Which three would you regard as the most important if you were going to make a case for non-evaluation? Use a 1 to indicate the most important reason, a 2 to indicate the second most important reason, and a 3 to indicate the third most important reason.*

The principal reason for not evaluating management training programs was that responsible officials didn't know what to evaluate because of foggy objectives. Also of importance were that some responsible officials did not know how to go about evaluating, and since evaluations were not required in some com-panies, no attempt was made to evaluate.

Results of this question are based on a weighted point system. The eight most important reasons are listed in the table below. The other five possible reasons listed on the questionnaire are: trainers are afraid of what they may find out, too few trainees in-volved to make it worthwhile, laziness on the part of the training staff, frightened to try evaluation because of complexity of the process, and previous evaluation results have been misused.

Reasons for Not Evaluating	Top 3 Industry Reasons Rank	Top 3 Personal Reasons Rank
Don't know what to evaluate because of foggy objectives*	1	1
Responsible officials do not know how to go about evaluating	2	3
It is not required, so why bother	3	—
It would probably cost too much	4	—
Training officials do not see the value of evaluation	5	—
Inability to secure necessary cooperation within the firm	—	2
There is probably not enough time	—	4–5
Inability to assemble essential expertise	—	4–5

Most probable reason

Study recommendations

A number of important points were brought to the surface as a result of the responses to the questionnaire used in this study. The following comments and recommendations may perhaps assist chief training officers in their decisions concerning evaluations of their management training courses.

■ Serious consideration should be given to management training program audits by someone outside the training department, if only on a periodic basis. The outside auditor could be an individual from another corporation, on an exchange basis, or a consultant who specializes in the field of management training. Periodic audits of this type will prevent an organization from becoming too "ingrown" and will instill fresh ideas to revitalize management training programs.

■ Each chief training officer should review his or her evaluation methods and attempt to adopt additional evaluation methods, where appropriate, for each type of in-house management training program. This would be particularly important in those companies which presently utilize less than the average of 48 percent of the evaluation methods that could be used.

■ Chief training officers should insist that management training programs not be conducted until the training program objectives have been established. If the person who requested a particular management training program is not able to develop training program objectives, it is possible that that particular management training program should not be conducted at that time.

■ Very seldom should training of any type be conducted solely for the sake of training itself. Chief training officers should insist that management training program sponsors review the need for a training program and that the sponsors are selective in who attends.

■ It is incumbent that chief training officers, or the individuals delegated the responsibility for evaluation of management training programs, be familiar with at least the fundamentals of basic evaluation techniques and research designs.

■ Many chief training officers apparently do not have sufficient time to devote to adequate evaluation of management training programs. If this is the case, a chief training officer should convince his or her superior to allot additional funds for this purpose, either through additional departmental personnel or through the use of outside agencies.

Measuring Training Results: Key to Managerial Commitment

Evaluations are important, even if they're not required. Here's why, with a case study of an effective evaluation program.

By JAMES D. BELL and DEBORAH L. KERR

Retraining beats out replacing workers as the best response to new job demands, technology, or management systems, according to current general consensus. But an individual business stuck in a phase of shrinking profits may quickly change its tune about training benefits. Although this is when the need for optimal employee performance and quality education programs is strongest, the skepticism of top-level management about training's effectiveness also picks up strength and, many times, weakens support for education efforts.

the organization and its workers? Do our programs result in the acquisition and use of new and internally marketable skills— the kind that this organization will need over the next three to five years?

Proof through evaluation

To demonstrate training and development's importance, trainers must not only present excellence programs but also must *prove* the programs get results: improved job performance, more efficient use of resources, and satisfactory returns on the training dollars invested. The proof is

> Management demands for a wider array of programs but not for a determination of their worth can mean danger in the training long run

Unless the recent growth in training and development is to be short-lived, those who plan, design, present, or market programs must start building a strong and lasting relationship with management by asking themselves some questions: Does the return on our training investment include increased productivity and cost containment? Does our training truly benefit

Bell is an associate professor in the Department of Computer Information Services and Administrative Science at Southwest Texas State University in San Marcos. Kerr is founder of Travis Management Consulting and Education in Austin, Texas.

revealed through program evaluation.

The concept of evaluation has received widespread recognition as beneficial, but the practice of evaluation has lagged behind. Few reports of actual program evaluation have been published; compared to the number of training programs, few evaluations have been conducted. According to the report of one study, less than 12 percent of 285 companies studied evaluated the results of supervisory training programs in management.[1] Another study, which surveyed 2,000 training professionals, predicted growth in the breadth and depth of training services provided and increased time spent on management activities such as planning and organizing,

while at the same time reporting a bleak outlook for evaluation.[2] Trainers spend time training, not evaluating.

A survey conducted at a 1984 meeting of a Texas chapter of the American Society for Training and Development supported these findings. While 90 percent of the trainers surveyed believed in evaluations, they did not use them because evaluations were not required by most firms for which they worked. Respondents reported that managers actually discouraged evaluation with "We wouldn't have hired you if we didn't think you were effective." Other studies report the same reluctance of management to "waste time" testing something that management has convinced itself is good.[3]

Management demands for a wider array of programs but not for a determination of their worth can mean danger in the training long run. Lack of demands for evaluation leaves open the potential for growth in training without accountability. This may lead to the continuation or even proliferation of ineffective programs or, in lean economic times, the perception by top management that training and development programs are superfluous and should be cut. If training and development functions are to eliminate the current roller-coaster approach to organizational support, effective evaluation must become a part of every program—whether or not top managers require it.

Effective evaluation in action

An evaluation design used recently at the University of Texas at Austin demonstrated that effective and economical evaluation *is* possible.

When the local job market is strong (such as in May 1985, when Austin unemployment was at 3.9 percent), the university has difficulty attracting skilled personnel to fill high-level support staff positions. To meet its needs the school periodically started offering four courses taught by College of Business faculty to improve the skills of support and secretarial people already on staff. Business Communication Skills was the program selected for evaluation during the 1984 spring and fall semesters. As in most organizations the program had previously been evaluated using only an end-of-course "happiness index." The 1984 evaluation was different in that it measured behavioral change. It was designed to determine whether trainees learned the skills presented in the program, whether trainees transferred the skills from the classroom to the work setting, and what effect the transfer had on use of bottom-line resources.

Business Communications Skills focused on improving the formal communication (writing, speaking, dictating) and supervisory skills of experienced support and secretarial staff. Participants met two hours per day, three times each week for five weeks. The program was held during normal working hours and participants were released from work to attend sessions. Workers participating in the program between 1982 and 1984 were selected for the evaluation.

Of the 96 participants, 86 were women. Their ages ranged from 20 to 61 years, with a mean of 31 years. Sixty percent of the participants were Anglo, 30 percent were Hispanic, 5 percent were black, and 5 percent were Oriental. Seventy-five percent were classified as Senior Secretary I and 25 percent were classified as Senior Secretary II (the higher level). Fifty-two percent of the participants were self-selected, 45 percent were assigned to the course by their supervisors, and 3 percent enrolled due to a colleague's recommendation.

The course covered principles and techniques of communication relating to writing (letters, memos, reports), speaking (public speaking, dictation, telephone use), self-management (listening, time management, planning, decision making), and managing others (selected supervisory skills). The goal of the program was not only to produce knowledge gains, but to enable participants to change their work behavior in these areas.

The course was evaluated with Del Gaizo's four-level evaluation model for employee training programs (see Figure 1).[4] Each level assesses an increasingly complex and difficult-to-measure aspect of training.

Level One: do they like it?

Level One is concerned with whether the participants liked the training program. An end-of-course questionnaire (a happiness index) was used to evaluate this aspect. Results indicated that the participants liked the course, felt it was appropriately instructed, and felt it related to their jobs. The mean ratings—based on a five-point scale with 5.00 as the highest rating—by participants for various aspects of the program were

■ the importance of the course in performing jobs—4.60;
■ the teaching methods and techniques—4.86;

■ the effectiveness of the course objectives, organization, materials, and instructor's response to participants—4.56.

Level Two: did they learn?

Level Two of the evaluation model looks at whether the participants learned the skills taught. An analysis of skills at this level indicated that participants improved their written communication, oral communication, and dictation skills. To measure the achievement in writing, two writing cases were constructed for use in one course group. The cases were checked for content validity by two business professors and one personnel specialist. They determined that the cases were equivalent in difficulty, representative of the skills taught in the course, and representative of the skills used in daily work.

Participants completed one writing sample with a 15-minute time limit during the first course session and the other sample during the last session. The course instructor and an outside evaluator rated the samples. High interrater reliability was found when a random sample of these cases was rated again by two professors not participating in the course or the evaluation study. The interrater responses identified every major and minor success or weakness in writing. Furthermore, an analysis of variance comparing pre- and post-tests indicated that writing on job-related tasks improved significantly.

Oral communication skill improvement was measured by an evaluation of in-class presentations completed before and after instruction in principles of oral communication. The post-instruction checklist rating indicated an average of 20 percent improvement. Dictation skills were

measured using a timed exercise. Analysis, based on a 20-item checklist, indicated that participants demonstrated performance comparable to college senior performance in a managerial communication course in which the average dictation score was B.

Level Three: do they use it?

Level Three considers whether participants used the newly learned skills on the job. A survey based on the course objectives and content was developed. It included a list of job tasks that a group of Senior Secretary I and II employees reported they perform regularly. Revisions were made after a content analysis was performed on all questions, and a pilot test of the survey was conducted as an additional check on reliability and validity.

The survey was then sent to 62 course participants from 3 to 23 months after they completed the program and to the 39 members of a matched, randomly selected control group. This control group matched the experimental group in specific job descriptions, age, years of work experience, and years of employment at the university. The response rate from the course participants was 92.5 percent and the response rate from the control group was 94.1 percent. Responses were mathematically equalized to adjust for the unequal size of the groups.

Program participants reported that they used the skills learned after they returned to their jobs. The participants indicated that they perform more tasks, perform different tasks, and have more job responsibilities than those in the control group (see Figure 2). The program participants reported writing letters 63.6 percent more

Figure 1—Del Gaizo's four-level evaluation model for employee training programs

Level	Index	Question
One	Happiness	Did participants *like* the program?
Two	Learning	Did participants *learn* the skills taught?
Three	Practical application	Did participants *use* the new skills on the job?
Four	Bottom line	Was the training *productive*, cost effective?

often, using oral communication skills as they gave directions and demonstrated equipment 88.2 percent more, using dictation skills 112.5 percent more, and supervising employees 107.1 percent more often than the control group.

This level of the evaluation also showed that 41 percent of the program participants made revisions to materials typed for others without conferring with their supervisors; only 34 percent of the control group reported this. Also, 32.6 percent of program participants reported composing reports, compared to 13.7 percent of the control group.

The survey revealed that 93 percent of the program participants continued to think the course was important to performing their jobs, and 92 percent believed that the techniques and principles presented and practiced during the program helped them perform their jobs more efficiently. In addition, 46 of 50 program participants indicated that they had kept their 60-page resource-reference book, and 37 indicated that they used the program materials at work. Half of the respondents reported using the materials more than once each week, while the other half reported monthly use.

Level Four: did it impact the bottom line?

Level Four considers whether the training program affected the bottom line. Measuring the bottom-line effectiveness of training programs is a difficult and elusive task. A direct approach to evaluation is suited to measuring training results in industrial or goods producing and supplying organizations (by monitoring sales or production performance figures before and after training). However, the direct approach breaks down when applied to service operations in which the input is heterogeneous and uneven, the product is intangible, and the assignment of dollar value to the product is difficult.

While the primary purpose of a university is not to generate monetary profits, any responsible manager must have some way of knowing what goal achievement has cost in terms of resources used. And staff time is the most expensive resource in any organization. In spite of difficulties, it was possible to identify training effects in areas directly related to the use of resources. The evaluated areas, primarily related to the use of work time by the employee and the supervisor, were as follows:

■ *Use of time.* Evaluation of the two pre- and post-test 15-minute writing samples indicated that writing improved after the course. Participants had learned to make better use of their time when composing messages.

■ *Number of revisions.* The survey results indicated that more program participants than control group members composed and revised routine written communications without direct supervision. This meant their supervisors had more time for planning and control functions and for responding to nonroutine information needs. When managers channel their knowledge and experience in this way rather than becoming personally involved in routine office work, the result is more efficient and effective use of personnel dollars.

■ *Duties performed.* Part of the survey was a list of standard duties, developed before this study from task lists submitted by a separate group of administrative support personnel. Program participants indicated that they performed these tasks more than did the matched control group. Program participants also performed tasks that the control group did not perform. The participants seemed to be using fewer personnel hours by accomplishing more and possibly saving the organization money that in the control group would be channeled to additional salaries and benefits. Survey results also indicated that the participants supervised other employees more than twice as often as the control group, and that after completing the program they did work that they had not done before. This translates to an increase in the work performed in each office because it enables managers to focus their attention on supervisory and managerial tasks.

■ *Improved relationships with supervisors.* When interpersonal communication is relatively conflict-free, the result is reduced misunderstanding of duties and assignments, reduced redundancy resulting from miscommunication, and more accurate perception of performance goals by the manager and the staff

Figure 2—Job tasks reported by the business communication skills program participants compared to job tasks reported by the control group

Duties	Program Participants a Percent	Control Group b Percent	Ratio (a–b)/b x 100 Percent
Bookkeeping/accounting	44.0	19.0	131.6
Transcribing from dictation	34.0	16.0	112.5
Making travel arrangements	34.0	16.0	112.5
Supervising employees	58.0	28.0	107.1
Giving directions, demonstrating equipment	64.0	34.0	88.2
Writing letters	72.0	44.0	63.6
Working on the computer (entering, retrieving)	76.0	50.0	52.0
Writing memos	72.0	50.0	44.0
Scheduling appointments	52.0	37.0	40.5
Attending meetings/recording minutes	26.0	19.0	36.8
Recording sick leave/absences	42.0	31.0	35.5
Collecting/checking employee time sheets	42.0	34.0	23.5
Distributing mail	66.0	56.0	17.9
Answering the phone	94.0	81.0	16.0
Maintaining office supplies	58.0	50.0	16.0
Filing	90.0	78.0	15.4
Photocopying	94.0	88.0	6.8
Performing receptionist duties	62.0	59.0	5.1
Typing	90.0	87.0	3.4
Other	34.0	41.0	– 17.1

member—all of which lead to increased efficiency. Sixty percent of the program participants indicated that the relationship with their supervisor improved as a result of the program. The remaining 40 percent reported that the relationship remained the same, which might mean that the relationship was satisfactory prior to enrolling in the course.

So what's a trainer to do?

This study demonstrated that evaluation can be incorporated into training programs by using pre- and post-program skill tests and self-report surveys. Trainers convinced of evaluation's importance and feasibility can strengthen management's support of the practice by doing the following:

■ Build into the design of each program an evaluation that goes beyond the happiness index. Get commitment to this as an integral part of the program. No manager would support a new product, computer system, or marketing plan without evaluation; if you expect your contribution to the organization to be seen as important, prepare to prove it! The evaluation can take the form of follow-up surveys or interviews based on the goals of the training program. The evaluation ideally should include the program participants, their managers, and a control group.

■ As you design programs and develop goals, involve managers and staff. For example, schedule a 15-minute meeting with a manager to identify three things that would help the operation run more smoothly.

■ Build your reputation on a solid foundation; always submit an empirical report after the program and again after the evaluation. These need not be long or technical, but must demonstrate specifically what has changed as a result of the training.

■ As your visibility and support grow, interview—both before and after programs—a sample of managers whose staff has been trained. Base your discussion on program goals and ask specific questions about benefits and the return on the human resource investment.

To quote Peter Drucker, "...few... [organizations] have any idea what they are getting for all the money and effort they spend on training, let alone what they should be getting."[5] If training is ever to be perceived as more than an expendable organizational frill, training and development professionals must insist that the same quality of analysis applied to marketing, production, and accounting systems be applied to training programs. Without this analysis, managers are playing a costly game of chance with their training budgets. And the odds are stacked against them.

References

1. Smeltzer, L.R. "Do You Really Evaluate, or Just Talk About It? *Training* (August 1979): 6-8.
2. Clement, R.W., and J.W. Walker. "Changing Demands on the Training Professional." *Training & Development Journal* (March 1979): 3-7.
3. Carlisle, K.E. "Why Your Training-Evaluation System Doesn't Work." *Training* (August 1984): 39.
4. Del Gaizo, E. "Proof That Supervisory Training Works!" *Training and Development Journal* (March 1984): 30-31.
5. Drucker, P.F. "A Growing Mismatch of Jobs and Job Seekers." *Wall Street Journal* (March 26, 1985): 34.

PART 3.
Reaction

At New England Telephone, training specialists use two quick, inexpensive methods for evaluating pilot sessions of new training programs.

Keeping Your Pilots on Course

BY JAN CHERNICK

You've developed a new training course and it's time to schedule the pilot session. A lot of effort has gone into the development; you think the program will meet the needs of your internal or external clients. But will it? How will you know?

Have you thought about how you will evaluate the pilot session? If not, now is the time to think about it. If the course isn't as effective as it could be, you should know that. Then you can make the necessary corrections before you release it for widespread delivery.

At New England Telephone, evaluation is considered during the development of every new course. But it's often impossible to conduct in-depth evaluations on every pilot; sometimes the specialists receive more requests for pilot evaluations than they can handle. Instead of time-consuming, in-depth evaluations, they use two quick methods that don't require the use of extensive resources.

An important part of the pilot evaluation process at New England Telephone is negotiation with the course designer or the person who requests the evaluation. That helps trainers determine whether to conduct an evaluation and what information it will provide. Several factors play a part in those decisions:

▶ the amount of corporate resources (money, staff, and time) used to develop the course
▶ the number of people who will attend the course
▶ major stakeholders' interest in the course
▶ time or resource constraints
▶ how the evaluation data will be used.

It doesn't make sense to spend resources on evaluation when you know that nothing will be done with the data you collect. For example, there may be constraints that prevent any changes from being made in a course design, such as a mandate that a course be implemented as is.

Once you've decided that you will evaluate your pilot, how do you do it? Two simple methods we've used successfully:

▶ "group debriefs" at the end of each day
▶ short surveys administered directly after specific modules are taught, and again at the end of the course.

Both methods provide valuable information and don't require major investments in time and resources.

What is a group debrief?

A group debrief is a directed group discussion with the participants of a short pilot course (generally three days or less). The best size for these discussion groups is usually 8 to 15 people. During the discussion, a moderator asks participants to rate individual modules of the course and then to respond to specific questions about the modules.

There is virtually no limit to the information you can obtain during a group debrief. You can solicit opinions about any areas of concern. We most frequently ask the participants for their opinions regarding course content. See the box, "What To Ask," for examples of the kinds of questions you might use.

Make sure that the person who conducts the group debrief observes the session first. A firsthand witness is much more likely to understand participants' comments about the course and to ask appropriate questions.

Select the moderator carefully. She or he is a critical link in the success of an evaluation. The moderator should have good facilitation skills as well as the ability to remain unbiased toward feedback as it is given.

The moderator should not be the course designer or instructor. A designer or instructor has a stake in the outcome of the evaluation. It would be difficult for such a person to remain objective and refrain from explaining why the course was designed or delivered the way it was.

Preparing for a group debrief

A group debrief requires some preparation. Make sure that participants in the pilot group are members of the audience being targeted in the training course. People who have another interest in the pilot (major stakeholders such as subject matter experts and supervisors of the target audience members) sometimes enroll as participants. But their perceptions may differ from those of the target audience.

The instructor, course designer, and participants should know in advance that there will be an evaluation at the end of each day of the pilot. This is critical for the participants. If the debriefing session will extend the day beyond normal work hours, people need to be able to plan for it. You may not get useful feedback if people are concerned about getting out late. Also, you may have to consider whether you'll need to pay participants for overtime.

We generally try not to debrief more than five or six modules per day. Otherwise, the session tends to become long and tedious. To make it easier for participants, we prepare evaluation worksheets. As the course progresses, the participants can use the worksheets to jot down their thoughts.

For example, a useful evaluation worksheet could start with the following general instructions.

> *Please note your general thoughts on the material presented. Consider the following: pace, flow, content, and amount of information presented. Please comment on the usefulness of the exercises, where appropriate. Do you know of any questions or issues that were not addressed?*

COURSE DESIGNERS NEED TO KNOW WHAT WORKED AND WHAT DIDN'T WORK

▬▬▬

The rest of the worksheet is divided into three sections, to correspond to the three modules of the course. Trainees can use each section to jot down pertinent comments, in order to have a record of their own perceptions to spark their memories during the group debrief.

At the end of each module, the instructor may want to stop and give participants a few minutes to note their thoughts about the module while it is fresh in their minds. This may disrupt the flow of a course, but it can also yield better feedback during the debrief.

If the modules take longer than anticipated and the group debrief

has to be shortened or eliminated (as occasionally happens even with the best of planning), at least you can refer to the worksheets for some written feedback.

During the pilot course

At the beginning of the pilot, the instructor should introduce the moderator to trainees. If you are the moderator, you should spend a few minutes talking to participants so that they will understand what is expected of them and feel more comfortable about giving feedback.

This discussion should include an explanation of why the course is being evaluated. For example, an evaluation may be conducted in order to determine people's reactions, to see if the course meets its objectives, or to find out if modifications are needed.

The moderator should let trainees know that a group discussion will be held at the end of each day of the course and that they will be asked to give their opinions about the material that has been presented during the day.

It is important for trainees to know why the moderator is present. Your role as moderator is to get participant feedback and to report it to the course designers. Trainees should understand that a moderator has no stake in the evaluation results.

Emphasize to participants that their feedback—positive and negative—is valued. Course designers need to know what worked and what didn't work in order to make improvements to the course.

At the beginning of the course, distribute the evaluation worksheets and explain their use. Let participants know that the worksheets will be collected, but that trainees do not have to put their names on them. This is also a good time to mention that trainees can use the worksheets for comments they don't want to share publicly.

Conducting the group debrief

The session is over and it is time to begin your group debriefing session. Taking breaks during such a discussion can be disruptive, so consider allowing a short break before you begin.

The only people who should be

End-of-Pilot Evaluation Survey
Accounting Coder Training—Trainee Questionnaire

Directions: Now that you have completed Accounting Coder Training, we would like to know your thoughts about the three-day program as an integrated whole. We would appreciate it if you would take a few minutes to respond to the following questions.

1. What is your overall impression of the training?
   - O Excellent
   - O Good
   - O Fair
   - O Poor

2. Listed below are objectives for the training course. Please indicate the extent to which each objective was met:

You will be able to use the Accounting Coder Guide.
   - O Successfully
   - O Somewhat successfully
   - O Not successfully

You will be able to select correct accounting system codes.
   - O Successfully
   - O Somewhat successfully
   - O Not successfully

You will be able to make the proper entries on cost-reporting forms such as the Employee's Personal Expense Voucher, Supplier Bill Payment forms, and so on.
   - O Successfully
   - O Somewhat successfully
   - O Not successfully

3. What is your opinion of the student guide? (Please check all that apply.)
   - O It was helpful in class.
   - O It was easy to use.
   - O I plan to use it on the job.

4. What is your opinion of the Coder's Manual? (Please check all that apply.)
   - O It was helpful in class.
   - O It was easy to use.
   - O I plan to use it on the job.

5. Please rate the effectiveness of the following:

Sample completed forms

Not effective				Effective
1	2	3	4	5

Wall displays

Not effective				Effective
1	2	3	4	5

Code Mania game

Not effective				Effective
1	2	3	4	5

End-of-module games

Not effective				Effective
1	2	3	4	5

Team activities

Not effective				Effective
1	2	3	4	5

If you rated an item 1 or 2, please explain why.

_____
_____
_____

6. Would you recommend this training course to others?
   - O Yes
   - O No
   - O I don't know

Please give a reason for your answer.

_____
_____
_____

7. What other comments do you have about this training course?

_____
_____
_____
_____
_____
_____

Thank you for taking the time to complete this questionnaire.

present during the debriefing are the moderator and the participants. Anybody else who has been observing the pilot should leave the room. At the very least, let such people know that only the moderator and participants should take part in the discussion.

To begin the group debrief, write the name of the first module at the top of a sheet of easel paper. Then draw a horizontal axis marked with a rating scale that ranges from 1 to 9. Explain to participants that 1 means "poor," "the worst," "you wish you'd been someplace else," or any appro-

tend to feel comfortable about giving positive feedback.

What would you change to make it better? What would have to change in order to persuade you to raise your rating to a "9"? This is a non-threatening way to find out what participants didn't like or what didn't work. Creative ideas are generated when people "piggyback" their own ideas onto each other's comments.

Continue the process for each module.

The hardest task for the moderator is to keep the discussion focused

Analyzing the data

So what do you do with all the data? To make the analysis easier, start by having somebody type up the information from the easel sheets and the evaluation worksheets.

The easel sheets will contain both subjective comments and numerical scores. It's best to type up any comments exactly as they are written.

To analyze the numerical easel sheet data, you'll need to compute an average rating for each module. Do that by multiplying the number of responses received for each rating by that rating's point value. Total the results and divide the sum by the total number of participants.

For example, say you have a group of 13 people at your debriefing session. Their responses look like this:
- One person gave the module a rating of 9. *1×9=9*
- Four people gave the module a rating of 8. *4×8=32*
- Five people gave the module a rating of 7. *5×7=35*
- Three people gave the module a rating of 6. *3×6=18*
- Nobody gave the module a rating of 5 or less.

The sum of those totals is 94. Divide that sum by the total number of participants to come up with the average score for the module. In this case, the average is 7.23, computed by dividing 94 by 13.

You can relate the average scores to verbal qualifiers if you wish, based on criteria you establish. People always want to know what the numbers mean and how they compare to the ratings of other pilots. We've established the following descriptors for our ratings:
- 1.0 to 3.9 is poor.
- 4.0 to 5.9 is fair.
- 6.0 to 7.9 is good.
- 8.0 to 9.0 is excellent.

Of course, you can use any descriptors that are appropriate for the particular program.

Compile worksheet comments into one "composite" set that groups together all participant comments for each module. This format makes common themes easily visible.

The survey method

What if you don't have the time or resources to do a group debrief after

THE HARDEST TASK IS TO KEEP
THE DISCUSSION FOCUSED ON POSITIVE
ACTIONS FOR IMPROVING THE COURSE

priate descriptor, and that 9 means "excellent," "the best it can be," or whatever.

Briefly review with participants what was covered in the module. To keep the discussion on a positive note, start with 9 and ask participants for a show of hands for each number on the scale. Under each rating, write the number of people who raised their hands for that rating. It's a good idea to double-check the number of responses to make sure that they match the number of participants.

Sometimes, participants are uncomfortable about giving their ratings verbally. A group may perceive a particular trainee to be a leader, an expert, or a dominant group member. People who do not agree with that person's rating may be reluctant to publicly voice their own ratings. Be sensitive to such feelings. If it appears that participants would prefer it, allow them to note their ratings privately on the evaluation worksheets.

When participants have rated the module, ask the following questions and write the responses as they are given:

What worked well? This provides feedback about the positive points of the module and usually stimulates group discussion, because people

on positive actions for improving the course. For example, if a participant makes a comment such as, "I didn't like that," then the moderator can ask, "What would you have liked?" If a participant says, "I didn't learn anything from that exercise," the moderator can ask, "What would the exercise need to look like in order for you to learn something?"

Sometimes it's necessary to ask additional questions during the debrief in order to clarify comments. Or you may need to poll participants when there is some doubt as to whether a comment reflects the opinion of one person or of the whole group.

Never make judgments about the feedback. As moderator, you should remain impartial. Write down the comments as they are given, even those that seem way off base. There is a natural tendency to respond to comments or questions, to advise participants of changes that may be planned, or to explain why the topics were presented as they were. It is critical not to do that. Such responses can make participants feel as if they are being put on the defensive and can inhibit discussion and feedback.

When you've covered all the modules in the debriefing, collect the worksheets and easel sheets for compilation and analysis.

your pilot program? You can compile some helpful information through the use of short, simple surveys.

Each pilot is different, with unique objectives, content, and activities, so customized surveys may be more useful than standardized forms. Of course, some basic questions can be asked for every course. The most effective surveys address both specific concerns and more general issues.

One approach that we've used successfully consists of short, end-of-module feedback worksheets and an end-of-pilot evaluation survey.

The worksheet for end-of-module feedback can be short and simple. One effective format gives the topic and purpose of the module and then lists three questions, each with a rating scale from 1 to 5. Here are some sample worksheet questions.

▶ *I understood the information: 1=with great difficulty; 5=with great ease.*

▶ *The information presented was: 1=not useful; 5=very useful.*

▶ *The pace of this unit was: 1=slow; 3=about right; 5=fast.*

The only other question is open-ended: *How would you improve this module?*

The module and rating descriptors will be specific to each course. Develop them from negotiations with the course designers about the information needed. Notice that the questions provide the opportunity for quantifiable as well as open-ended responses. You can use a separate sheet of paper for each module to facilitate distribution and collection.

An example of an end-of-pilot evaluation survey is shown in the figure on page 71. It includes some standard questions to gauge participant reactions. Questions on objectives, student materials, and activities are customized in order to provide course designers with information about certain components of a particular course.

Usually, the primary concerns in evaluating pilot courses focus on course content, not instructor performance. If instructor ability or performance may influence the success of the pilot, you may want to consider asking some additional questions to address those issues.

Once you have created the format for such a questionnaire in your wordprocessor, you can use the format over and over again. All you need to change is the customized information.

Remember that the number of participants in pilot sessions is usually small; the information you gather is probably not statistically reliable.

If the participants' opinions about the pilot course are markedly different—or if you are concerned that the participants are not all members of your target audience—then you should gather more data.

Jan Chernick *is staff manager for training evaluation at New England Telephone, 280 Locke Drive, Marlboro, MA 01752.*

PART 4.
Learning

Evaluating Test Questions

By David Blair and Steve Giles

More Than Meets the Eye

It seemed a routine mission for the crack team of training evaluators at Lockheed Martin's Center for Continuing Education. They were in for a surprise.

It began as a seemingly straightforward mission: To evaluate the relative difficulty of questions on a written exam for radiological workers who had completed a safety training course.

But for the group in charge of the test item evaluation, the mission proved to be anything but routine.

The Evaluation and Performance Measurement (EPM) group, formed to oversee training evaluation at Lockheed Martin Energy System's Center for Continuing Education (CCE) based in Oak Ridge, Tennessee, soon discovered it had its work cut out for it. An evaluation of the radiological test

---
In This Story

▼ evaluation
▼ program design and development
▼ program delivery
▼ tests

questions, which trainees argued were overly difficult, uncovered far broader issues in evaluating post-training performance.

In the end, the EPM team—a group of three experienced training evaluators with backgrounds in instructional design—determined not only the cause of worker complaints over the exam, but other factors that called into question the validity of the trainee performance evaluation. The team also mapped out a series of steps to ensure that future testing would be fair and bias-free.

COMPLAINTS OVER TESTING

In early 1995, the Health, Safety, and Environmental Management (HSEM) Institute at CCE was grappling with concerns over in-class testing and performance

reviews of radiological workers who had completed an advanced-level safety training course. After completing training, several trainees provided feedback critical of the written exam's degree of difficulty.

Students also were critical of a post-test shop floor performance review that evaluated their knowledge of 32 so-called "practical factors"—individual performance measures used to evaluate worker competency. "The practical factors evaluations are not placing the same weight on each of the test items," was one common complaint among students. "There are unwarranted failures," was another. Complaints about the performance review component raised eyebrows over possible subjectivity among instructors during that phase of the testing.

When representatives of the local union bargaining group began voicing concerns over the test, the institute's management pledged to evaluate the validity of the radiological test. As a result, the EPM group was asked to conduct a traditional test item analysis of the written test, as well as an analysis of the practical factors exam results to see if failures were above expected rates. Some 116 participants who underwent radiological training at the company's environmental restoration and waste management site in Oak Ridge in early 1995 became the study sample for the EPM evaluation.

WRONG ANSWERS

After an initial analysis of items on the written test, the EPM group determined that three questions were being missed at a statistically significant rate. Frequency distributions—used to identify test questions that were missed by 50 percent of the trainees more than 50 percent of the time—fingered three identical questions on two different versions of the exam.

The group initially assumed that certain instructors were not covering the material adequately. However, incorrect answers were found to be evenly distributed among all instructors, so the EPM researchers began scrutinizing the course's training materials. It became clear that information relating to the questions missed was not being covered adequately in the lesson plan.

The first order of business was to revise the lesson plan to augment those topics frequently missed on exams. In addition, several questions identified by the analysis as unclear were rewritten to remove ambiguity. But the initial complaints that the test was too difficult were not validated by the statistical analysis.

Meanwhile, the performance evaluation phase of the testing had generated its

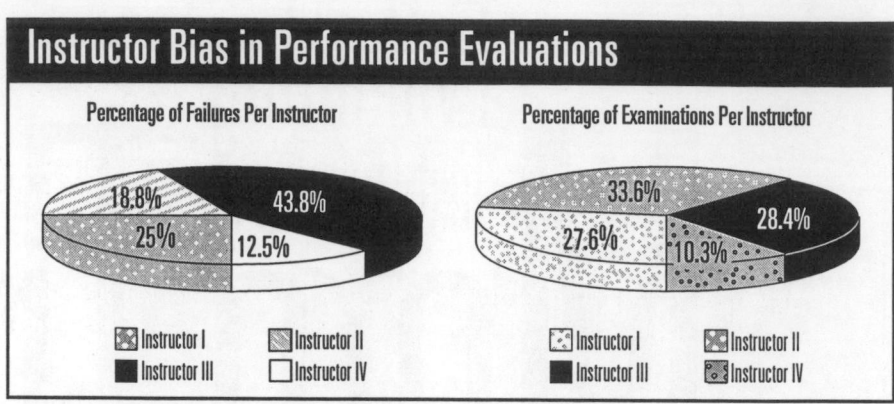

own set of complaints, including student claims of subjectivity and bias among different instructors on the 32 practical factors. The EPM group did, indeed, discover disproportionate pass/fail ratios among instructors performing the practical factor evaluation. (See charts above.) One instructor in particular (Instructor III) had scored a much higher percentage of student failures compared to the total number of trainees evaluated (43.8 percent of failures among only 28.4 percent of examinations.)

In examining this glaring discrepancy more carefully, the group discovered that the instructor was scoring errors in a different manner than the others. Contrary to instructions, the instructor had stopped students when they did not perform a certain task in a prescribed manner—before they could complete the task.

Corrections were made to ensure that all instructors would handle future practical evaluations in the same manner. And all failures were reviewed to make sure that students who might have failed under flawed testing circumstances (even though those who had failed retested and passed) had no record of a failure in their file.

LESSONS LEARNED

In the final analysis, the EPM group's test item evaluation restored confidence in both the Health, Safety, and Environmental Management Institute's testing practices, as well as those of the Center for Continuing Education.

The group's work helped the HSEM Institute in many ways. Union concerns were successfully addressed. In-class instructor evaluations of the delivery of course material, which were already part of the review process for HSEM instructors, also were augmented to provide more measures of instructor quality.

The effort taught the group's members a few lessons, too. As a result, the group added a statistical software package to its toolbox to aid in future test item evaluations. The group members also began performing similar evaluations for other courses taught at the institute. And, most importantly, they began scrutinizing test items earlier—at the pilot stage of new training courses—to avoid similar repercussions.

Mission accomplished.

David Blair is program manager and **Steve Giles** is an institute leader with Lockheed Martin Energy System's Center for Continuing Education (701 SCA Building, 701 Scarboro Road, Oak Ridge, TN 37831; 423/576-7759).

Analyzing Knowledge-Based Tests

Analyze criterion-referenced tests for more effective design and delivery of technical training.

O n the fourth day of a five-day program, the trainer administered a knowledge-based, criterion-referenced test (CRT). First thing next morning, the trainer reviewed the results of the test with participants. Reaching Item 12 the trainer said, "I'm not sure what happened here, but almost everyone missed this item." The participants immediately responded:

"I don't remember covering some of this information!" one trainee said.

"If most of us missed the question, is it a good question?" another asked.

"Are you going to drop this item from the test?" pleaded a third.

The trainer uttered something about looking at the test later, then finished the review. Participants left feeling frustrated and wondering about their test scores. The trainer returned to his office, looked at the test, and thought:

Did the test measure what it was designed to measure? Do test items need to be revised before the test is used again? Which analysis procedures are appropriate for a CRT? Should content, instructional methods, or the time devoted to teaching the objectives be changed?

Analyzing a CRT can provide valuable information for trainers interested in improving both test quality and training delivery. This article offers trainers a useful guide for analyzing CRTs. It is a companion piece to two articles published previously in *Technical & Skills Training*, "Writing Knowledge-Based Tests" (April 1991) and "No-Sweat Tests" (May/June 1991).

By Richard L. Sullivan, Jerry L. Wircenski, and Mary Jo Major

Norm-Referenced vs. CRT Tests

The purpose and results of norm-referenced and criterion-referenced tests differ, with technical trainers often preferring CRTs. Norm-referenced or standardized tests, which are typically administered to large groups of people, are not extensively used in technical training programs. They measure how well each test-taker scores compared to other test-takers. Most norm-referenced tests produce a reliable ranking of the test-takers and, as a result, emphasize individual differences.

But when the test's purpose is to determine if a learner has mastered one or more training objectives, criterion-referenced tests are preferred. Technical trainers use CRTs extensively to determine when trainees have mastered specific knowledge and skills.

The results of norm-referenced tests and CRTs are viewed differently. If you want to determine how your test-takers compare to other test-takers, a norm-referenced test is best. But it will not necessarily reveal how well the training program was designed, delivered, or evaluated. CRT results, on the other hand, record how well your trainees mastered the material or skills. If the majority of trainees achieved the objectives and scored well on the CRT, the training program appears to have been effective. But if the majority of trainees scored poorly, questions arise regarding both training and the test.

The process of developing and analyzing a knowledge-based CRT is presented in Figure 1. Although the focus of this article is on the analysis of CRT results, a review of the development steps can be helpful:

In This Story

▼ evaluation
▼ subject matter experts
▼ program design and development

Test Development and Analysis Process

Figure 1

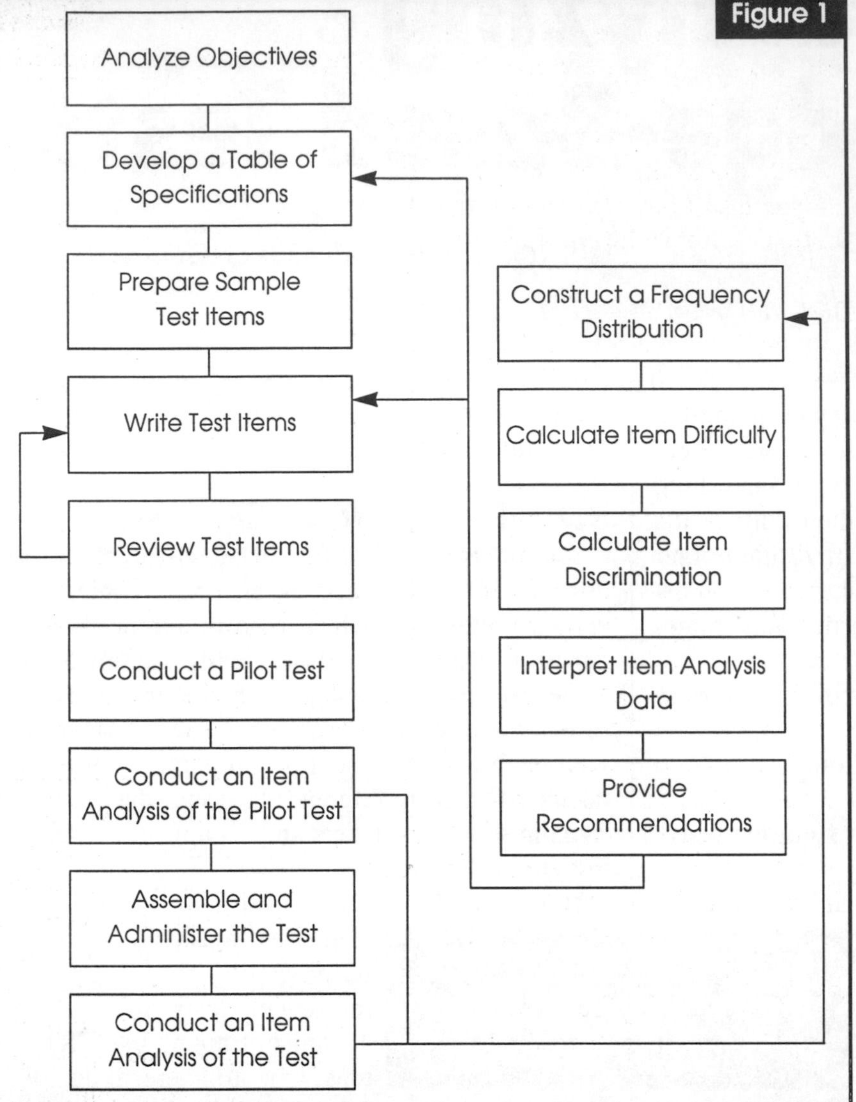

Analyze Objectives

↓

Develop a Table of Specifications

↓

Prepare Sample Test Items

↓

Write Test Items

↓

Review Test Items

↓

Conduct a Pilot Test

↓

Conduct an Item Analysis of the Pilot Test

↓

Assemble and Administer the Test

↓

Conduct an Item Analysis of the Test

Construct a Frequency Distribution

↓

Calculate Item Difficulty

↓

Calculate Item Discrimination

↓

Interpret Item Analysis Data

↓

Provide Recommendations

Frequency Distribution Graph

Figure 2

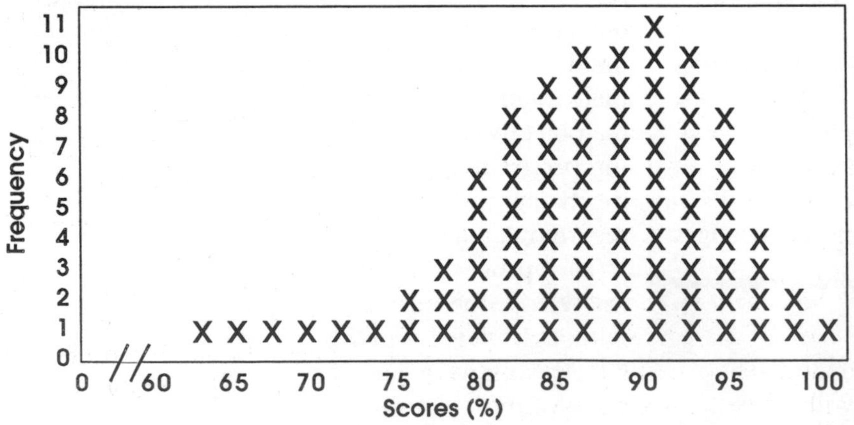

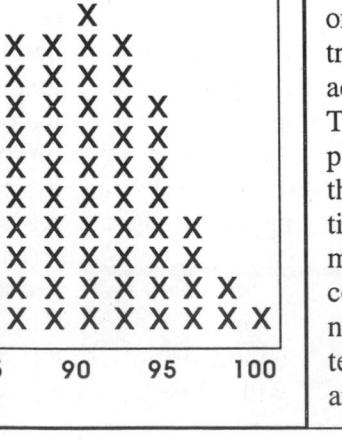

▼ Analyze objectives. Ensure that all training objectives describe the specific knowledge or skill levels required of the trainee and that test items are designed to measure learning.

▼ Develop a table of specifications. A table of specifications or test plan helps determine how many test items will be written at each level of the cognitive domain for each objective. The cognitive or knowledge-based learning domain describes how we acquire information. It is divided into six levels. Essentially these levels represent simple to complex training content or information. The table of specifications ensures that test items reflect the levels present in the training content for each training objective.

▼ Prepare sample test items. Write sample or model test items as guides for writing other test questions. Four multiple-choice test items are shown in Table 2.

▼ Write test items. Develop test items using the sample items as guides.

▼ Review test items. Review items (use a panel of subject matter experts) to ensure that they are based on the objectives and are written to standards. Revise test items accordingly.

▼ Conduct a pilot test. When possible, administer test items to a group similar to the population to be tested. Ideally, participants in the pilot group should be tested only after they have completed the training course. This provides an accurate assessment of the test. Trainers may include practice or pilot items on existing tests and then should analyze item effectiveness. Another option is to administer practice tests or quizzes consisting of new test items. This not only lets trainees take practice tests, it also allows the trainer to analyze the new test items.

▼ Conduct an item analysis of the pilot test. When a pilot test is administered, use the item analysis process described in this article to assess the quality of the test.

▼ Assemble and administer the test. Randomly select test items from an item pool, prepare and duplicate the test. Administer the test. When the same test is administered each time a training course is conducted, the test answers eventually get out. But numerous computer software packages exist that allow trainers to create a bank or pool of test items. Items may be classified according to item type (e.g., multiple-choice, true-false); the objective on which the item is based; item difficulty or complexity; and other variables. Items can then be randomly selected so that the same objectives are always tested but the specific test items change.

▼ Conduct an item analysis of the test. After administering the test, assess test quality with an item analysis that examines item difficulty, discrimination, and distractor effectiveness. Item difficulty is the percentage of trainees who correctly answer a question. Item discrimination is how well an item distinguishes between those trainees who know the information and those who do not. Distractor effectiveness is the ability of each incorrect response (distractor) in a multiple-choice item to distract those who are not quite sure of the correct answer.

The test analysis process is the right-hand portion of Figure 1. Use this five-step process to analyze pilot and final tests:

1. Construct a Frequency Distribution

A frequency distribution graphically displays how trainees did on the test and provides a prelim-

inary view of the effectiveness of the test. Frequency distributions may appear as tables (see Table 1) or as graphs (see Figure 2). The frequency distributions illustrated show the scores of 90 technicians on a trainer-developed test that contained 50 multiple-choice test items. The passing or cutoff score (minimum performance standard) for this example test was 70 percent.

Examine the data in Table 1 and Figure 2. The frequency distributions may look alarming from a norm-referenced viewpoint. The number of high scores may make the test appear to be too easy.

But remember that trainees taking the CRT were given the training objectives (and subsequently the test objectives) from the onset of training. They had an opportunity to learn the required information, and were tested over their acquisition of the information and application. So it is not surprising that the majority of them scored well, as shown in Table 1. Had this been a norm-referenced test, the range or spread of scores would presumably be greater and the shape of the distribution more "bell-shaped."

When a CRT test frequency distribution is not bell-shaped, it is "skewed." The graph in Figure 2 is skewed in a negative direction, with scores ranging from 64 percent to 100 percent, because 87 out of 90 technicians passed the test. A negative skew means that most scores were high, while a positive skew means that most scores were low. The technical trainer analyzing a CRT should expect a negatively skewed frequency distribution.

You can conduct an item analysis of a CRT containing multiple-choice and true-false test items with or without a computer pro-

Frequency Distribution

		Table 1
Score	Percentage	Frequency
50	100	1
49	98	2
48	96	4
47	94	8
46	92	10
45	90	11
44	88	10
43	86	10
42	84	9
41	82	8
40	80	6
39	78	3
38	76	2
37	74	1
36	72	1
35	70	1
——— Passing Level ———		
34	68	1
33	66	1
32	64	1

gram. This article shows how to conduct an analysis without a computer. Most test analysis programs may be used in conjunction with a computer and scanning device. Once the answer sheets are scanned, the computer prepares a test analysis report. When a computer analysis is not available and you want to know the percentage of trainees correctly answering an item (difficulty), who correctly responded to each item (discrimination), and the effectiveness of incorrect responses in distracting those who do not know the answer (distractor effectiveness), use the procedure presented here.

An item analysis to calculate item difficulty involve four steps:

▼ Score the tests by a scanning machine or by hand.

▼ Order the test papers from the highest to the lowest score.

▼ Compare the high scorers and low scorers. The analysis uses

Item Analysis Table

Table 2

Item	Group	N	Alternatives a	b	c	d	Correct Response	Number Correct	Difficulty	Discrimination
1	High	24	0	16	4	4		16		
	Low	24	7	9	3	5	b	9	52%	+0.29
2	High	24	0	0	0	24		24		
	Low	24	0	0	0	24	d	24	100%	0.0
3	High	24	17	4	3	0		3		
	Low	24	2	8	14	0	c	14	35%	-0.46
4	High	24	7	4	7	6		7		
	Low	24	3	6	9	6	a	3	21%	+0.17

Questions

Item 1: The chassis electrical system

A. produces a spark that ignites the air/fuel mixture.

B. powers safety and warning devices, convenience accessories, and lights.

C. produces, controls, and monitors electrical current needed while the vehicle is running.

D. powers the electrical motor used to start the engine.

Item 2: Combustion occurs within the engine. Which of the following is the definition of combustion?

A. Resistance to motion between two parts in contact

B. Heat resulting from friction between two parts in contact

C. Process of heat transfer

D. Process of burning

Item 3: In terms of the engine, friction is considered

A. an undesirable element.

B. a source of heat.

C. resistance to motion.

D. a desirable element.

Item 4: The purpose of the engine cooling system is to

A. circulate air and liquid to transfer heat of combustion and friction away from the engine to prevent engine damage.

B. promote clean, efficient burning of the fuel mixture within the combustion chamber.

C. minimize the escape of gasoline vapor from the carburetor, fuel tank, fuel cap, and crankcase.

D. mix fuel with air in proper proportions for efficient burning in the engine.

roughly the top and bottom 27 percent of the scores. For example, with 90 technicians taking a test, the number in the high group would be 27% × 90 = 24.3 or 24 trainees. The same number would be in the low group. Note that when the number of trainees is relatively small (fewer than 20), the trainer may divide the group in half and at least analyze distractor effectiveness.

▼ Starting with the top score, count off the number of tests in the high group. Count the same number of tests for the low group,

beginning with the lowest score and counting the selected number. These are the tests used in the analysis process.

At this point, complete a table as shown in Table 2. Before calculating item difficulty and discrimination, a trainer could make a preliminary analysis of the results. Using the data in Table 2 as an example, the following conclusions are possible:

▼ Item 1 appears to be a good, yet difficult, test item. The distractors (alternatives A, C, and D) are working. More trainees in the

high group answered the question or item correctly (16) than did those in the low group (9).

▼ Item 2 may either be too easy or perhaps it relates to a critical point that has been stressed.

▼ Item 3 shows a problem. Many more trainees in the low group answered the item correctly (14) than did those in the high group (3). The item may be somewhat ambiguous and the high scorers are reading something into response A. Also, no one selected alternative D, so it is not a functioning distractor.

▼ Item 4 appears to be very difficult, with all distractors functioning extremely well.

2. Calculate Item Difficulty

Item difficulty is represented by the percentage of trainees answering the item correctly. Assume we are analyzing the responses to Item 1 as shown in Table 2. Item difficulty would be the percentage of test-takers correctly answering this item.

The following formula calculates item difficulty:

$$\text{Discrimination} = \frac{\text{high group correct} + \text{low group correct}}{\text{total in both groups}}$$

The difficulty for Item 1 in Table 2 has this solution:

$$\text{Difficulty} = \frac{16 + 9}{48} = \frac{25}{48} = 0.52 = 52\%$$

This indicates that 52 percent of the trainees answered Item 1 correctly.

3. Calculate Item Discrimination

Item discrimination is represented by the correct items and tells us how well an item distinguishes between those trainees who know the information and those who do not.

The following formula calculates item discrimination:

$$\text{Discrimination} = \frac{\text{high group correct} - \text{low group correct}}{\text{number in one group}}$$

The discrimination for Item 1 in Table 2 has this solution:

$$\text{Discrimination} = \frac{16 - 9}{24} = \frac{7}{24} = +0.29$$

This value of +0.29 indicates that the item is discriminating in a positive direction and that more of those in the high group responded correctly than those in the low group.

4. Interpret Item Analysis Data

Remember that item difficulty and discrimination are more often associated with norm-referenced testing. When test items have difficulty levels around 50 percent and high discrimination values (such as +0.5 and above), a bell-shaped curve with a range of scores is more likely. Difficulty and discrimination in criterion-referenced testing are used to signal possible concerns with a specific test item.

Item difficulty ranges from 0 percent to 100 percent. A difficulty level of 82 percent indicates that 82 percent of the trainees answered the item correctly. What is an acceptable difficulty level? This will depend on the type of training program, the training objectives, the importance of the information, and the emphasis placed on the information during training. For criterion-referenced tests, item difficulty often ranges from 70 percent to 100 percent. Once again, trainees are aware of the objectives, they participate in training sessions that focus on the objectives, and they take tests based on the objectives. Thus, many trainees score well on the test, resulting in difficulty levels above 70 percent.

Some training organizations use a value of 50 percent as the difficulty level that signals concern. Values below 50 percent for a test item may indicate one or more of the following:

▼ The test item stem needs revision.

▼ The test item distractors need revision.

▼ Test items related to the same objective need examination. If related items appear to be effective, the item may need revision. If there are problems with other items measuring the same objec-

tive, these concerns may involve methods of instruction, training materials, or the time devoted to teaching that objective.

Using the data in Table 2, both Item 3 and Item 4 would need examination and revision.

Item discrimination ranges from -1.0 to +1.0, with +1.0 indicating that all trainees answering the item correctly were in the high group and all answering incorrectly were in the low group. A value of -1.0 indicates the opposite. A value near zero indicates no discrimination between those in the high and low groups for that item. Trainers should look for positive discrimination values. For a CRT, discrimination values will often be near zero, as many trainees score well on the test. In general, any item with a negative discrimination value needs revision before reuse. The distractors should be examined and revised accordingly. In Table 2, both Item 2 and Item 3 need examination and revision.

When analyzing the results of a trainer-developed, criterion-referenced test, trainers can expect the following results:

▼ Most scores will range from 70 percent to 100 percent.

▼ A graph of the frequency distribution will have a negative skew.

▼ Item difficulty values will range from 50 percent to 100 percent, with an average difficulty in the 85 percent to 90 percent range.

▼ Item discrimination will have positive values.

▼ Some items will have distractors or incorrect responses that are not selected due to relatively high difficulty levels.

Limitations of item analyses include sample size and appropriateness for timed and essay tests. Most item analyses involve a relatively small number of trainees. Consider item analysis information for small groups as only an estimate. Other data and personal judgment should play key roles in item retention or revision.

5. Provide Recommendations

An item analysis gives information for improving the quality of both the test and the training program. When you have concerns about the test, consider one or more of these actions:

▼ Revise individual test items.

▼ Delete individual test items.
▼ Check that test items are based on the training objectives.
▼Check the table of specifications or test plan to ensure that an appropriate number of test items are asked per objective and per level of learning domain for each objective.
▼ Check the time and topic schedule to ensure adequate time is being devoted to conduct training related to each objective.
▼ Ensure that the methods of instruction used are appropriate for each training objective.

Richard L. Sullivan is a professor in the Occupational and Technology Education department at the University of Central Oklahoma (Edmond, OK 73034-0120; 405/341-2980, x 5741).

Jerry L. Wircenski is a professor in the Occupational and Vocational Education department at the University of North Texas (P.O. Box 13857, Denton, TX 76203; 817/565-2714).

Mary Jo Major is a trainer and consultant for CommuniSkills (8224 N.W. 99 St., Oklahoma City, OK 73162; 405/ 722-1371).

Does the Trainee Know Best?

By Jerry L. Harbour

What can a large group of trainees in a technical field reveal about their perceptions of your instructional methods? To what extent does the perceived effectiveness of an instructional method affect actual learning?

To answer the second question, technical trainers must draw conclusions from the answers to the first. Recognizing this, I developed an instructional methods inventory (IMI). The IMI's objectives are to clarify and document the perceived learning effectiveness of a number of different group and individual instructional techniques, from lectures to closed-circuit TV. While its initial findings do not try to measure the relationship between a trainee's perception and reality, they provide surprising data on trainees' individual, personal opinions.

The audience

The IMI was initially directed to an audience of geologists and geophysicists working for large, integrated oil companies. My assumption was that these mature, degreed professionals had been exposed to numerous instructional methods and media during their academic and professional careers and therefore represented a valuable source of information. I hoped that a survey of the opinions of such a discerning audience would yield detailed information and prove to be reasonably objective.

In the initial pilot study, the IMI was given to 200 geologists and geophysicists working for a major U.S. oil company. One hundred and seventeen IMIs were completed by participants who had just finished a one- or two-week

Harbour is a private training consultant based in Stillwater, Oklahoma.

Advocates of technological tutors may be mystified at the low ratings

corporate technical seminar. The remaining 83 were distributed randomly to professionals not currently in a training program.

The majority of the participants were 20 to 39 years old; just over half of the whole group were 30 to 39. Ten percent were in their forties, but only two of the participants were over 50.

The distribution of the participants by degree reveals that 37 percent had earned a BS, 49 percent an MS, and 14 percent a PhD.

The survey

The IMI, shown in Figure 1, is in two parts. The first part asks participants to rate 24 instructional techniques on a scale from 1 (ineffective) to 9 (highly effective). Note that the scale reflects only relative, not absolute, values. The cumulative ratings of the instructional techniques are shown in Figure 2, and I will discuss the results later in this article.

The second portion of the IMI involves a series of isolated questions dealing with other facets of training. The first question lists eight words and asks the trainees to assign two of them to the methods they rated highly effective. My choice of these eight descriptive words—active, creative, relevant, doing, structured, observing, logical, and experimental—is based partially on a learning style inventory model de-

veloped by D. A. Kolb in *The Learning Style Inventory*. A majority of the trainees used *active*, *creative*, and *doing* to describe what they considered highly effective techniques, which suggests that they are inspired by involvement in the learning process. This conclusion, however, should not surprise those familiar with Malcolm Knowles's andragogical concept of adult learning put forth in his book, *The Modern Practice of Adult Education*. He maintains that learning is increased when the learner is actively and creatively involved in the process.

Brochures advertising technical training attach importance to the stature, reputation, and public record of the trainer. In order to measure the effect of such advertisements on individual trainees, the next question on the IMI asked participants to rank what they thought were the most important characteristics of a good instructor. These characteristics included:
- competency in the subject matter;
- publication record;
- ability to create a learning environment;
- reputation and stature in the profession.

The ability to create a learning environment was judged the most important characteristic, with competency in the subject matter coming in a close second. Reputation and publication record had little importance.

The next question addressed the oldest and still one of the most frequently used instructional methods: the lecture. How much lecture can an individual withstand during a normal training day? In an effort to calibrate trainees' lecture tolerance, IMI participants were asked to state how many hours per day of lecture they thought they could absorb effectively. Interest-

Figure 1 — The Instructional Methods Inventory

Age _____   Highest Degree _____

PART 1

From a learning viewpoint, how effective do *you* perceive the following individual and group instruction techniques? Please use a rating scale from 1 (ineffective) to 9 (highly effective).

Instructional Techniques	Ineffective	Moderately Effective	Highly Effective
1. Lectures (not illustrated)	1 2 3	4 5 6	7 8 9
2. Group dicussions	1 2 3	4 5 6	7 8 9
3. Computer-assisted instruction	1 2 3	4 5 6	7 8 9
4. Annotated reading lists	1 2 3	4 5 6	7 8 9
5. Lectures (illustrated)	1 2 3	4 5 6	7 8 9
6. Group brainstoming sessions	1 2 3	4 5 6	7 8 9
7. Interactive videodisc	1 2 3	4 5 6	7 8 9
8. Reading a book	1 2 3	4 5 6	7 8 9
9. Videotapes (viewed in a group setting)	1 2 3	4 5 6	7 8 9
10. Small-group, structured problem-solving exercises	1 2 3	4 5 6	7 8 9
11. Self-paced, individualized learning packets	1 2 3	4 5 6	7 8 9
12. Videotapes (viewed alone)	1 2 3	4 5 6	7 8 9
13. Instructor-presented case studies	1 2 3	4 5 6	7 8 9
14. Field trips	1 2 3	4 5 6	7 8 9
15. Workbooks (worked alone)	1 2 3	4 5 6	7 8 9
16. Closed-Circuit TV (viewed alone)	1 2 3	4 5 6	7 8 9
17. Demonstrations by an instructor	1 2 3	4 5 6	7 8 9
18. Small-group, practical simulation exercises	1 2 3	4 5 6	7 8 9
19. Self-discovery exercises (conducted alone)	1 2 3	4 5 6	7 8 9
20. Thinking alone	1 2 3	4 5 6	7 8 9
21. Closed-circuit TV (viewed in a group setting)	1 2 3	4 5 6	7 8 9
22. Small-group, self-discovery exercises	1 2 3	4 5 6	7 8 9
23. Writing a paper	1 2 3	4 5 6	7 8 9
24. Mentally reviewing a subject	1 2 3	4 5 6	7 8 9

PART 2

25. Of those instructional techniques that you rated highly effective (7–9), which *two* words best describe why you ranked them as such?

_____ Active	_____ Relevant	_____ Logical	_____ Structured
_____ Observing	_____ Creative	_____ Experimental	_____ Doing

26. What do you feel is the ideal length of a seminar?   _____ days

27. How many hours of lecturing do you feel that you can tolerate and *absorb* during an all-day training session?   _____ hour(s)

28. Please rank (4-highest to 1-lowest) the most important
_____ Competency in the subject matter.
_____ Publication record.
_____ Ability to create a learning environment.
_____ Reputation and stature in their field of speciality.

29. In a training seminar, do you prefer multiple instructors?   _____ No   _____ Yes   If yes, how many? _____

30. What do you feel is the ideal size for a training seminar devoted to group interaction and learning?   _____ participants

---

ingly, professionals who completed the IMI at the end of a one- or two-week training seminar wanted to sit through fewer lecture hours than those who randomly received the IMI. It appears that, on average, approximately three hours of lecture, with appropriate breaks, in an all-day training course is sufficient.

When asked whether they preferred one or more instructors in a seminar, 68 percent of the trainees favored only a single instructor. Results also show that they think a training seminar should have no more than 12 trainees and last approximately four and a half days.

The learning versus the medium

As we can see from Figure 2, the

Figure 2 — Perceived effectiveness of instructional techniques

Instructional Techniques	Ineffective	Moderately Effective	Highly Effective
1. Field trip	1%	16%	83%
2. Lectures (illustrated)	0%	23%	77%
3. Small-group, practical simulation exercises	2%	22%	76%
4. Small-group, structured problem-solving exercises	1%	27%	61%
5. Instructor-presented case studies	1%	31%	68%
6. Writing a paper	7%	33%	60%
7. Group brainstorming sessions	5%	40%	55%
8. Group discussions	6%	38%	56%
9. Demonstrations by an instructor	1%	49%	50%
10. Small-group, self-discovery exercises	9%	48%	43%
11. Thinking alone	4%	68%	28%
12. Self-paced, individualized learning packets	8%	65%	27%
13. Reading a book	10%	65%	25%
14. Individual, self-discovery exercises	11%	61%	28%
15. Mentally reviewing subject	14%	64%	22%
16. Interactive videodisc	15%	66%	19%
17. Workbooks (worked alone)	17%	64%	22%
18. Lectures (not illustrated)	25%	55%	20%
19. Computer-assisted instruction	23%	65%	12%
20. Videotapes (viewed alone)	23%	65%	12%
21. Annotated reading lists	26%	62%	12%
22. Closed-circuit television	24%	68%	8%
23. Videotapes (viewed in a group stting)	25%	67%	8%
24. Closed-circuit television (viewed alone)	28%	68%	5%

most favorably rated instructional methods involved active group participation or graphic presentations made by an instructor. Again, we can see the importance to the trainees of being part of the learning process.

Strong advocates of technological tutors such as computers, videos, and the like may be somewhat mystified at the low ratings given computer-based learning (CBL) and various video-based systems. Attempting to discover the ideal instructional machine has preoccupied educational researchers since 1912, when E.L. Thorndike recommended pictures as a labor-saving device in instruction. This research has continued through the radio era of the 1950s and the television era in the 1960s until today, with computer and advanced video interactive systems now predominant in so much of the research.

Research and summary articles by R.E. Clark, however, seriously challenge the concept that one particular learning medium is more effective than another. According to Clark's article, "Reconsidering Research on Learning from Media," appearing in *Review of Educational Research*, analysis of media studies clearly suggests that in-

structional media do not influence learning under any conditions. He notes that "the best current evidence is that media are mere vehicles that deliver instruction but do not influence student achievement any more than the truck that delivers our groceries causes changes in our nutrition."

Learning is more important than the machine that delivers it

Differences in learning outcomes using various media are more likely attributed to independent variables—especially differing instructional program designs and, to a lesser extent, the novelty of high-tech tools—than to the specific medium itself. Studies indicating increased learning effectiveness through CBL, for example, simply may be measuring differences in the way the program is produced or the increased attention given to a novel situation.

But the IMI is not to be used to iden-

tify the actual effectiveness of one instructional medium over another, because it measures only perceptions of media effectiveness. These perceptions may affect learning outcomes by directly affecting the learner's persistence and perseverance. The learning, however, seems more important than the machine that delivers it. Perhaps more effort is needed to motivate adults to learn, rather than simply to repackage the learning in another medium.

What it all means

Preliminary findings of this instructional methods inventory suggest the following:

■ When polled collectively, a mature and well-educated group of professionals can reach a consensus concerning the perceived learning effectiveness of certain instructional techniques.

■ Techniques perceived as being highly effective include field trips; illustrated lectures; small-group (two to four people), relevant simulation exercises; small-group, structured problem-solving exercises; and instructor-presented case studies.

■ The most popular training seminar

is one that lasts four and a half days, is composed of 12 people, has a single instructor who can create an effective learning environment and is highly competent in the subject matter, and employs a creative and active mix of instructional techniques.

■ Training courseware must be creative, active, and participatory. Adults seem to learn best when they are active in the learning process.

■ While the overwhelming preference is to learn in an interactive group setting, trainees have some need for effective instructional methods such as reading a book or writing a paper that can meet an immediate, individual training need.

■ Surprisingly, most technological tutors are not perceived to be highly effective training tools. A possible explanation is that this particular study group was technologically sophisticated, and such instructional techniques did not have the value of novelty.

■ This initial study did not investigate the relationship between learner preference and actual learning outcomes. Do trainees know best? If they think a training approach is effective, does that mean that they actually learn more? And if they do, how much more? Certainly, that is an issue deserving further study.

The conclusions drawn from this study are based on a highly educated, specialized sample population. I don't know how well such findings can be generalized to a broader training segment. I do know, however, that as the potential contribution of human resources is increasingly realized, a greater effort will be placed on adult learning. Improving and understanding our perceptions of learning effectiveness will not only have an immediate and positive impact on the individual adult learner, but will also improve our corporate training programs and, ultimately, the productivity of our society as well.

IBM Takes the Guessing Out of Testing

By George M. Alliger and Harold M. Horowitz

An innovative new way to assess training courses can measure not only trainees' knowledge but also their confidence in their answers.

Measuring how effectively training satisfies corporate goals is a major challenge for training and development specialists. Many training programs have no assessment mechanisms to determine their quantitative effectiveness. Instructors typically solicit the opinions of the students. They ask, "Did this course meet its stated objectives?" or "Did you enjoy this course?" But "happiness-sheet" feedback hardly answers the crucial question, "Are the students learning anything?" To answer that question, trainers must measure knowledge gain and retention.

In 1985, IBM's internal education organization began a programmatic research effort to find out whether new, internally developed training technologies increased knowledge transfer and retention. In an experimental course, the organization successfully developed and implemented a unique method of knowledge testing, employing pre- and post-tests. The testing goes beyond reaction measurement (Kirkpatrick's Step One) and also provides more information than that gained from knowledge testing (Step Two). (Donald L. Kirkpatrick discusses his four steps for training evaluation—measuring reaction, learning, behavior, and results—in "Evaluation of Training," *Training and Development Handbook,* edited by Robert L. Craig.)

Alliger is an assistant professor of industrial and organizational psychology at the State University of New York, Albany, NY 12222. Horowitz was program director for instructional applications research at IBM, and now is president of HMH Associates, a firm specializing in interactive classroom technologies.

Instrument construction

A six-hour experimental course, designed to make maximum use of new training technologies, was the initial target for the new program of knowledge measurement. The course designers separated the course objectives into 10 main segments; each segment encompassed, on average, 10 key learning points. From that material, 100 content-valid, true-or-false questions were written. The course designers took great care to write questions that were challenging but fair, and relevant to the learning points and course objectives. In fact, the designers' goal was to create, when possible, questions that were "mini-case-studies"—that is, questions that required some situational analysis as well as knowledge of a principle or idea. In addition, the designers compared the learning points with the questions to ensure that the questions thoroughly covered the course's 10 knowledge areas and the learning points within each area.

After extensive experiments with the questions to obtain estimates of difficulty, the designers created two non-overlapping, parallel families of tests (four tests per family). Each test contained 25 questions, balanced according to the knowledge areas. Because of the particular circumstances, the designers considered that multiple forms of the test were important, not only to ensure differences between pre- and post-test questions, but also to minimize the ability of instructors to "teach to the test" instead of teaching the course as designed. In the same

---

Figure 1 — Means and standard deviations for different scoring methods

	Mean	SD
Pre-Test		
S1 (score on "reasonably sure" answers only)	52.9	17.4
S2 (score on "reasonably sure" and "not sure" answers)	70.8	10.7
CI (confidence index)	65.4	21.6
IS (intuitive score—correct answers marked "not sure")	55.2	29.6
Post-test		
S1	79.9	9.9
S2	82.9	8.7
CI	94.4	8.2
Gain (difference between pre-test and post-test S1)	27.4	18.4
Reaction (students' overall satisfaction: 1=high; 5=low)	1.5	.7

Sample sizes for pre-test variables, 2,210; for post-test variables, 2,226; for gain, 1,812; and for reaction, 1,511. Intuitive score was not calculated for post-test.

way, particular circumstances (such as time constraints) dictated the choice of the true/false format and the number of questions.

Trainees could answer each question, of course, either correctly or incorrectly. The designers, however, developed a true/false format that they hoped would be more interesting to those taking the test and at the same time yield more information than a standard true/false test. For each question, students could answer that they were "reasonably sure" of a true or false answer, or that they were "not sure but probably" the answer was true or false, or that they didn't know the answer. That way, the tests measured not only what students knew but also their confidence in that knowledge.

Administration and scoring

Before presenting the tests, instructors gathered data on student demographics, such as intracompany organization, occupation, time in profession, and time in company. That data would be used, for example, to analyze which internal organizations did particularly good jobs of preparing their students.

Students took the pre-tests before taking the course and the post-tests several days after it. The tests were administered on optical mark-read forms, and the programs that computed the responses were designed to score the tests in several different ways, to take advantage of the unique test format. Descriptions of the scores follow.

■ S1. First, each test was scored according to "reasonably sure" responses only. That score (S1) represented the test's primary score and was the first one the students received as part of their feedback. One may argue that S1 results underestimate the true score, because guesses (those answers in the "not sure" category) are not included in its computation. The logic was that because the subject matter was critical to a student's job, being unsure about a principle or fact did not meet the criterion of knowing. In any case, results showed that the following, more lenient computation might have served as the primary test score without altering conclusions.

■ S2. The second method of scoring the tests was to measure both "sure" and "not sure" answers; in that manner, S2 resembled traditional test scoring, where guessing is included. For

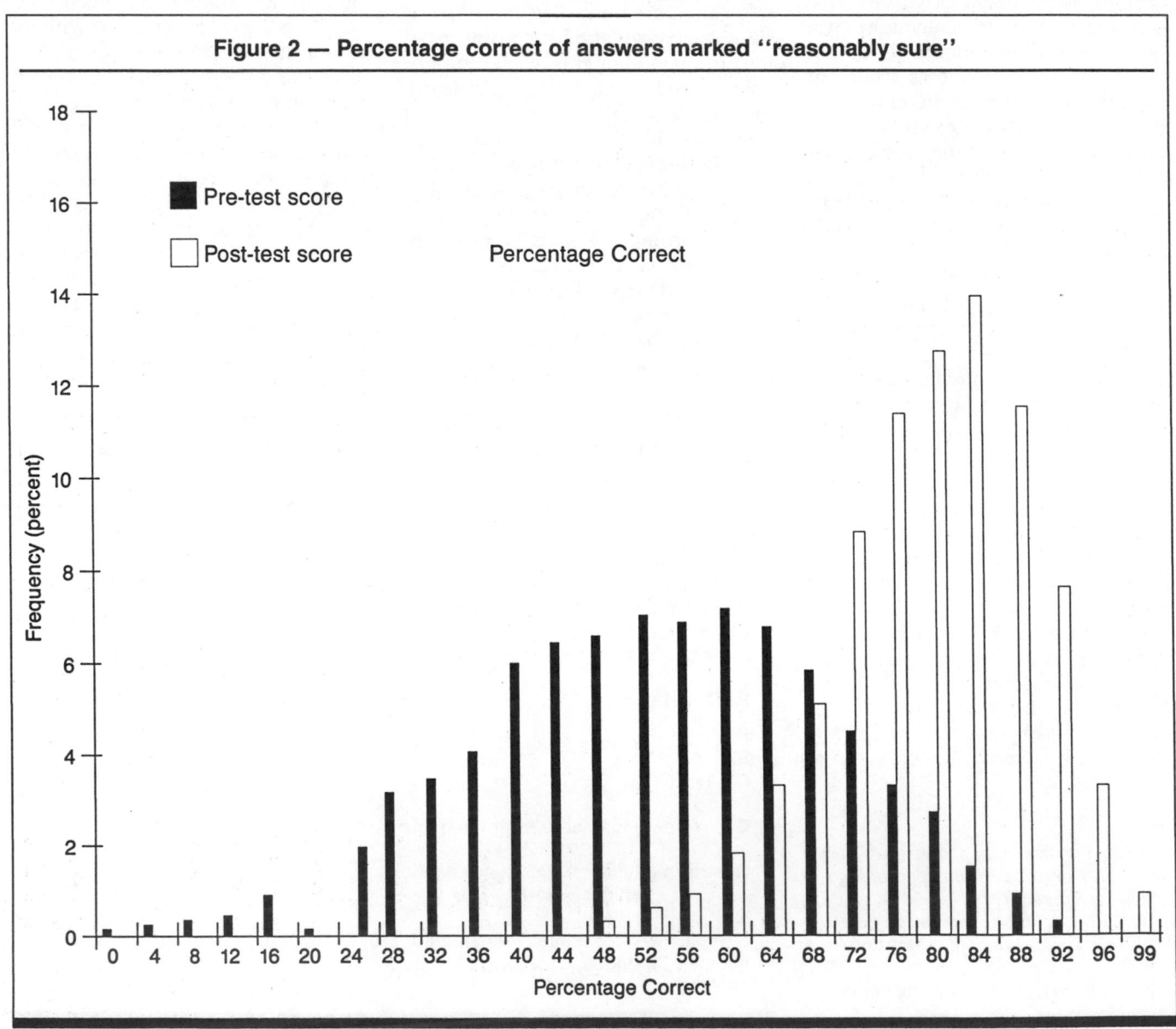

Figure 2 — Percentage correct of answers marked "reasonably sure"

■ Pre-test score
□ Post-test score

Percentage Correct

Frequency (percent) / Percentage Correct

anyone who gets at least one correct "not sure" answer, S2 will be higher, obviously, than S1.

■ **CI.** The confidence index represented the percentage of answers that the students felt "sure" about—the higher the percentage of "sure" answers, the greater confidence the student indicated. A 100 percent CI, for example, meant that the student was sure about all of his or her responses. On the other hand, the score did not reflect the actual correctness of those responses.

■ **IS.** The intuitive score was the percentage of correct answers in the category of "not sure." IS was used only for pre-test data, because students generally use intuition, rather than knowledge, prior to learning; IS answered the question, "Are the students guessing correctly?" The "don't know" category served as an overflow—in order to keep the "not sure" category unaffected, it did not enter into the computation. In fact, the "don't know" category was rarely used by the students.

For reasons of interpretability and reliability, an IS was not computed for a student if he or she provided fewer than three "not sure" responses.

■ **Gain.** Gain is the difference between the pre-test S1 and post-test S1.

■ **Reaction.** At the end of the test, the students rated their overall satisfaction with the course on a scale of one (high) to five (low).

Means and correlations

Figure 1 shows the means and standard deviations for the four variables in the experimental testing. As expected, S2 showed a higher mean value than S1. That difference is more pronounced in pre- than in post-test data, probably because only a small percentage of answers were "not sure" on the post-test, as the higher post-test CI reflects. One can note in passing that the training had a strong effect on each variable.

Figures 2 and 3 illustrate the relationship between S1 and CI across the range of students. In both sets of

scores, increases in means and decreases in variances occurred between the tests, but one can see that the shape of the distributions differs from one to the other, CI piling up at the 100-percent ceiling.

Figure 4 shows the correlations between the pre- and post-test variables. The strongest relationships are between S1 and CI for pre-test data, which indicates that the surer the students were about their answers, the higher their scores were.

On the post-test, the correlation drops to .58, due to the decreased standard deviation (see Figure 1). Strong negative correlations between both pre-test S1 and CI and the gain highlight the fact that lower scores in correctness and confidence on the pre-test meant greater increases in post-test knowledge; that slant toward low initial scores is a typical problem of gain scores. The correlation between S1 and S2 on the pre-test was only .58, indicating the distinction between the two types of scoring. On the post-test, the

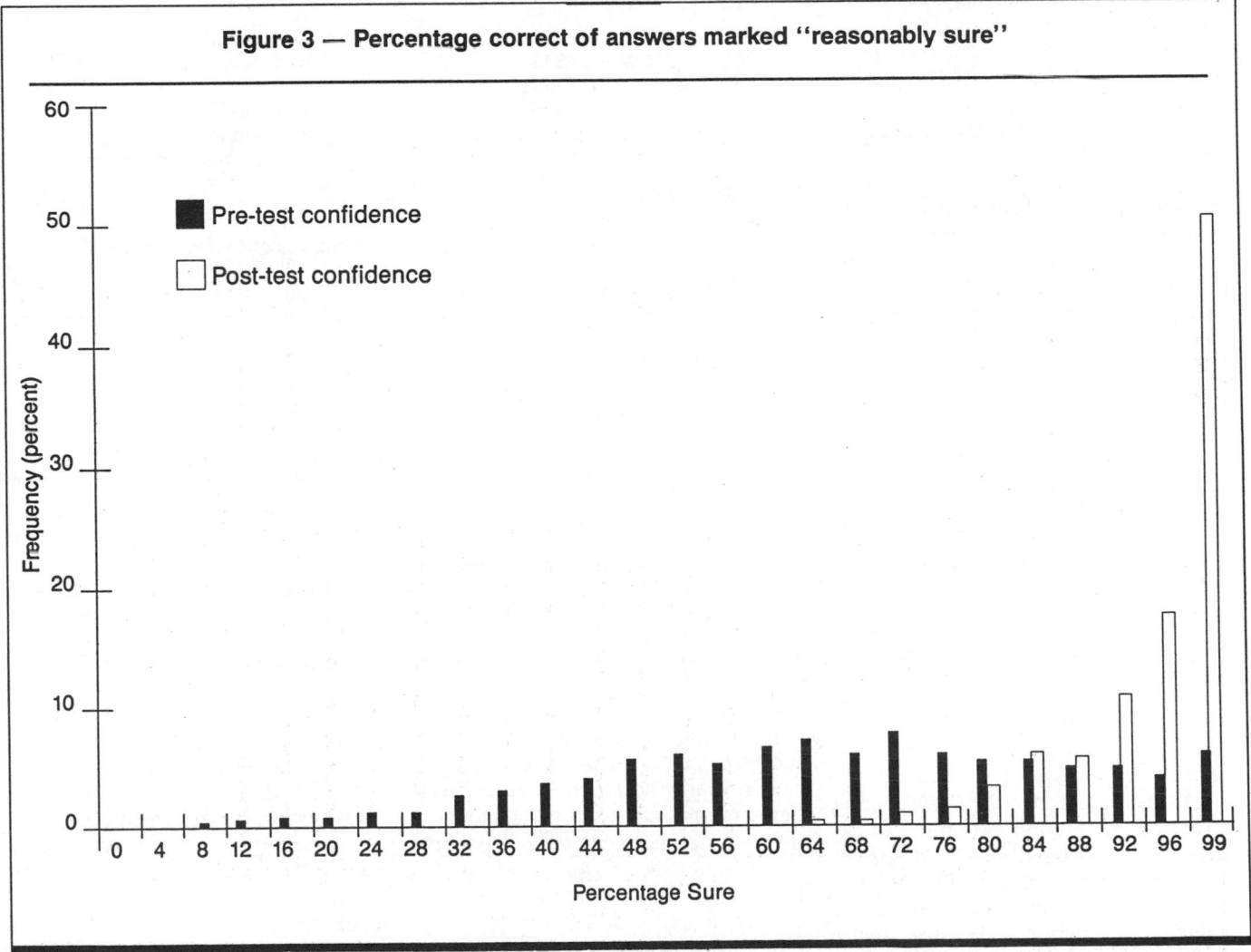

Figure 3 — Percentage correct of answers marked "reasonably sure"

■ Pre-test confidence

□ Post-test confidence

Frequency (percent)

Percentage Sure

scores correlate strongly, due to fewer "not sure" responses.

It is interesting to note that reaction had a correlation of virtually zero with measures of pre- or post-knowledge (.02 and − .02 respectively) and knowledge gain (− .03). Although those low correlations may be due in part to the restricted range of the generally favorable reaction ratings, they also reflect the probability that course ratings and actual learning are different measurements. That distinction may have implications for the use of Kirkpatrick's Step One (reaction) measures to assess the impact of training programs.

Using the results

The indices obtained from the tests aided the organization in several ways:

■ Course administrators analyzed each question and used the results to help redesign the course.

■ A larger study used the results to measure course effectiveness.

■ The demographic analysis of the results aided in an organizational examination.

■ The measurement results served to rate the effectiveness of the course instructors.

■ In the step that took most advantage of the new test format, students received feedback on the results. Figure 5 shows a fictional example of typical feedback to a student. The form, generated entirely by computer, was accompanied by a cover letter that explained the nature of the various scores.

Discussion

The introduction of the CI, or confidence index, as reflected in the CI data and indirectly through the other scores, may be the biggest success of the test format. Behavior in the workplace is probably a function not only of what a student knows but also of how certain he or she is of that knowledge. Moreover, the fact that students did not have problems understanding the format could indicate that separating knowing into "sure" and "not sure" is a practical measurement approach.

Further research in the design of the questions is certainly possible. For example, a multiple-choice test format could present questions and options followed by a secondary scale on which the student could indicate his or her level of confidence. Of course, the

most interesting aspect of the test format would be a validation that related scores to performance: does the level of confidence add to the prediction of performance supplied by the standard, unitary measurement of knowledge?

This method of testing measures knowledge and confidence simultaneously; to put it another way, it provides a way to take the guessing out of

testing. The test format and its results can be interpreted easily, and many students expressed appreciation for the feedback they received. In industrial training and development, few methods can accurately measure knowledge gain and retention—perhaps IBM's experimental measurement method could be valuable for other companies as well.

Figure 4 — Major variable intercorrelations

		Pre-test				Post-test				
		S1	S2	CI	IS	S1	S2	CI	Gain	Reaction
Pre-test	S1		.58	.92	−.44	.19	<u>.03</u>	.31	−.85	.02
	S2			.36	.13	.21	.18	.12	−.44	.01
	CI				−.48	.13	<u>−.05</u>	.34	−.81	.02
	IS					<u>.00</u>	.10	−.15	.41	.00
Post-test	S1						.84	.58	.36	−.02
	S2							.13	.43	−.01
	C1								<u>−.02</u>	−.02
	Gain									−.03

Sample sizes for the above correlations vary from 1,511 to 1,259. However, only underlined correlations are not significant at at least the $p < .05$ level.

Figure 5 — Student feedback form

**Personal and Confidential
Course Score Card**

A

Report Date 4-12-89
Class Number: 554a
Name: John Q. Student
Years in Company: 5
Months in Job: 1

Organization: Marketing
Function: Administration

Avg. Years in Company: 9.4
Avg. Years in Position: 2.1

B

Score (%)

	Your Score	All Marketing	All Administration	All Students
Pre-test	28	47	47	47
Post-test	72	79	76	80

C

Confidence Index (%)

	Your Score	All Marketing	All Administration	All Students
Pre-test	40	61	59	58
Post-test	88	96	91	94

D

Intuitive Score (%)

	Your Score	Number Answered "Not Sure"	All Marketing	All Administration	All Students
Pre-test	67	12	54	55	57

By Vivian Marshall
and Rob Schriver

USING EVALUATION TO IMPROVE PERFORMANCE

At Martin Marietta Energy Systems, a five-level model evaluates knowledge and skills separately.

Traditionally, training professionals and managers have focused on training as an end result: Training was measured by the number of employees trained, hours spent in training, and number of training sessions held.

In fact, employers were so focused on training itself that they did not worry about whether the trainees used the training on the job or if the training actually met job needs. Employers simply passed excessive labor costs along to the customer.

Moreover, because employers were worried about training—and not job performance—evaluation was usually based on asking the participant if he or she "liked" the training, or if he or she "felt" that the training was good. And while trainers created paper and pencil tests to evaluate the participant's knowledge, those tests often only evaluated one's ability to regurgitate information given in the training session.

In This Story

▼ evaluation
▼ performance-based training
▼ transfer of training

This focus on training as the end result meant that few if any companies tried to evaluate training beyond the first two levels—which examine attitudes and knowledge—of the four-step model developed in 1959 by Donald Kirkpatrick (see Figure 1).

PERFORMANCE-BASED TRAINING

But during the 1960s, a new approach to training and evaluation emerged. Robert Mager, Joe Harless, and others took the instructional systems development (ISD) model from the military and adapted it into a new systematic approach called performance-based training (PBT). PBT contains the following five phases:
▼ analysis
▼ design
▼ development
▼ implementation
▼ evaluation.

Developing PBT requires us to invest time in front end analysis, which provides needed information for developing training and performing all levels of evaluation. This information includes the following points:
▼ performance problem
▼ performance indicators (for long-term summative evaluation)

KIRKPATRICK'S MODEL FIGURE 1

Level 1	**Level 2**	**Level 3**	**Level 4**
Measures attitudes and feelings	Measures knowledge	Measures trainee's use of skills on the job	Measures business results

▼ target audience
▼ job role and responsibilities
▼ job tasks
▼ steps of tasks.

But despite the development of PBT, few employers during the 1960s and 1970s were willing to spend the money necessary for front end analysis. They thought they could buy enough technology to take the human element out of the work processes.

Then, as foreign competition eroded U.S. market shares during the 1980s, employers slowly realized that buying new technology would not keep them competitive. And in the late 1980s, a skill deficiency problem emerged, as did the awareness that America's educational institutions were providing industry with graduates who lacked the necessary skills and knowledge to meet job requirements. Employers began to realize that they had to do something to fill this gap for their employees.

As a result, training departments started trying to improve the quality of training. But again, the focus was on training as the end result. Training improvements were made to the formative side of training, and an evaluation in best-case scenarios was performed, as shown in Figure 2.

NEW EVALUATION MODEL

In 1992, at Martin Marietta Energy Systems, we developed evaluation guidelines for PBT. First, we researched the literature, including Kirkpatrick's model, and interviewed people associated with the training process. As a result, we discovered that some people misinterpreted Kirkpatrick's model. They believed that evaluating for knowledge was the same as testing for skills. Because skills and knowledge were lumped into the same category (level 2 of Kirkpatrick's model), people assumed that they were testing "skills" when all they were testing was knowledge—whether the trainee "knew" the material.

But PBT requires that the trainee demonstrate *both* skills and knowledge before leaving training. This means the trainee can perform under job-relevant conditions to the standards required for the task.

We determined that a new five-step model was needed (see Figure 3) to separate evaluation for knowledge *and* skills:

▼ Level 1 in our model also measures attitudes or feelings about aspects such as the instruction, instructor, and training environment. We recommend a "smile" sheet to gather this data.

FORMATIVE EVALUATION FIGURE 2

	Internal Reviews	Subject Matter Expert Reviews	Pilot Test
Target audience selection	✔	✔	
Quality assurance	✔	✔	✔
Instructional validity	✔		✔
Content validity		✔	✔
Learner acceptance			✔
Learner mastery of knowledge			✔
Learner demonstration of skills and knowledge			✔
Course and instructor critique			
Postcourse trainee critique			
Postcourse trainee supervisor critique			
Skills transfer to job			
Organizational impact			

| MODEL FOR EVALUATING KNOWLEDGE AND SKILLS | FIGURE 3 |

MODEL FOR EVALUATING KNOWLEDGE AND SKILLS — FIGURE 3

FORMATIVE | **SUMMATIVE**

1	2	3	4	5
Smile sheet measures attitudes and feelings	Paper and pencil measures knowledge	Performance demonstration measures skills and knowledge	Skills transfer, behavior modification measured by job observations	Organization impact, including ROI, measured by cost savings, problems corrected, etc.

▼ Level 2 measures knowledge by itself, using a paper and pencil test.

▼ Level 3 requires the trainee to demonstrate both skill and knowledge by performing the task to job standards and under job-relevant conditions. In other words, the trainee must demonstrate that he or she can do the task. There is no need to test for knowledge itself if the trainee can already demonstrate his or her knowledge through the actual performance of the task or skill.

▼ Level 4 measures skills transfer, or behavior modification. Evaluating this level must include job observations in the work environment to ensure transfer of the skill from the training session. Such evaluation occurs at some specified interval after the training course, such as 60 days, 90 days, six months, or one year.

But beware: Some individuals think that they are performing level 4 evaluation when, at some interval after the training was conducted, they use a questionnaire that measures feelings and attitudes. This is still level 1 evaluation, even though it occurs after the training.

▼ Level 5 measures organizational impact and the return on investment. To perform this level of evaluation, you need to establish the performance indicators during the analysis phase. But even after doing that, performing level 5 evaluation is tricky because of the many uncontrollable variables that can affect the results.

Training itself can only be measured using the first three levels. Levels 4 and 5 measure more than just training because, as stated earlier, many variables can affect the trainee's performance. In other words, if the work environment does not support the new skill learned, the trainee will not use the skill or the skill will not transfer.

THE REAL END RESULT

To set up an evaluation program successfully, training must be viewed as a means to the end; not the end itself. The goal of a training program must be improving performance in the workplace.

We do not mean to imply that evaluation on the formative side is not important. All levels of evaluation are vital to ensure the quality of training and to help ensure organizational improvement and competent workers. The evaluation matrix in Figure 4 highlights the importance of both formative and summative evaluation. So if we can focus our efforts and dollars on performance improvement, we will see a significant return on dollars spent for training.

Vivian Marshall and **Rob Schriver** are supervisor of performance systems development and manager of performance support systems, respectively, for Martin Marietta Energy Systems, Inc. (701 SCA Bldg., 701 Scarboro Rd., MS 8240, Oak Ridge, TN 37831; 615/574-2450).

	FORMATIVE EVALUATION			SUMMATIVE EVALUATION	
	Internal Reviews	Subject Matter Expert Reviews	Pilot Test	End-Of-Course Evaluation	Follow-up Evaluation
Target audience selection	✔	✔			
Quality assurance	✔	✔	✔		✔
Instructional validity	✔		✔		
Content validity		✔	✔		
Learner acceptance			✔	✔	
Learner mastery of knowledge			✔	✔	
Learner demonstration of skills and knowledge			✔	✔	
Course and instructor critique				✔	
Postcourse trainee critique					✔
Postcourse trainee supervisor critique					✔
Skills transfer to job					✔
Organizational impact					✔

Using Competency Exams for Evaluating Training

A competency exam can help you decide whether a training program has been effective, but only if the exam can pass the test of content validity. Here are seven steps for ensuring that your exams are valid.

By Jack E. Smith and Sharon Merchant

Evaluating training is a critical aspect of the training process. One widely accepted approach now in practice is the use of competency exams, which are administered at the completion of training. Competency exams attempt to measure how well knowledge and skills are transferred in training.

Competency exams can be written (paper and pencil) or practical (hands on) and are an excellent way to measure a training program's success. They benefit organizations by requiring employees to demonstrate actively their understanding of new knowledge and new skills.

Properly developed and administered, competency exams can answer various questions:
■ Can the person who has been trained to weld, for example, actually do the kinds of welds that are required on the job?
■ Can the customer information representative provide customers with the correct information or refer them to the proper agency?
■ Can the manufacturing supervisor successfully use a new computer-based manufacturing resource planning system introduced into the plant?

By retraining those who have not acquired the knowledge and skills needed to do their jobs, an organization can improve employees' performances and the effectiveness of the organization as a whole.

Accountability for learning

Competency exams can also increase trainees' motivation levels, because a "testing hurdle" is required at the end of training. When trainees know they will be tested when training is completed, they are more likely to attend sessions, actively participate, concentrate, and study course materials. In this way, trainees are made accountable for their learning.

Trainers are also held accountable for their instruction. Competency exams provide trainers with valuable feedback for improving training. For

Competency exams increase trainees' motivation levels, because a "testing hurdle" is required at the end of training

example, consistently low scores by trainees on certain parts of an exam may indicate that the training should be revised. More information may be needed, exercises may need clarification, or more time may be required to cover a particular topic. Confusion or misunderstanding on the part of trainees can provide trainers with direction for improving subsequent training efforts.

Smith *is a principal of Jack E. Smith and Associates, a consulting firm at 34770 Whittaker Court, Farmington, MI 48024.* **Merchant** *is the MRP Training and Education Coordinator for Land Systems Division at General Dynamics, 38500 Mound Road, Sterling Heights, MI 48310-3286.*

Seven steps to content validity

Trainers who develop competency-based training and use competency exams must ensure that the content of each of their exams is valid.

Content validity asks the question: "Does the exam adequately measure a trainee's performance on a job-relevant body of trained knowledge, skills, and behaviors?" With a content valid approach, those who know the requirements of the job try to ensure relevancy in training. They then check exam items against content.

Subject matter experts can help ensure content validity. Experts know about job requirements and training programs. Experts may include supervisors, experienced incumbents, trainers, and other knowledgeable personnel. Subject matter experts can help ensure exam content that is compatible with instructional objectives, training, and job content.

For best results, competency exams and statistical analysis should be monitored by expert testing professionals, particularly with respect to the construction of test items.

Content validity is demonstrated by performing a series of systematic steps, beginning with a thorough examination of job and training content and ending with the administration of an exam and analysis of the results. The following seven important steps can facilitate the development and administration of valid competency exams, both written and practical.

Step 1: Develop learning objectives

Construct meaningful, well-written learning objectives that describe what a successful learner is able to do after receiving training.

Learning objectives typically are derived from two sources. First, they are developed from job analysis information, which carefully examines important job tasks along with the knowledge, skills, and abilities needed to perform those tasks. Second, they come from new methods or procedures being introduced into the workplace. It is important to identify carefully new tasks, methods, knowledge, skills, and abilities to use them for creating learning objectives.

Figure 1—Examination content outline

CONTENT AREAS			Written items (paper and pencil)	Practical items (work samples and simulations)
General categories	Sub-categories	Weight (percent)		
I. MRP—II overview		5	1, 2, 3, 4, 5	
II. MRP—II		50		
	A. Concepts		9, 10, 11, 13	1, 2, 4
	B. Netting logic and application		8, 12, 14, 15 16, 18, 19, 25 26, 27	3, 5, 6, 7, 11 13
	C. BOM		6, 7, 17, 28	8, 9, 10, 12 14, 15, 16
III. Maintenance transactions		25		
	A. Creating an order		20, 21, 22	17, 18, 21
	B. Allocating material		23, 24, 29	19, 20
	C. Releasing order		30, 31	22, 23, 24 25, 26

Step 2: Outline exam content areas

Once job-related learning objectives are developed, construct a thorough outline of the exam content. An outline ensures that relevant subject matter is neither omitted nor inappropriately emphasized on an exam. Figure 1 shows a sample outline for a generated manufacturing orders training module.

Step 3: Establish exam format

Once the content of the exam has been constructed, decide on its format (written or practical) and choose which types of items—multiple-choice, true/false, or matching—to use. Exam items should closely conform to actual job tasks. The closer the link between test items and required job behaviors, the easier it is to ensure and demonstrate content validity.

Step 4: Construct exam items

Develop and write the test items for the exam. Competency exams fall into two broad categories: written and practical. Most people are familiar with written exams. Practical exams require trainees to actively demonstrate skills in situations that best simulate the job environment. For example, a welder may be required to execute a particular type of weld according to certain specifications.

Put it in writing. Here are some general guidelines for written exams.
■ Construct questions and answers that are definite, precise, and objective. Types of questions include true/false, multiple-choice, matching, and fill-in-the-blank.
■ Construct test items in language that is appropriate for the reading level of the trainees taking the test, unless the test is measuring reading ability or vocabulary.
■ Do not replicate statements from the training manual. Exam items should test comprehension of the training material rather than an ability to memorize.
■ Begin the exam with a few easy, yet relevant, questions. Taking an exam can be an intimidating experience for some trainees. A few easily answered questions at the beginning often relieves test anxiety.
■ Make sure that exam items do not provide clues to the answers of any other exam items.
■ Avoid the use of interlocking or interdependent items, where answering one item is contingent upon correctly answering another.
■ Avoid trick questions. Trick questions tend to put test takers on the defensive. They also give trainees who are unfamiliar with the material an advantage; they can randomly select an answer, while prepared individuals tend to make the wrong response.
■ Avoid ambiguous and subjective questions. One way to ensure that questions are clearly written is to have other instructors or subject matter experts review the exam beforehand.
■ Beware of questions that deal with trivia. Stick with questions that promote problem solving and reasoning and inference, even though they are more difficult and time-consuming to construct. Keep course objectives in mind so as not to fall into the common trap of constructing easy-to-write, irrelevant questions.

Hands-on testing. Practical exams have several advantages over written ones. Employees can be tested in situations that clearly resemble their jobs and can apply training concepts using the same skills used on their jobs. Performance results are immediately visible to both trainer and trainee. Also, trainees often perceive practical exams as being fairer than written exams.

Here are a few guidelines for constructing practical exams.
■ Use work samples or simulations to measure skills by recreating certain aspects of a job under controlled conditions.
■ Work samples should duplicate the conditions and challenges faced in training and on the job, including considerations such as the typical length of a task, the complexity of the task, and the availability of references or work aids such as manuals, desk procedures, and computers. Ideally, an exercise should duplicate as closely as possible the normal duties and conditions expected of the employee.
■ Develop enough exercises to constitute a reliable sample of a trainee's behavior. Generally, the more hands-on exercises on the exam, the better.
■ Develop a set of scoring criteria and an objective scoring procedure. This step may be difficult for some trainers who mistakenly believe that they will know good performance when they see it. The problem is that what one trainer considers unacceptable, another may deem satisfactory.

Common criteria for testing might involve measurements such as the number of units processed, conformity to quality standards, the amount of reworking or repeated trials allowed, and the speed of completion.

Construct questions and answers that are definite, precise, and objective

Scoring methods are often checklists of acceptable or unacceptable performance as rated by experienced judges. If scoring depends on individual judgment, then two or more subject matter experts should independently score each exercise, and there should be common agreement between experts as to what constitutes acceptable or unacceptable performance. Areas of disagreement may suggest that changes are needed on the disputed items, or these items might have to be excluded altogether from the final exam.

New employees should be allowed more time to complete an exam than experienced employees.
■ Interdependent items should be avoided, as on written exams. For example, successfully passing practical exercise number 5 should not be contingent upon successfully obtaining the correct data or information from exercises 1 through 4.
■ Pilot-test exercises with a representative group of examinees or with subject matter experts who have never seen the exam. A trial run should duplicate the intended use and conditions of the work sample. Pilot-testing

work samples is crucial because creating realistic job replicas is complex; many problems can arise that even an experienced test developer cannot predict.

■ Revise or change exercises to reflect the results of trial administration. In some cases, a new or different job simulation may be necessary. Other times, some fine-tuning of directions, scenarios, or scoring methods may be the solution.

Administer the revised exercises to a different group of examinees or experts. When that is not practical, carefully observe and take notes during the first administration. Minor problems can often be corrected before a second administration.

Step 5: Evaluate content validity

When the exam is complete, the next major step is to systematically evaluate each of its aspects. The following questions must be answered affirmatively in order to prove content validity:

Are all exam items job-related? And does success on the exam translate to satisfactory performance on the job?

Each exam item should be rated in terms of its relationship to successful job performance. Knowledge or skill that is not needed on the job has no place on a competency exam.

Are the knowledge and skills that are being tested adequately covered during training?

Each item should be examined to determine the degree to which the skill or concept being measured is adequately addressed during training. This review should occur prior to administration, but additional information can be collected when training and testing are completed. For example, a 50 percent success rate on an item may signal to the trainer that this

Competency Testing at General Dynamics

Beginning in 1984, the Land Systems Division of General Dynamics began a new, division-wide program for planning and organizing its manufacturing process. This new computer-based operating system, called Manufacturing Resource Planning (MRP II), integrates all major manufacturing functions. MRP II is a new way of directing the distribution and use of resources to meet General Dynamics Land Systems's manufacturing requirements.

Successful implementation of MRP II requires functional-specific users to be thoroughly trained and competent before they are allowed to go on line and access the database. Indeed, MRP II experts cite inadequate training as the number-one reason for system failure.

A special division-wide task force representing key functions and plants was created to implement MRP II throughout the Land Systems Division.

The task force included a Training and Education Coordinator who had responsibility for coordinating all training development, user qualification, and education activities that support MRP II implementation. More than twenty training courses were developed for hundreds of employees. The task force and top management mandated that no employee would be allowed access to the database without proper training. Employees would have to demonstrate that they could meet requirements for the appropriate functional-specific training courses. As a result, competency exams were constructed to demonstrate user qualification.

Here is how General Dynamics conducted its competency testing.

■ 1. General Dynamics contracted a consultant expert in competency testing to ensure the systematic validation of all written and practical exams.

■ 2. An internal, division-wide MRP II Training and Education Coordinator coordinated the validation efforts and ensured consistency in approach and content.

■ 3. The MRP II training and testing target group included some union-represented employees. The consultant and Training and Education Coordinator met with union officials to explain the user-qualification process, the justification for using competency exams, and how the exams would be used.

This session also addressed various issues and concerns raised by union representatives, including re-testing, failure in training, and seniority. Union officials were assured that employees would have an opportunity to re-test and that failure would result in reassignment with no reduction in pay or loss of seniority.

■ 4. A content validity manual, *MRP II Qualification Testing: A Manual for Trainers,* was developed and given to each trainer and subject matter expert. This manual was the basis for a workshop on content validity and served as a reference document during the validation process.

■ 5. Subject matter experts and designated trainers from each of the functional areas attended a one-day workshop on content validation. Before the training, experts were asked to read the content validity manual. During the workshop session, the experts engaged in a series of exercises that took them step by step through the content validation process.

■ 6. The external consultant and the MRP II Training and Education Coordinator held individual meetings with the subject matter experts for each of the functional areas. The purpose of the meeting was to clarify roles, schedule sessions, and reinforce material covered in the manual and in the training.

■ 7. Work sessions were held with subject matter experts to systematically develop and validate written and practical exams for each of the functional areas. Work sessions began with a review of training, content, and learning objectives. They continued through the seven steps of the content validity process until the exam had been administered and analyzed and appropriate feedback given to the trainer. The content area should be reviewed, and adjusted if necessary.

Do the separate items sufficiently reflect the entire content of the exam? And, do they fairly represent all aspects of the training?

The key to getting an accurate sample of a trainee's skills is to construct an outline of the exam content areas. The purpose of the outline is to ensure that the exam is comprehensive and does not weight certain areas too heavily at the expense of other equally important ones.

Are enough items included on the exam to ensure the reliabilty of a

development of competency exams was integrated with the development of the training material and content whenever possible.

The results

The use of competency testing at General Dynamics has had positive side effects. First, the validation process helped many subject matter experts to systematically examine their training. Because the experts had little or no background in training or testing, developing content valid exams helped them verify that their training was relevant and that only essential information was included in training.

Second, competency testing kept trainers on their toes during the actual training process. Success on a competency exam is as much a reflection of trainers as it is of trainees.

Thus, trainers were careful to prepare for and cover all relevant information during training and to avoid the common temptation to say to trainees, "Don't worry if you don't understand it now, you'll pick it up back on the job." When trainers knew they were being judged by exam results, they were more determined to deliver successful training.

Finally, our preliminary results indicate that competency testing has helped maintain training and testing consistency and integrity regardless of plant location.

trainee's exam results?

Reliability tells us how much confidence a measuring device merits. If a test is highly reliable, then trainees can be expected to obtain comparable scores on similar tests covering the same content area.

The primary reliability issue for competency testing is the length of the exam. If every possible question covering all the material presented in the course could be asked, reliability for how well each trainee mastered the training material would be 100 percent. That is not possible because of limited time and resources, so the trainer must instead select a sample of representative items to be given as the competency exam.

The more questions asked (assuming all items have content validity), the more adequate the sample of all possible questions. By asking more questions, a trainer comes closer and closer to estimating a trainee's true level of knowledge and skills after training.

Are the instructions and items well-written and in a form that is understandable to trainees?

The exam process is a form of communication between the technician who develops the test and the trainee who takes the test. Like other forms of communication, it can be effective or ineffective. Consistency and "understandability" provide the framework for good communication. Commom terminology adds to consistency. For example, the terms used on an exam should closely reflect the terms used during training and on the job.

"Understandability" means simply that trainees should comprehend the language of the exam. Exam developers should keep the vocabulary of their trainees in mind when writing test items.

Has the exam been pre-tested for content, clarity, completion time, and coverage?

Under ideal circumstances, the completed exam should be pre-tested on a group of employees similar to

those who have completed the training program. This is generally not possible for most competency testing, but there are two other approaches for previewing exams.

One way is to set aside the exam questions for a few hours, or even days, and then give them a final review. This hiatus allows a trainer to look at the items from a fresh perspective. Questions or hands-on exercises that seemed clear at first may appear vague or inadequate at second glance.

The other approach is for experts or peers to review content or take the exam. They can then provide feedback on clarity, coverage, completion time, and other features. Too often, the trainer has in mind what the question and answer should be. Another expert who looks at the same question and answer may not know what is expected; therefore, the test-taker

Keep trainees' vocabulary in mind when writing test items

may not know, either. Pre-testing can occur at the same time that subject matter experts are reviewing and rating exam items with respect to job relevance and adequate coverage in training.

Will the exam have "face validity?"

In other words, will the text look correct and fair to the trainees? Test takers often become upset if an exam appears to have little or no relationship to their jobs. "Face validity" can be enhanced by ensuring that all parts of the exam (written and practical) closely reflect job requirements. While "face validity" is not a replacement for demonstrated content validity, it does create more favorable reactions.

Will the exam be administered and scored under standardized conditions?

Standardization specifically refers to the requirements of standard content and standard procedures for administration and scoring. Employees should receive identical exams that are administered under the same conditions and scored or evaluated against the same performance standard.

Figure 2 provides trainers with a

useful checklist for evaluating the content validity of their exams.

Step 6: Exam administration

The goal of a content valid exam is to obtain an accurate estimate of a test taker's performance on the job. Everything should be done to ensure that reliable results are obtained from each trainee. Standardization is one way of obtaining that reliability.

Another way to increase the likelihood of reliable results is to provide a positive test environment—meaning adequate space, a comfortable and well-lit room, needed resource materials, and a lack of distractions and interruptions.

A third way to get reliable exam administration is to have a knowledgeable and prepared test administrator who is familiar with the exam content and who understands the concepts being tested. Ideally, this should be the trainer, not a proctor.

Additionally, good exam administration means giving exam performance feedback to trainees as quickly as possible. Immediate feedback on exam results not only provides trainees with a more positive view of the training process, but it has also been shown to enhance trainee learning.

Step 7: Analysis and feedback

The job of training, test building, and content validation is an ongoing process. Training and exams should be reviewed at intervals in the light of current needs and changes in the workplace. An exam that is satisfactory today may need to be revised or replaced tomorrow due to changing technologies and job requirements.

Once an exam has been administered to a group of employees, it is possible to generate valuable information on trainee performance, instructor effectiveness, and test item characteristics. A statistical analysis of responses made by trainees on the exam can provide such data. Feedback information can then be used to improve both the content of the exam and the content and presentation of the training program.

Test designers can calculate such test and item statistics as exam means and standard deviations, item difficulties, item discrimination indices, and item test correlations.

The most useful statistic for competency exams is the difficulty level of each exam item. This statistic shows how many trainees failed an item or work sample, and it is easy to calculate. If an exam item has been properly developed, then a high difficulty rating typically indicates problems with the training content or process. Changes such as increasing the time and emphasis spent on that topic may be necessary.

The competency exam in use

There are three important issues concerning what happens when a competency exam is given.

The first issue concerns preparing the test takers. The goal is to provide employees with ongoing performance feedback during training so that trainees are adequately prepared for the exam and so that the exam contains no surprises in content or format.

Questions and exercises interspersed throughout training can provide feedback, or trainees can be given a set of practice questions or exercises. For example, if a trainee is going to be asked on a test to call up and interpret computer transaction screens, then computer-based exercises should be part of training.

The second issue is re-testing. What happens if an employee does not pass an exam? Generally speaking, the policy should be to allow employees the opportunity to retake exams. The goal of competency testing is not to screen out trainees but to ensure that each employee masters the targeted learning objectives. Those who fail should receive remedial training and be allowed to re-test. Remedial training may include additional formal or classroom training, on-the-job coaching or training, or self-study.

The final issue concerns personnel decisions. Some organizations use competency exams to make hiring and promotion decisions and to upgrade employee classifications. If a competency exam can affect an employee's future status, the exam may come under scrutiny to determine if it complies with equal employment opportunity regulations. Following the steps outlined here then become even more important, and each step should be fully documented in writing. Properly developed and administered exams can be insurance against discrimination charges.

FIGURE 2—Content Validity Checklist

Question	Yes	No
1. Are exam items job-related?		
2. Are the skills being tested adequately covered during training?		
3. Do the exam items sufficiently reflect the entire content domain?		
4. Are enough items included on the exam to assure the reliability of training exam results?		
5. Are the instructions and items well-written and in a form that is understandable to the trainees?		
6. Has the exam been pre-tested for content, clarity, length (time), and coverage?		
7. Will the exam have face validity for those employees taking it?		
8. Will the exam be administered and scored under standardized conditions?		

PART 5.
Behavior

Training 101

Are you looking for new ways to collect confidential employee opinions and make use of them? Take another look at a traditional method. That old employee survey has some new twists.

What's new with the old concept of surveying employees for their ideas? Plenty! Recent innovations may make the employee survey the centerpiece for modern concepts of empowerment, feedback, and participative management. Let's take another look at employee surveys and discover how they meet the new needs of the 1990s.

The purpose behind employee surveys is to solicit worker perspectives about how a company is doing. Employees often have viewpoints that can move the organization to new heights and counterbalance the all-too-frequent "business-as-usual" comfort of many managers.

Why don't employees tell management what they think in the course of day-to day operations? Some do. But they may not be taken seriously, partly because managers have no way of knowing if a few assertive voices actually represent most employees. Many employees are afraid to speak out, because they assume their comments are unwanted or will come back to haunt them.

For example, in recent surveys of employees at several organizations, only 29 percent to 41 percent of workers agreed with the statement, "We say what's on our minds without fear of attack or reprisal."

Employee surveys were devised to help management discover what employees really think. Typically, they are written questionnaires with mostly multiple-choice items. Most surveys are completed anonymously by all employees or a "representative sample" of employees, and then are tabulated by demographic groups.

Surveys have provided important information to management for some time, but several factors account for their recent popularity in organizations. More sophisticated strategic planning, customer satisfaction, and quality-improvement techniques now stress internal as well as external data-gathering. Also, the growing mobility of workers and the tight labor market (while lessened during the recession) have sparked managers' interest in employee satisfaction. Widespread downsizing has raised issues of employee morale and loyalty.

Several major problems have emerged with the administration and analysis of employee surveys:

▶ When a survey is anonymous, it's difficult to get a high return rate. When it is not anonymous, the reliability of answers is questionable because employees may be afraid to tell the truth.

▶ "Check-mark" answers sometimes raise more questions, rather than providing useful data. But narrative answers may be too cumbersome to categorize and tabulate.

▶ Too often, management does nothing with the results of a survey; at least, employees perceive no change.

▶ Long time lapses between the completion of surveys by employees and the implementation of changes by management weaken the tool and any reactions taken in response to its findings.

▶ Traditional questionnaires are one-time measurements of employee feelings about specific factors such as their jobs, supervisors, and pay—rather than perceptions about the broader business developments that are important to companies.

Several recent innovations have reduced the problems associated with traditional questionnaires, and have increased the benefits of employee surveys to organizations.

An ongoing process. The days of the "once-and-done" survey are over. Why solicit employee ideas only once? What about employees who join the company after a survey? How do we know if progress has been made since the last survey? Periodic, regular communication with employees makes more sense.

Surveys are increasingly part of management's overall planning and development strategy, providing regular information for decision making—like financial and customer reports. The new major players in the survey game—line supervisors and managers—want to know how the business is doing from the employees' perspective.

Involvement of line management. Employee surveys once were the province of industrial psychologists and management consultants. More and more, internal human resource or organization development specialists are spearheading the activity. And increasingly, they are involving frontline supervisors and managers throughout the process, from initial design of the instrument to action planning based on the results.

External consultants may also play a part, as advisors throughout the survey process.

Greater supervisory and management participation helps develop line ownership of the process. After all, line supervisors and managers are likely to be held accountable for implementation of any changes made as a result of the survey. Their most critical role may be in the leadership of employee feedback sessions, described below.

Broader business issues. Traditionally, employee surveys were used to gauge satisfaction with narrow issues such as pay and immediate supervisors.

The trend now is to ask employees to rate the organization's progress more broadly. For example, an article in the September 1987 issue of *Training* magazine said that IBM, which has used surveys for more than three decades, now asks employees how it is doing with innovation, efficiency, and the use of information systems. Xerox asks employees to rate its progress in becoming a "total-quality" organization. Johnson and Johnson employees evaluate the company's success in meeting commitments to customers, employees, stockholders, and the community.

This refocusing of the process heightens the interest of managers and supervisors, especially those who are evaluated on performance in those areas. And it helps employees achieve alignment with broader organizational goals and objectives. In today's fiercely competitive business environment, that alignment is a strong side benefit of an employee survey—if not a direct objective.

Customization. Companies once purchased standard survey instruments from national sources. That practice saved in-house staff from laborious development, tabulation, and analysis of data. It allowed for comparison of responses with related industry groups. In general, it was considered to be an economical way to run an employee survey.

Those same advantages still apply, but HR people should keep in mind that standard surveys may not capture what many organizations really need—data on key issues that are critical to current operations. When deciding whether to develop a survey in-house or purchase an off-the-shelf instrument, also take into consideration the specific products or services a firm provides, the firm's geographic locale, and the company culture. Such factors can make it difficult to draw meaningful comparisons with the results of other organizations.

Also, in-house analysis of the results may be easier than it used to be. Computerized data analysis now allows firms to speed up the process of tabulating data in-house.

Outside assistance from an expert in employee surveys can be helpful in many steps of the process:
- designing surveys
- offering advice at critical points
- assuring anonymity or confidentiality
- providing complex tabulations

But external consultants should be prepared to tailor instruments to client needs, rather than relying on off-the-shelf questionnaires and procedures.

Expanded feedback sessions. After the results are in, feedback sessions are essential. In these sessions, managers meet with employees to share survey results and plans for improvements. A firm that does not hold such sessions can reduce employee satisfaction and willingness to

---

Asking the Right Kinds of Questions

Questionnaires can use one or more of the following types of question formats.

Multiple choice. Multiple-choice questions contain two or more mutually exclusive answers. The respondent must pick one of them.

Use multiple-choice questions when all the possible responses to a question can be included, when those responses can be worded so that they are mutually exclusive, and when the forced selection does not result in bias. Advantages of the multiple-choice format include easy tabulation and interpretation, and short response time.

Multiple answer. This format is similar to the multiple-choice format, except that the respondent is allowed to choose more than one statement. Also, multiple-choice responses are exclusive; multiple-answer responses are not. Multiple-answer questions are used to help respondents remember and to ensure that they consider all viable options. Researchers may choose to list only those answers that are of special interest to them, and can provide space for respondents to write in other answers.

Ranked questions. These questions ask respondents to indicate, in order, their personal preferences or perceptions of the relative importance of the answers. A respondent may be asked to choose a first choice, or to rank some or all of the possible responses by numbering them in order of preference.

Open-ended questions. These allow respondents to answer with no prompting. Use open-ended questions when individual responses are important, when the range of responses can't be predicted, or when free expression is needed to clarify a multiple-choice answer.

Scaled questions. These questions are used to determine opinions or attitudes. A scaled question measures direction (positive to negative) and intensity (strongly positive to strongly negative). Survey developers commonly use three types of scales:
- Semantic-differential scales allow respondents to indicate how they feel about a specific item by selecting a position on a bipolar scale (such as agree/disagree or good/bad).
- Diagrammatic scales are grids or diagrams on which respondents indicate their position with respect to a statement. Words or numbers are usually not included on the grid, with the scale expressed instead by an abstract category or continuum—for instance, with symbols or pictures.
- Likert scales provide standard sets of words and ask respondents to indicate agreement or disagreement with a statement. Possible answers are assigned weights, which are used to compute individual and group ratings. For example, on a 5-point scale, 1 could indicate "strongly agree," 2 could indicate "agree," and so forth, with 5 representing "strongly disagree."

— *adapted from* Info-Line 9008, "How To Collect Data"

share ideas with management.

Feedback sessions have evolved into much more than reporting results and planned actions. Many companies now use them as a second phase of the process. In these firms, groups of employees discuss with managers the meaning of the survey results. For example, they may talk about why they feel less content with the company than they did three years ago. Or they may offer ideas on what could be done to improve corporate innovation.

Typically, managers or supervisors lead these meetings with each division or department. They may hold several sessions with the same team of employees. Once the sessions have clarified the issues, the group or the feedback session leader develops an action plan, which may be considered the third phase of the employee-feedback process.

Of course, the feedback sessions and group problem solving fit in nicely with the current business trends toward empowering employees and pushing responsibility downward.

High-tech embellishments. As with everything, computers have given employee surveys a big boost. Obviously, they can speed the tabulation of data, but there's more—much more!

Imagine this scene: Forty-seven employees of a manufacturing firm are seated in front of keypads, responding to questions about their company—the standard questions normally asked on mail-in surveys. Once they've all answered the question of the moment, their group results are flashed on a big screen in front of the room. Smiles, gasps, and looks of surprise flash across faces as they discover the meaning of the latest tally.

They follow the voting with discussion—sometimes in small groups and sometimes with the whole group. They talk about what the results mean.

"Why do you think 20 percent voted 'very satisfied' while another 20 percent voted 'very dissatisfied' on that question?" the facilitator queries.

"Well, each group is treated differently here," answers a young man in the back row.

Even quieter employees are likely to join the discussion, buoyed perhaps by the strength-in-numbers concept, after they see that others voted the same way they did.

"Decision technology" is growing in America. The process allows employees in groups of 10 to 150 to react to organizational issues with less likelihood of being swayed by the biases of the loudest or most aggressive group members.

The voting is anonymous, as with the feedback given on most written surveys. But unlike other methods of surveying, decision technology allows both managers and employees to receive instant results. The software's instant tallying makes it possible for the group to clarify questions about what the responses mean, on the spot. Finally, the computer and the group do the analysis work at the session, rather than tying up support staff for days or weeks.

The process offers anonymity, objectivity, speed, and clarification—a combination of advantages that are not available through other means. Growing numbers of corporations are setting up conference rooms specially equipped for decision technology. And some management-consulting firms can transport such systems to company locations for facilitated workshops.

Changes in attitudes. The new trends affecting employee surveys reflect changes in the way we do business. They support organization cultures that are increasingly geared toward business issues, that are intent on funnelling responsibility downward, and that mix the latest technology with more traditional methods.

The employee survey is becoming less of a stand-alone project. Instead, it is an integral component of an ongoing planning and decision-making process for collecting information. Managers who want to update their surveys should consider recent innovations. And those who once berated employee surveys for their traditional deficiencies may want to take another look.

— *H. John Johnson*
consultant
336 East Chestnut Street
Lancaster, PA 17602

Tabulating and Analyzing the Results

After you collect the responses to an in-house employee survey, you'll need to tabulate and analyze them. Here are some general guidelines:

▶ Be prepared for incomplete surveys and incomplete responses. Decide whether you're going to disregard all of a respondent's answers if he or she didn't complete the survey.

▶ If you are using a computer to tabulate results, check for data-entry errors, especially when the operator is entering the first responses. This will prevent the recurring errors that can be created when an operator does not understand a task.

▶ If staff members are tabulating results by hand, make sure everybody is using the same tabulation system.

▶ Be sure that anyone tabulating results understands the criteria for making decisions about questionable responses.

▶ Paraphrase carefully the answers to open-ended questions; don't change the meaning of a response.

▶ Use charts and graphs to make the results of each question evident at a glance.

▶ Use a cross-tab table for a pictorial comparison of the results of two or more questions. Computers can be very useful for doing this. Cross-tab tables can help you analyze cause-and-effect and complementary relationships. For example, the cross-tab between a question about age and one about professional development might reveal that 20 percent of employees over age 50 want development opportunities.

Several books offer more detailed instructions for tabulating results. They include *Handbook in Research and Evaluation* by S. Isaac and W.B. Michael, and *Survey Research Methods* by E.I. Babbie.

adapted from Info-Line 8612, "Surveys From Start to Finish"

Collecting Data the E-Mail Way

A TEAM AT AT&T
FOUND AN EFFICIENT
ELECTRONIC WAY TO
SOLICIT RESPONSES TO
AN EMPLOYEE
QUESTIONNAIRE. IT
REQUIRES NO COMPLEX
TECHNOLOGY—JUST
SOME COMMON-SENSE
EXTENSIONS TO
ORDINARY USES OF
ELECTRONIC MAIL.

BY LORRAINE PARKER

The team at AT&T faced a special challenge. We were a Process Quality Management and Improvement (PQMI) team, charged with investigating the expatriation and repatriation processes. The people we needed information from were employees working abroad, in places from Montego Bay to Manila, and everywhere in between. Of course, everyone on the team generously volunteered to travel to the target population for face-to-face fact-finding, but we had no budget for that. In these days of corporate cost-cutting and travel-cutting, not many companies do.

So how else could we effectively and efficiently gather data from a target population spread throughout the world?

Besides the special challenges of soliciting information from a global population that spanned most of the earth's time zones, the PQMI team faced an unfortunate reality: too many corporate citizens receive too many questionnaires too many times a year. We had no reason to assume that employees in the target population would place any special importance on our questionnaire. We acknowledged that a paper instrument would not be likely to yield the kind of response rate we wanted.

After some discussion, the PQMI team found a way to overcome the geographic and time-zone drawbacks, as well as the proliferating-paper syndrome. We decided to use electronic mail, or e-mail, to collect the information.

Some alternatives

The decision to use electronic mail to transmit the data-gathering instrument was not made lightly. The PQMI team considered various alternatives, including regular company mail, the company's international pouch, and facsimile.

Team members decided that regular company mail was too slow and that its delivery was too uncertain. In addition, company mail involved paper. The company's international mail pouch offered reasonably sure delivery, but it, too, would mean a paper questionnaire. A faxed questionnaire would certainly be delivered to the targeted employees' offices—but not necessarily to the intended recipients. And sending by fax is fast, but it also involves paper.

Because of the size of the population—140 expatriates—telephone was never a consideration. It would be too time-consuming for the three-member team.

After discussing such alternatives, PQMI team members became enthusiastic and confident about the merits of e-mail for the project at hand.

E-mail advantages

Frankly, e-mail's merits are considerable. One of its major advantages is the obliteration of time-zone hassles. With e-mail, the PQMI team would be able to communicate during its normal working hours; respondents all over the world would be able to answer during their normal working hours.

The use of electronic mail would eliminate the team's serious concerns about sending a paper instrument. The team members were pleased that they would not be adding to anyone's inundation by paper in general—and by paper questionnaires in particular. They acknowledged that in their own work lives, they often discard paper questionnaires, unanswered. They had a hunch that the target population might have similar habits.

Electronic mail can be deleted, but at least it does not wind up in a wastebasket.

E-mail is fast and easy to use. The delivery of an e-mail document is sure. The use of the "form" feature of our e-mail system allows for interactivity, but the interactivity can take place within everyone's respective workday, despite the disparate time zones. The cost is reasonable; with an electronic-mail interface, designers complete off-line all writing, editing, and creating of forms. The user goes on-line only to transmit; charges are based only on transmission time.

Another real attraction of an electronic-mail instrument is flexibility. Respondents can choose either to answer the questionnaire on their screens and return it electronically, or to print out paper copies, fill them out, and deliver them by mail or fax.

All of those qualities seemed to make electronic mail the right choice for the PQMI project.

E-mail constraints

We did see some drawbacks to using e-mail as our delivery system. Obviously, all of the message receivers must be e-mail subscribers. Of the PQMI team's 140-member target population, only 100 had electronic-mail addresses. The other 40 received paper copies by company pouch.

Another constraint on conducting a survey by e-mail is that the target population has to be finite. In our case, it was.

E-mail systems tend to have rigid keying requirements. For our survey, this meant that directions to respondents on how to answer and how to return the questionnaire had to be clear, simple, and absolutely correct.

System compatibility is another potential problem. If respondents use a variety of electronic-mail systems, their systems have to be compatible. The industry is working on this problem and is making substantial progress; soon, most large electronic-mail systems will be able to "talk" to each other. Luckily, no managers in our telecommunications corporation, in any location, use a competitor's electronic-mail system. So the PQMI team found no system compatibility problems.

Initially, it seemed that the most serious system constraint was that the form configuration on our system limited a document to just over two hundred lines. We needed more lines for our survey, and finally overcame the problem by sending the document to respondents in two segments. The team decided that the solution was regrettable, but necessary.

Some potential constraints deal with human, rather than technical, factors. Potential respondents might harbor negative feelings about electronic mail. They might be uncomfortable about their abilities to respond on-line. Some might simply delete the entire questionnaire rather than figure out how to print a paper copy. The team thought it could accommodate these constraints, and kept them in mind as it started to develop the data-collection instrument.

Developing the instrument

The PQMI team developed the instrument systematically. The number of questions was limited—not only because of the system's 200-line limit, but also to ensure that the length of the questionnaire wouldn't irritate or inconvenience the target population.

The questionnaire contained 32 questions, in five sections. Experience had shown that people are less reluctant to answer when it is clear that they are being asked for their personal opinions. So questions were asked in the first-person singular (for example, "I believe I was chosen for my overseas assignment by...."). Other questions were included to verify the consistency of responses.

The team deliberately made demographic questions the final section. This way, respondents answered all topical questions before they were asked for personal data, which can stimulate some reserve about answering. The team formatted the response scales to contain consistent wording; scales always went from positive to negative.

Some questions were open, allowing respondents to compose their own answers. Others were closed, giving respondents a selection among specified choices, with most of the closed items containing "other—please specify," as one of the choices.

After determining the questions' content and format, the team turned its attention to the directions for answering the questionnaire and returning it, electronically or on paper.

In some ways, writing the directions was the most challenging task of all. Everyone on the team had been exposed at some time to directions that were so complex that the tasks in question did not seem worth the trouble. Computer documentation fails frequently and monumentally, because of its convoluted and complex prose.

But our e-mail system and its "form" feature are unforgiving of incorrect keying, so the team had to pay special attention to ensuring that the directions were crystal clear. In total, only nine lines of very simple physical keying directions were necessary for the entire document.

Satisfied with its efforts on the questionnaire, the team now had the file translated from our wordprocessing system into ASCII. Then it was downloaded into our electronic mail system. Next, a form was created on the e-mail's "form" feature. The form allowed for on-line responses and electronic return of the completed questionnaire.

It was time to test the instrument on eight members of the target population. The results of the trial led to some minor revisions.

Transmittal and results

Finally, the questionnaire was transmitted electronically to everyone in the target population with an e-mail address. The very next day, the team received its first completed questionnaire.

Responses continued to arrive at a steady rate for the next two weeks. One week before the specified deadline, the team sent the questionnaire to nonresponders, by a fax that could be accessed from the e-mail screen. In other words, on our system, the team typed the recipient's fax number at the "to:" prompt, where the e-mail address was. Then our e-mail sent the questionnaire via fax. Of course, features and procedures vary among different electronic-mail systems.

Sixty-eight of the 100 e-mail questionnaires were returned, for a 68 percent return rate. Of the 40 questionnaires sent by company pouch, only 15 were returned, for a modest 38 percent return rate. Why did we find such a difference in response rate, with an identical questionnaire? Empirical evidence indicates that the delivery medium accounts for the difference.

People in my corporation have conditioned themselves to throw out what they perceive to be the company equivalent of junk mail—such items as ads, questionnaires, and personnel announcements. Employees do it without a single qualm or backward glance. The electronic-mail medium does not deliver a disagreeable volume of junk mail—at least not yet. Certainly it does not convey many questionnaires to subscribers.

The limited volume of e-mail junk mail is probably one reason for the satisfying response rate to the PQMI e-mail questionnaire.

Another reason for the success of e-mail as a questionnaire delivery method is an intangible quality associated with electronic communications. This is a subjective, personal element, rather slippery and elusive of definition. There is a mystique in high tech, in computers, and in digitization of written words. It has to do with being part of the cutting edge in electronics. Although 28 percent of the e-mail respondents chose to send back paper copies of their answers, they still responded when their screens presented them with the questions.

That is not to discount any of e-mail's inherent and very real virtues as a delivery medium. The team's hard work in creating a user-friendly instrument certainly made the product attractive, but I think the allure and the novelty of the medium clinched the sale.

The expatriation and repatriation PQMI team profited from what some people call a passing phenomenon, but it did profit. Instruments transmitted via e-mail may not continue to elicit such high response rates; that window of opportunity may close as the mystique wears off and e-mail junk mail increases. Sooner or later, e-mail subscribers may become calloused pressers of the "delete" key, who pitch your questionnaire, unread, into the electronic world's equivalent of the circular file.

But for now—especially if you have to gather data from a target population scattered around the world—electronic mail is still an attractive delivery option. ■

Lorraine Parker is a member of the education and training team of AT&T's International Operations Division, 412 Mount Kemble, Room C300, Morristown, NJ 07960.

This article expands on Parker's earlier piece, "Front-End Analysis Via Telecommunications," which appeared in Performance and Instruction *in February 1990.*

By Kenneth M. Nowack

Getting Them Out and Getting Them Back

Here are some guidelines for getting your employee questionnaires to a random, representative group of workers— and 10 tips for getting employees to complete and return them.

You've decided to use a questionnaire to complement your interview and focus-group training needs assessments. You carefully construct the questionnaire, involve everyone you can think of in the development phase, pilot test it, revise it, and send it out to your specified target group.

But you don't get back as many as you would have liked; you're not even sure that they are representative of the organization. You hope that your interviews and focus groups will do a more complete job of delineating training needs.

Sound familiar?

Not all attempts to administer and collect paper-and-pencil information have to follow that pattern. By using some important guidelines, you can dramatically improve the quantity and quality of information gathered by surveys, questionnaires, and other paper-and-pencil assessments.

Several major issues can make or break the relevance of information gathered through questionnaires. If your questionnaires don't go to a representative sample of your target audience, or if the sample that receives them won't complete them, then most of your important development and design effort has been wasted. Three questions are key:
■ How can you ensure a representative sample when administering organizational questionnaires?
■ How can you determine what sample size is appropriate for your data-gathering procedure?
■ How can you increase compliance to complete and return questionnaires?

Random and representative

Once you have developed your questionnaire, you must define your target audience. That often involves an agonizing—and sometimes mysterious—process The first critical decision is whether to send the survey to everyone in the target audience or to a representative sample.

In survey research, a general rule is "the more data, the merrier." Each time you administer a questionnaire to an employee group, you create a "soft" intervention. It may be worthwhile to include all employees in your target audience; that sends the message that everyone's feedback is important and will be considered in organizational change efforts.

Nowack is with Organizational Performance Dimensions, 20950-38 Oxnard Street, Woodland Hills, CA 91367.

If it is not feasible or necessary to administer the questionnaire to all employees of your target audience, consider using a sample. A representative sample of a target audience allows you to generalize the results effectively to apply to the entire target audience. Sampling procedures can be more economical in terms of time and expense for such items as data analysis, keypunching, and postage.

In making the decision to use a sample rather than the entire target audience, remember two important concepts—it must be random and it must be representative.

A random sample means that everyone in the target audience has an equal opportunity to be selected to complete your organizational questionnaire. If the employees you use are not randomly selected from the target audience, the sample you select can strongly influence your results. That could make it hard to accurately generalize your questionnaire findings to apply to the entire organization.

Let's look at an example with a training needs-analysis questionnaire

---

If it is not necessary to send a questionnaire to everyone in the target population, how many should you send out?

---

involving a company with three large divisions of 500 employees each. You decide that it is not feasible to send out 1,500 questionnaires, so you will sample from the three divisions.

First, you hand-pick 50 employees whom you know personally, because you know you can count on them to complete the questionnaire. Next, you go to your organizational phone book and take the first 100 names. You send out 150 questionnaires and are happy to receive 100 back.

You quickly analyze the data and brief the division managers on your findings. But one of the division managers refuses to accept the results, because less than 10 percent of the sample from her division was included in the findings. She disregards all of your questionnaire findings and interpretations.

The problem is that not every em-

ployee in the three divisions had an equal chance of being selected. The well-meaning attempt to select 50 employees initially because you knew they would comply was not an example of random selection. In fact, those employees may be different because they know you. Their particular training needs may or may not be useful to generalize to the rest of the employees in all three divisions. You thought that using the phone book would be a clever way to determine who should be included in the administration of the questionnaire, but it didn't work out that way.

Because the sample was not truly representative of the type and number of employees in each of the three divisions, the results of the survey were discounted.

How many is enough?

If it is not necessary to send a questionnaire to everyone in the target population, how many should you send out? What is the minimum number of respondents needed for you to be confident that the sample size will

reflect the sentiments of the entire target population?

The minimum number depends on several statistical factors:
- the expected response rate of the questionnaire (a 50 percent response rate is generally considered good);
- the precision of the population estimate (for example, within plus or minus 5 percent);
- the confidence level (for example, a 95 percent confidence level means that 95 out of 100 times a sample will provide the desired precision level).

You can calculate the minimum sample required using this formula: Minimum sample size = (population size)(.96)/(.0025(population size) + .96).

For example, to make valid inferences from a population of 700 supervisors and managers, you would need to have at least 248 questionnaires returned to have a 95 percent confidence level that the results were within plus or minus .05 accuracy for the entire target population. The equation you would use to calculate that minimum sample size would be $(700)(.96)/(.0025(700) + .96) = 248$. If

---

An Example

You have a target population of 1,500 that you want to sample for a training needs assessment study. The population is made up of the following:
- 10 percent supervisors and managers;
- 25 percent administrative and clerical staff;
- 25 percent professional staff;
- 30 percent technical personnel;
- 10 percent manufacturing and production workers.

1. How many questionnaires will you need to send out in order to meet the minimum sample size required to make reliable inferences about the questionnaire data, assuming a 50 percent return rate?

2. How many questionnaires will you need to send out to receive a representative sample of professional staff? Of supervisors and managers?

Answers

1. Use the formula to find that the minimum sample size needed is 306 for a population of 1,500. Assuming a 50 percent response

rate, you would need to send out at least 612 questionnaires.

2. To make sure that the sample was representative of the target audience, you would need to distribute about 153 questionnaires to the professional staff (.25 X 612) and about 61 to supervisors and managers (.10 X 612). Again, assume that the response rate is 50 percent and the sample is randomly selected.

The example shows how to get a sample that is representative of the job classifications of the target population. You should try to ensure a representative sample with all demographic variables of interest to the project (such as job level, gender, ethnicity, geographic location, division, and product line). Of course, that assumes that you have thought about how you might want to analyze your questionnaire—before you distribute it. Remember, poor representation of your target audience will diminish the usefulness and acceptance of questionnaire findings.

Ten Ways To Increase Response Rates

The following 10 techniques may help to increase the compliance and response rates for questionnaires you use in your organization.

1. Make sure that participation is voluntary and either anonymous or confidential. Employees who feel coerced into participating may comply, but may provide incomplete or biased answers.

Determine whether you need to identify employees (for example, you may require a post-intervention follow-up using a questionnaire). If so, inform employees of the rationale for identifying them and of what you will do to make sure that their identities are protected. If employees believe that their identities will be revealed (or that their results will be posted in the bathrooms or elevators), they may think twice about returning the questionnaire.

2. Provide a complete cover letter addressed personally to each employee. The cover letter should be signed by the highest-level executive you can find, and should include the following:
- purpose of the questionnaire;
- whether it is anonymous or confidential;
- how the survey will be used;
- how and to whom respondents should return the questionnaires;
- when respondents should return the questionnaires.

3. Make it easy for employees to return the questionnaires. Provide a postage-paid envelope (external mail) or pre-addressed mailer (company internal mail) to help employees.

4. Make the questionnaire look professional. With today's technology, including desktop publishing with laser printers and typeset-quality fonts, there is simply no excuse for producing a questionnaire whose appearance could minimize employee participation. You can greatly increase employee compliance by spending a few more dollars to have a graphic artist or desktop-publishing specialist help with the layout, graphic design, typesetting, and reproduction.

Remember, each questionnaire leaves an impression on the employee who receives it. As in all organizational interventions, you want to create a professional, positive, and competent image with your internal customers.

5. In general, as the length of any questionnaire increases, its reliability increases and the compliance rate decreases. That rule creates a challenge in questionnaire design and administration within organizations. Make sure that your questionnaire is long enough to be reliable (or to have internal consistency or integrity) but not so long that is discourages employees from completing it.

The "compliance/reliability" issue is often resolved by clarifying exactly what you wish to measure, eliminating redundancy, and using brevity as an absolute editor. Compliance can be increased with a longer questionnaire by ensuring that it is professional looking, with a pleasing graphic layout that is easy to read.

6. Include a separate demographics page with the questionnaire. It can come either at the beginning of the questionnaire or at the end; there is no agreement on which placement has a better effect on compliance. Limit the demographics to variables that you feel are relevant to the project and that you will analyze statistically. Remember, the more demographics you ask, the easier it is to identify individual employees. As demographic variables increase, the compliance rate will tend to decrease slightly.

Controversial or potentially upsetting questions (such as those on substance abuse or discrimination complaints) should be placed toward the end of the questionnaire. Research has shown that compliance on controversial questions increases when they are placed in the middle or at the end of the questionnaire or survey.

7. Alert managers and employees ahead of time that a questionnaire is being developed and will be sent out. Make special presentations to managers alerting them to the purpose of the questionnaire. Use organizational communication channels such as company newsletters or announcements to describe the importance of the questionnaire and the anticipated use of the results—before it is mailed out.

Encourage supervisors and managers to make announcements encouraging employees to fill out questionnaires and to return completed ones during staff and team meetings.

8. Provide incentives to reinforce employee compliance. Employees can be motivated to comply when some form of incentive is tied to the return of the questionnaire. Your organization may not be able to raffle off a new car, but it can provide some inexpensive incentives that most employees will value. Some examples might include discount movie tickets, books, training programs, meals in the executive dining room, preferred parking spaces for a period of time, discounts in the company store, and organizational paraphernalia such as coffee mugs, briefcases, and pen sets.

Such incentives can be used even when the questionnaires are completely anonymous. Randomly assigned "double raffle tickets" with numbers can be attached to each questionnaire sent to an employee. Employees are instructed to keep one of the tickets and leave the other attached to the questionnaire when it is returned. One or more of the tickets is selected from the pool of returned questionnaires, and the winners can claim their prizes.

They may sound "too cute" for your organizational culture, but such incentives can dramatically increase employee participation. Remember, without an adequate response rate from your employees, you may not be able to make valid inferences about the questionnaire results.

9. To increase the response rate, follow up with employees after the questionnaire has been distributed. If the questionnaire is not anonymous, follow-up phone calls are effective for increasing compliance. When anonymity is preferred, follow up through a memo encouraging cooperation in completing the questionnaire. Postcard size reminders and the use of organizational communications, such as a company newsletter, can also be used to encourage employees to return the questionnaires by the target date.

For a large organizational questionnaire, it may be valuable to conduct an evaluation of those who did not comply (assuming that the individuals can be identified). Such a study may provide important information about the organization's current climate, as well as delineating employees' reasons for not providing feedback.

For example, one study revealed that employees thought that the organizational questionnaire was too long, that questions were too personal, and that respondents could be easily identified despite assurances of confidentiality.

10. Provide employees with feedback about the questionnaire results. This is one of the most powerful ways to ensure participation in a survey. Employees are curious; they want to know the major questionnaire findings and, more important, what is going to be done about their concerns and recommendations.

Organizations that break the "feedback rule" will pay later—the next time they ask employees to share their observations, attitudes, and feelings by using any data-collection method. Employees who do not feel as if their opinions are heard and valued will simply stop providing them.

Feedback can occur in a variety of forms and can even be targeted to special audiences. Management and employee briefings, although time consuming, can be useful. They provide an opportunity to share questionnaire results and to discuss possible organization interventions. Written reports, executive summaries, and newsletter articles are also useful feedback tools to consider after your questionnaire has been analyzed.

you anticipated a 50 percent return rate, you would need to distribute about 500 questionnaires in order to secure the minimum sample size required.

The accompanying figure summarizes the minimum sample sizes required for different size target populations, calculated from the formula. The minimum sample sizes given in the figure will provide a 95 percent confidence level that the sample is within plus or minus 5 percent of the actual population estimate. If your return rate is at least 50 percent, you can feel confident that your sample is providing you with information that is truly reflective of the target population.

Now that you have a statistical method for determining the minimum sample size required from your target population, you need to know how to guarantee that the sample is truly

---

Minimum sample sizes

Table for determining the minimum sample size required to generalize results to apply to an entire population. Level of confidence is 95 percent that the sample proportion will be within plus or minus .05 of the population proportion. Sample size refers to questionnaires returned, not those sent out. Assuming a 50 percent return rate, send out at least twice the minimum sample size.

Population Size	Sample Size
10	10
50	44
100	80
200	132
300	169
400	196
500	217
600	234
700	248
800	260
900	269
1,000	278
2,000	322
3,000	341
4,000	351
5,000	357
6,000	361
7,000	364
8,000	367
9,000	368
10,000	370
20,000	377
30,000	379
40,000	380
50,000	381
100,000	384

---

representative of the population. That will assist you in "selling" both the analysis and the interpretation of the information gathered. It may also help you to get managers to accept organization interventions and training programs that may be recommended as a result of the questionnaire findings.

The box (on page 146) gives an example of a target population to sample, and some questions and answers for making the sample representative of the employee groups and large enough to be reliable.

Getting employees to comply

Now that you have figured out how many questionnaires you have to get back to make worthwhile inferences about the target audience, go ahead and distribute them to a random, representative sample. Then, hope that employees will comply and return them to you. Low response rates raise important concerns.

For example, if significantly fewer than 50 percent of people in your sample return the questionnaires, the attitudes of the respondents may be very different from those of the non-respondents. Making inferences about employee attitudes based on the information gathered may not be in the best interests of the organization. When response rates are low, interpret questionnaire results with caution; respondents may not be accurate barometers of the target audience.

Worth the effort

It is not enough to design effective questionnaires, surveys, and other paper-and-pencil tools to assess employee attitudes, observations, and opinions. The sampling procedure must be random, representative, and large enough to make valid inferences about the organizational population.

Getting employees motivated to respond to your survey will ensure that all the hard work that went into questionnaire design, pilot testing, revision, and administration is well worth the effort. With the techniques outlined above, you are well on your way to successfully getting out your employee questionnaires—and, more important, getting them back.

Everything You Always Wanted To Know About Employee Surveys

BY KAREN B. PAUL AND DAVID W. BRACKEN

Here's a way to design effective employee surveys by answering some commonly asked questions.

More than 70 percent of organizations survey their employees either annually or biannually. Yet, there doesn't seem to be a lot of specific, practical, and documented information about creating employee surveys. It isn't easy to construct surveys that elicit accurate data from employees to determine organizational needs. The following strategy can help you design winning surveys.

Q & A

Here are answers to 12 of the most commonly asked questions about designing surveys, based on our own research and consulting experience.

Question 1. Is there a best time to conduct surveys? It depends on the required response rates. Obviously, it's best not to conduct surveys that require high response rates during times when many employees are unavailable—such as vacations and holidays.

The timing also depends on how the survey data will be used. Survey administrators may want to schedule surveys during regular business-planning cycles so that the survey results can be used in strategic planning and budgeting. Or, the survey's administrator may want to schedule a survey so that the results will be published just before performance appraisals are conducted.

Quarterly sample surveys that target only a small segment of the employee population can do spot checks on employee attitudes. The results can be compared over time with results from other employee surveys and measurement instruments. The quarterly surveys may include one-time items on issues currently affecting employees.

Overall, survey administrators should ensure that employees view participation in surveys as an important business event.

Question 2. Which are best: sample surveys or census surveys? Surveys of a small, selected segment of the employee population sometimes generate results that are just as reliable as results acquired through more expensive census surveys, which involve the entire organization. But when the employee population is small, sample surveys aren't as reliable as census surveys.

When using a sample approach, the survey administrator must know how to calculate the required sample size to represent the population adequately. To calculate sample size, the administrator must have an accurate list of the demographic groups that will participate. Using results from past surveys, the administrator may want to predict response rates so that the survey design is more likely to produce results that are reliable statistically.

The survey administrator also must be able to accept the large margins of error (confidence intervals) typically generated by sample surveys. Such intervals can have a direct effect on various comparisons among groups—such as demographic and trend comparisons.

Sample sizes are calculated according to a margin of error that the organization deems acceptable, typically plus or minus 5 percent. The smaller the margin of error is, the larger the required sample size. If the sample becomes too large, the survey may be less cost-effective than a census survey.

Reports on survey results must note the margin of error or confidence interval to indicate the degree of confidence with which one could say that the results accurately reflect the truth. It's important to report the confidence interval, to avoid misinterpretation or misapplication of the data. When the confidence interval is not taken into account, differences that appear statistically significant may, in fact, be due only to chance.

As a communication tool, sample surveys provide only some employees with the opportunity to give input. Consequently, sampling may dilute organization-wide communication and decrease employees' commitment to accepting the survey results and any ensuing actions.

Employees selected to take part in a sample survey may wonder why they were chosen, and they may worry about the confidentiality of their responses. They may also think—out of loyalty—that their responses should reflect the views of co-workers who aren't participating. Employees not selected may worry about being excluded.

Sampling places some limitations on reporting. Reports should include only the groups specified in the sample. In interpreting survey results, one shouldn't assume that the data apply to similar groups within the employee population or to groups that are new to the population since the survey.

Some surveys select participants based on job classification or other organizational structures. The results of such surveys may not be valid if the organization restructures after the survey and respondents are assigned to other work areas. Still, people will request survey reports on work groups and trends, to fit new organizational charts.

To make results adaptable, surveys should include demographic coding for group functions, departments, and locations—even though it's difficult to design coding that will ease respondents' suspicions about confidentiality and accuracy.

Question 3. How long should surveys be? This question is open to debate. Some studies show that

AN EVEN NUMBER OF RESPONSE OPTIONS, WITH NO NEUTRAL MIDPOINT, TENDS TO FORCE RESPONDENTS TO TAKE A STAND

shorter surveys get higher response rates than longer ones. Other studies suggest that the number of questions is irrelevant, as long as respondents are interested in or committed to the aims and outcomes of the survey. Interestingly, different demographic groups may prefer different-length surveys. For example, managers tend to like surveys of three to five pages; nonmanagerial employees tend to prefer surveys of 10 to 12 pages.

Generally, experts say that a survey should take about 30 minutes to complete. Given time and cost constraints, 80 to 100 questions—excluding questions about demographics—is an ideal length for most surveys.

Question 4. Does it matter how many response-scale points are used? Different surveys use different response formats. The typical format is a rating scale with response points

arranged from low to high, along equal intervals—for example, a scale of 1-2-3-4-5. The five-point Likert scale probably is the most widely used rating scale.

The number of points on the scale depends on the kind of information required. Some rating scales are designed to assess people's opinions on certain topics in a positive or negative direction—for example, "agree or disagree" and "satisfied or dissatisfied." On such scales, an odd number of response points may encourage equivocal responses, depending on the label at the midpoint. A neutral midpoint label such as "not sure" lends itself to uncertainty. Our study shows that 20 percent of respondents choose a neutral option if one is offered.

An even number of response options with no neutral midpoint tends to force respondents to take a stand, though people with strong attitudes typically give the same ratings with or without a neutral midpoint. In fact, overall response rates are sometimes lower on surveys that don't have neutral midpoints. The choice of whether to include a midpoint also depends on the specific content of the question. Sometimes, it just doesn't make sense to allow respondents to be uncertain.

To help interpret respondents' use of the neutral midpoint to express uncertainty, the survey should ask a prescreening or "filter" question: "Do you know enough about the topic to have an opinion?" Or, the survey can offer a "don't know" option outside of the range of the rating scale, especially to identify respondents' levels of knowledge about various organizational issues.

If the survey author determines that the target audience is able to make fine distinctions between various options, more response points may yield results that can be used to interpret or track data across time to detect trends.

Question 5. Does it matter what the rating-scale anchors are or how many scales are used? It's usually best to label a scale's high end with the most positive anchor. That's because most organizational surveys measure positive concepts such as effective communication, competitive edge, and teamwork. When analysts calculate means or summary scores, high numbers will immediately identify positive attributes or attitudes. This approach helps simplify analysis.

An example of a scale with the most positive anchor at the high end is "5—highly satisfied," "4—satisfied," "3—neither satisfied nor dissatisfied," "2—dissatisfied," and "1—highly dissatisfied."

It's important to choose an accurate label to attach to the midpoints of three-, five-, and seven-point scales. Respondents might apply different meanings to the same options—particularly when the rating scale is supposed to represent different degrees of agreement or satisfaction. The midpoint should be part of a continuum, not a completely different concept.

The number of response points on the rating scale helps determine the anchors or labels. On surveys with more than six response numbers, the survey author might label only the end points—for example, "strongly agree" and "strongly disagree" or "always" and "never." But when the survey results will be reported as averages, survey designers should label each point.

The labels should correspond to the psychological values respondents are likely to assign. Depending on what those values are, the type of rating scale can skew results. One study of an upward-feedback instrument compared results from identical questions using two different scales—a frequency scale and a satisfaction scale. The different rating scales generated very different response patterns on several of the same items.

It's best to use only one type of rating scale in a survey. Switching rating scales tends to confuse respondents and cause errors and frustration. Changing scales may hamper the reporting of results. And too many scales can make a survey too lengthy.

Sometimes the wording of items dictates which type of scale to use. Here are some examples.

▸ "Our benefits are as good as or better than those offered by other companies in this area." (agreement)
▸ How satisfied are you with your benefits plan? (satisfaction)
▸ How would you rate the benefits plan? (grade)
▸ To what extent does the benefits plan provide you and your family with adequate coverage? (extent)
▸ "My benefits cover my medical requirements." (frequency)

Last, the survey instructions should tell respondents to leave an item blank if they think they don't know enough about it to have an opinion.

Question 6. Do the demographics belong at the beginning or end of a survey? Surveys with the demographics at the end tend to produce higher response rates. Once respondents have spent time answering the other survey questions, they're likely to complete the demographic section at the end because of a "completion tendency." In addition, respondents may be tired by the end of a survey. They may be more likely to complete the survey if the final items are the typically unchallenging demographic questions.

Generally, demographic questions about age, gender, and education aren't threatening. But such questions may intimidate certain respondents (such as some members of minority groups) or may seem threatening when combined with other demographic questions. When controversial demographic questions are placed at the beginning of a survey, respondents may react negatively and not complete the other questions.

In organizations in which trust is an issue, achieving full participation on a survey is more important than getting demographic information. In such situations, it's best to keep the demographic questions to a minimum. Fortunately, survey information without the demographic information is still usable.

Another way to reduce respondents' anxiety about demographic questions is to make such questions optional or to explain how the information will be used. Many people worry about being identified. They may not know that most survey administrators don't have the time, energy, or resources to track people's opinions and identities.

For example, the survey can read: "In this section, we are asking for some information about you to help us understand the survey data. We will not use the information to identify you. We will include your answers with those of other employees so that your responses remain anonymous."

Question 7. Is it better to combine items from various topics or to group all of the items about a single topic in the same place on a survey?

Studies show that responses fall into "sets" when survey authors cluster certain items under category headings. Response sets imply that respondents are answering all of the items similarly, based on their attitudes toward the category heading. But categories don't completely determine the types of responses received. A well-designed survey clusters items that have a common basis, sometimes derived from statistical analysis.

Evidence suggests that survey developers can avoid response-set bias by carefully selecting the category headings. For example, using "shaping excellence" instead of "rewards and recognition," may remove negative implications for some people.

Surveys that group items by categories can be effective in communicating an organization's priorities—especially when the headings are based on the organization's vision or values and when the items define desired behaviors. Category-type surveys also help employees see a direct relationship between the survey instruments and the survey reports.

Question 8. Is it important to have some negatively worded items on a survey? A form of response bias occurs when respondents consistently choose one response option regardless of the question, even drawing a line through one answer column. In such cases, one has to wonder whether the respondents even read the questions. Including negatively worded items helps deflect this blatant type of response bias by forcing the respondents to choose different options to express their opinions.

An example of a negatively worded item: "Our benefits are worse than those offered in most companies."

But negatively worded items don't always work as intended. Respondents may view such items has having certain meanings due solely to their negative phraseology. And negative wording doesn't work well with some types of scales—such as satisfaction scales.

When the survey is to be a two-way communication tool, negatively worded questions might imply that top-level managers think that employees are harboring negative attitudes. Respondents may feel as if they're being tricked, and that the survey is some kind of test. Such feelings can reduce response rates. Negatively worded items may cause some people to resist management actions that are based on survey results.

The survey administrator should monitor negatively worded items closely during data collection, analysis, reporting, and interpretation. If survey analysts are using summary scoring—which aggregates or adds up the ratings of similar items within a category—negatively worded items may require them to reverse the scores of negative items before adding those scores to the scores of positively worded items. Some administrators find this kind of scoring and analysis difficult to do.

When a survey contains negatively worded items, the survey administrator should
▶ decide early on how results will be presented in reports
▶ request a sample report of survey results and ask someone outside of the survey process to interpret it
▶ request item frequencies on the raw data from negatively worded items and from some positively worded items surrounding the negative ones
▶ ensure that the users of the survey results handle data according to the survey administrator's instructions and that reports on the results explain how to interpret negatively worded items.

One way to minimize response bias is to keep surveys short, explain to respondents why surveys are important, and share information about how the organization will use the data. It makes sense that when people understand how survey data will be used—and when they aren't fatigued by an over-long survey—they will respond more carefully and completely.

Question 9. Are norms useful? Norms are descriptive statistics that enable survey administrators to compare mean survey results with the scores of people in a defined population. A major consideration in using norms on surveys is the quality of the information on which the norms are based. Do the norms represent the types of companies with which the

TO REDUCE ANXIETY ABOUT DEMOGRAPHIC QUESTIONS, MAKE THEM OPTIONAL OR EXPLAIN HOW THE INFORMATION WILL BE USED

survey organization wants to be compared? Is the data base updated and purged of old information at least every two years?

By definition, a norm is an average. Using norms for the purpose of comparison may imply that an organization is satisfied with average parameters. Many survey administrators think that norms give managers a false sense of security when employees achieve or exceed averages. A more proactive approach is to establish a benchmark for continuous improvement, based on past survey results. The ultimate goal should be to achieve the organization's vision, as determined through past surveys.

Sometimes, the external suppliers of survey norms require that the norms be used without modification. In such cases, survey authors may have to word items in specified formats that don't meet their needs. They may also have to use whole sets of normative items, including some that are irrelevant to the objectives of their organizations.

It's best to use the survey organization's own vision to define and establish benchmarks. If normative data are still required, the data provider should describe and ensure the high quality of the normative data base.

Question 10. Which is better: custom surveys or standardized surveys? Many standardized surveys yield reliable and valid information; customized surveys can yield invaluable organization-specific information. The primary considerations are fit, cost, and the way survey results will be used.

Fit refers to the extent a standardized or off-the-shelf survey is consistent with an organization's priorities. Many standardized surveys require a company to buy into someone else's definition of an effective organization.

When a survey is used mainly as a communication tool, its message and results should convey the organization's own values. Some standardized surveys can be modified. But if the survey aims to communicate an organization's vision, then a customized version is best. If the organization is concerned about job stress or job satisfaction, a standardized survey may be more useful.

With standardized surveys, users pay only a part of the development costs, compared with paying in full for the development of customized surveys. Small organizations often find the development of custom surveys to be too expensive, unless the surveys are used repeatedly.

Question 11. What are the differences between mail-in and on-site surveys? Generally, the benefits of on-site administration outweigh the seemingly high cost. Still, opponents of on-site surveys point to the cost of taking employees off their jobs. They may mistakenly assume that employees don't complete mail-in surveys during work hours.

The main consideration against using mail-in administration is a typically lower response rate—about 60 to 75 percent lower than the response rate for surveys administered on site, which tend to produce about a 90 percent response rate. Yet, it may be impractical to administer surveys on site in organizations that have several sites or that lack meeting rooms large enough to accommodate all respondents.

Remember: Employees look for management's commitment in the survey process. Asking employees to fill out surveys on their own time is hardly a sign of management support. On-site administration presents an opportunity to communicate the commitment of senior-level managers. One company showed employees a video featuring the CEO giving his commitment to the survey and promising to take action on the results.

Another benefit of on-site administration is a faster turnaround time. Sessions can take place in just a few days, with survey administrators present to answer respondents' questions.

Question 12. Is a minimum number of respondents required to produce reliable results? On surveys of a particular work group—employees who report to the same supervisor—reliability is partly determined by the size of the group. Large groups tend to generate more reliable and consistent results. The results from small groups are more affected by the answers of individual members. As group membership or personal attitudes change in small groups, results can vary widely.

Confidentiality is a big concern, especially in organizations with limited experience in conducting surveys. In such situations, it's best that an organization initially survey a large group until employees trust that their confidentiality will be protected in future surveys.

Traditionally, survey experts consider 10 respondents to be the minimum size for reliable reporting of a particular work or functional group. One drawback is that most organizations have some work groups of fewer than 10 employees. Problems can arise when the supervisors of such groups don't get survey results—especially in the areas of career development and performance appraisal. Consequently, an unofficial industry norm has emerged: a minimum of five respondents.

There are many decisions to make along the way when constructing employee surveys. What works well is often learned the hard way. Knowing some answers to commonly asked questions about surveys can help ease the process.

Karen Paul *is an industrial and organizational psychologist in organizational learning services at 3M Human Resources, 3M Center 225-1N-13, St. Paul, MN 55144-1000; 612/733-9925.* **David Bracken** *is director of consulting services, NCS Information Services Division, 4401 W. 76th Street, Edina, MN 55435; 612/830-7692.*

Measuring Skills and Behavior

By Kate Ludeman

In addition to improving productivity, a well-developed skills assessment program can help you measure and demonstrate the worth of your HR department.

Remember the Continuous Process Improvement (CPI) tenet: Nothing improves until it is measured. The CPI corollary, "as soon as something is measured, it automatically begins to improve" is certainly true when you measure job behaviors and skills. Once measured, employees begin to improve automatically, simply because people are paying attention to their skills and behaviors and giving them feedback.

Skills assessments offer many other benefits. They accelerate people's learning, job performance, and professional development by offering the reliable feedback necessary for continuous improvement. They also provide useful feedback to self-managed teams and to facilitators of quality programs.

These assessments provide a significant way for HR departments to increase productivity during downturns by showing employees how to work more effectively.

In a 1985 survey of working Americans by the Public Agenda Foundation, only 23 percent said they work at full potential. Gary Berger, vice-president of International Survey Research in Chicago, reported on 1990 attitude surveys of 6,193 managers and 25,465 employees. He said that feedback about a person's performance was rated as the fourth most important factor in job satisfaction, but that fewer than 40 percent of respondents are satisfied with the feedback they receive. Closing this feedback gap is the key to performance improvement.

The role human resources plays in motivating and educating employees has become more important as companies have shed layers of manage-ment, giving managers more direct reports. The *Wall Street Journal* reports that, in the last 10 years, U.S. companies have eliminated the jobs of three million managers.

Removing a tier of management most often means more direct reports and less time for one-on-one coaching. A human resource department armed with a customized skills assessment program can provide the feedback lost when the span of management increases.

In addition, assessments provide measurable proof of behavior changes, legitimizing the HR department's role in improving employee productivity by rendering it quantifiable. Especially when the economy is soft, the HR department may need to demonstrate its worth. Skills assessments are one way to improve effectiveness and to measure and demonstrate how the human resource department is doing.

Continuous improvement

Customized assessments provide a high-impact way to tie HR into a company's quality program.

Just as people search for quick fixes to lose weight with fad diets rather than make lifestyle changes to achieve lasting results, companies often start total quality control programs and temporarily succeed in trimming some fat, but fail to make company-wide lifestyle changes. Employees fall back into old modes of behavior, and true continuous improvement is never realized.

The human resources department is the guardian of the lifestyle or culture of the corporation. It is up to human resources, then, to make possible the companywide cultural changes necessary for employee behavior to improve permanently and continuously.

Such change requires focus. It is impossible for an employee or manager to be outstanding at every aspect of his

Ludeman *is president of the Worth Ethic Training Company, 240 La Cuesta Drive, Portola Valley, CA 94028.*

Put Quality To Work
TRAIN AMERICA'S WORKFORCE

or her job. People must know what areas are critical to their performance and how much, if at all, they need to improve. The notion of not encouraging employees to excel at all aspects of their jobs but to concentrate instead on the most crucial areas is more important to productivity than it seems at first glance. As Mimi Bluestone notes in "The Push for Quality" in the June 8, 1987, *Business Week*, it is akin to the approach to production quality that has made Japanese manufacturers so competitive.

The statistical quality consultant Genichi Taguchi has shown that greater uniformity around a target results in lower losses per production unit, even if some of the units are out of spec. In comparing a San Diego television manufacturer to a Japanese television manufacturer, Taguchi

developed within the company and the feedback is uniquely relevant to the company culture. The best type of skills assessment allows you to select categories based on job competencies and on company values, with questions customized in each category based on employees' actual job contents and levels.

As a result, customized assessments allow comparisons that are entirely rooted in a company's particular values, business goals, and development needs. That is possible because the company is the source of the data.

Quality programs

Today, quality experts know a lot about measuring changes in paper flow or manufacturing processes, and developing quality programs for such processes. When the programs are

that are necessary for changes in processes to take root.

Training assessment

There is only one concrete way to know if your training programs are actually making a difference: pre- and post-measurements of the skills you are teaching.

"The pressure to prove the value of training is greater than ever," say Anthony Nathan and Michael Stanleigh in the January 1991 issue of *Training & Development Journal*. They recommend comparing baseline performance data with post-training performance data to know if a training program really has improved employee performance.

Scott Parry, in the December 1990 issue of *Training and Development Journal*, emphasizes the importance of providing trainers with feedback about learners' use of new job skills. Without feedback, it is impossible to know how to revise training programs for maximum skill development.

It is important to build in a method to evaluate the effects of training, especially if your firm is pursuing the Malcolm Baldrige National Quality Award.

For example, the company should be able to describe how it decides what quality education and training is needed by employees. It should be able to describe the key indicators used to evaluate training, and to explain how the indicators improve the quality of training. Several companies that came close to winning the 1990 Baldrige award were criticized for their failure to evaluate training in a systematic way.

Customized skills assessments meet these needs. Questions can be developed so that each category of skills aligns with a training module. The pre-training feedback is provided as part of the program and helps focus employees on those areas they need to improve. Six months later, a post-training assessment lets them (and you) know that their efforts to improve have worked.

Knowing ahead of time that they will receive post-training feedback increases employees' motivation to implement their development plans and to put to work the new skills they learned in the training program.

In the November 1990 issue of *Training and Development Journal*,

Quality program participants perceived a gap between the company's quality talk and actual actions

found that the U.S. company was careful to keep all units produced within specifications, while the Japanese company focused on getting as many units as possible to the ideal target, even at the expense of shipping some parts out of specification.

The net result was that the San Diego plant's loss per unit was $1.33, while the Japanese plant's was only $0.44 per unit. Pinpointing areas of importance and taking measurements results in more efficient performance from people as well as products.

In the same way, customized skills assessments focus managers and employees on the critical areas that require improvement. Customized assessments use today's computer technology to meet the need for performance measurements that can promote continuous improvement—which could be more aptly named "continuous people improvement."

The type of feedback a customized assessment offers is more valuable than that of traditional assessments. That is because the questions are

broadened to include white-collar and management jobs, people often must make specific changes in their behavior to make the programs succeed.

Unfortunately, these individual behavioral changes often do not occur unless the HR function focuses on helping employees make them. As a result, many quality programs fail to stimulate the culture changes required for long-term results.

A 1990 Gallop survey for the American Society for Quality Control polled 1,237 employees and found that 36 percent of employees in companies with quality programs don't actually participate in the programs. Among quality program participants, one major complaint was a perceived gap between the company's quality talk and its actual actions.

Continuously improving the performance of people within an organization is the only way to achieve the expanding goals of total quality control. A customized skills assessment can help meet those goals by measuring the changes in people's behavior

Debra Cohen cites extensive research that found a significant correlation between learning and goal setting in the training process. Pre-training assessments make it easy for participants to see the areas in which they need to improve and help them to set appropriate goals.

Career development

Customized skills assessments provide a powerful foundation for companywide career development programs. Employees often under-assess both their strengths and their problems. As a result, they don't develop needed skills. An accurate assessment can serve as a private personal trainer, particularly if the process is repeated at regular intervals so participants can gauge their improvement.

A company whose growth has been flat may want to offer employees and managers the opportunity to improve skills so people will be better positioned when the company begins to grow again. Employees in flat or downsizing organizations may well understand that their individual performances over time are their only real job security—the only investments that walk out the door with them every night. Participating in an assessment process gives them tools with which to monitor the development of those investments.

From the company's perspective, according to research by the American Society for Training and Development and the Department of Labor, U.S. employers spend between $90 billion and $180 billion on informal on-the-job training each year, but only $30 billion on formal training programs.

Customized assessments clarify for employees the areas that need attention and further development.

Reinforcing company values

Many companies publicize the values they consider crucial to their success, but have no way of ascertaining whether employees really work in alignment with those values. In 1990, several companies in the final selection for the Baldrige award were criticized because they failed to measure the implementation of their company's values and connect management training with their published values.

With a customized assessment, you can build categories of questions around your firm's values, such as customer service, innovation, risk-taking, and concern for your employ-

Five Assessment Stages

There are five basic stages in using a customized assessment:
- designing the questionnaire
- collecting the data
- reviewing the findings
- helping the person being assessed to use the report to stimulate growth and make important changes
- reassessing the same skills and behaviors 6 to 12 months later.

Look for a program that allows you to tailor it to your company at each stage.

1: Designing the questionnaire

Designing the questions for a customized assessment can be done several ways.

If you have a good data base of questions to work from, you can select questions by yourself or with the group's manager. Both methods are adequate, but they are not as effective as a method that involves more people. Involving more people will help you achieve the buy-in that you need in order to convince people that the ultimate report focuses on key factors that are necessary for success.

At this initial stage, I have found that the most powerful way to design a customized assessment is to work in small groups.

Pull together a series of groups from different levels and different departments. For each category of people being assessed (for example, product market managers, software engineers, and data entry clerks), provide key competencies, skill categories, success factors, or values around which to group the questions.

Ask people to think of the most successful people they have worked with at their level, at the level above them, and at the level below. If this is a group of managers who manage first-level supervisors, for instance, have each manager think of the best supervisor who has reported to him or her over the last two years, the best managing peer, and the best director at the next level up.

Each group will develop a list of behaviors or skills in each category that seem to have made these people successful. Do this in groups of six to eight people and post the information gathered on the wall.

Have a writer cluster each group's items into a usable set of questions for each of the three levels of management. Continue this process until you have in-

volved representatives from each level in each department that will participate in the assessment process.

The final stage is extremely important. The questions the groups have developed must be meticulously edited. Each question must be clear and easy to understand so that everyone who reads it will interpret it the same way. Equally important, each question must focus on only one behavior or skill. For example, "Do you set long- and short-term goals?" is not a useful question, because a person would not know what area needed improvement—long-range planning or short-term work goals.

It is important that questions are focused on behaviors, not qualities. That can be tricky. For example, "Do you provide a good role model for the work group to follow?" is not a helpful question because a person would have no idea what problem was indicated by a low score.

After you have identified questions for each of the various questionnaires you plan to use in a particular intervention, you should ask a group of people to fill out the questionnaire. Debrief them immediately afterward. Find out if ees. Under each value, you can develop different behaviors and skills, customized not only by level (such as managerial, technical, and administrative) but also by business unit (such as finance or sales).

High-potential programs

Customized assessments allow you to tailor assessments to individual employees. This is particularly useful for high-potential and executive development programs.

For example, Gary Z., an aggressive fast-track director, always knew he was better at talking than listening. An assessment, with questions focused on communications and the need for control and attention, shocked Gary. It showed him that his team members uniformly hated the way he monopolized meetings and dominated one-to-one conversations. A six-month follow-up assessment showed he had stopped interrupting others, but his nonverbal behavior indicated that he was still not really listening to what other people were saying. He was absorbed in his own ideas when he should have been listening.

A third assessment, one year after the first, showed a dramatic improvement in listening and especially in sharing the stage. In fact, he had begun to develop solid skills in facilitation and had become far more participative in his management style.

Bottoms-up performance reviews

Customized assessments offer a way around one flaw of most performance evaluation systems, in which feedback is gathered only from above. Offering someone feedback only from his or her manager encourages the employee to search for ways to impress that person that may not be completely relevant to providing the best possible job performance. Critical problems the employee may have with staff or peers that are not perceived by the manager will go unsolved. Or the person's behavior may change, but only in relation to the manager.

Most people are very good at managing *some* relationships. A person may manage *up* well with his or her manager and the manager's manager. Perhaps he or she manages employees well but can't sell ideas effectively to management. A 360-degree circle of feedback from management, peers, staff, and internal and external customers can show precisely who

---

any questions seemed vague or ambiguous, or if they found they could interpret a single question in different ways.

2: Collecting data

Ask each person being assessed to identify respondents to participate in the survey. Provide guidelines to ensure that a person does not suggest only those people with whom he or she has an excellent relationship.

For example, you might suggest that a manager identify key people such as those who directly report to him or her, all department peers, major internal customers, his or her manager, and perhaps the manager's manager. Strive for a well-rounded group of respondents that will offer a 360-degree portrait of the employee.

The instructions sent to respondents with the survey should emphasize that their responses will be anonymous and presented in terms of averages and ranges, with the exception of the manager's data, which will stand alone. Use a code sheet with numbers in place of respondents' names to allow you to track completed assessments and still assure respondents of anonymity.

3: Reviewing the data

The summary reports will show the actual development needs that you should address in the training program. You will also know the specific development needs of each person in the room with you, so you will be able to tailor your comments and small group discussions to ensure that needed areas are addressed.

4: Presenting the findings

The least time-consuming way to present the report is to give the employee the assessment along with a self-paced workbook that will help him or her study the data and develop a plan of action.

This may not be the most effective way for an employee to review the data, but it may be the only practical approach, given lean human resource departments and tight budgets. With employees in the field, it is often the only solution that works logistically.

The most effective way to present assessment data and to provide guidance in building development plans is in small groups. These can be half-day or day-long programs focused entirely on using the assessment data to build development plans.

Another alternative is to couple the assessment process with appropriate training programs to help people develop the skills they need. Peer support can go a long way toward firming a commitment toward continuous improvement of skills.

Whatever feedback process is used, it is important to have people examine their strengths as well as the areas in which they need to improve. That way, they can learn to leverage their strengths and build careers that allow them to capitalize on their innate abilities.

5: Following up

Knowing ahead of time they will receive post-training feedback increases people's motivation to implement their development plans and put the new skills they've learned to work.

Six to 12 months later, give the assessment again so that people can see specifically where they've improved and identify areas that still need attention. This allows you to evaluate the program's effectiveness in stimulating managers and employees to grow and develop.

needs which changes to occur.

Clustering responses by category of respondent provides a breadth of feedback that is impossible with the typical performance evaluation. Problems that have escaped notice during the annual performance appraisal may become glaring.

For example, Janice W.'s assessment report showed many strengths, except in integrity. Her manager thought she was reliable and trustworthy in the broadest sense, but integrity ratings from her peers and direct reports were dismal. They believed they couldn't count on her, that she didn't honor her commitments to them, and that she was out for herself. They saw her

edge and traditional selling skills.

A customized assessment can be tailored to focus salespeople on skills that increase the likelihood of a sales approach that builds revenue and long-term, satisfied customers. A well-informed salesforce will not sacrifice the development of a loyal customer for a shortsighted focus on racking up sales.

Customized assessments can become the backbone for a concentrated focus on customer service. Customers can be clustered by sales regions with non-anonymous feedback. If there is a problem in a region or with a customer, it is easy to pinpoint the cause of the problem and

that employees' needs and expectations are met. This means you should address areas in which the assessment shows a large gap between the current reality and the desired situation.

Essential criteria

The assessment should be easy to use and practical to implement. Your assessment program should allow you to do the following:

■ select from a pool of hundreds of questions and edit them or even create new ones

■ provide guidelines to use in creating your own questions

■ organize specific questions into categories based on job competencies, corporate values, or training modules

■ input data using a scanner to ensure accuracy and save administrative time

■ show the importance or priority of each skill

■ compare current skill levels with the levels you expect

■ provide easy-to-understand bar graphs showing the importance of various skills and the level of change desired

■ show the distribution or range of responses

■ use bar graphs to compare each employee's scores over time, so that changes can be plotted

■ compare employees' individual scores to department, company, or national norms

■ identify and pull out from the rest of the report a person's primary strengths and development areas in order to avoid data overload

■ couple the assessment with a development process that can be easily customized for your work environment.

Overall impact

Customized skills assessments will allow you to provide department and company summaries showing productive changes in your employees' skills and job behaviors, year after year. This provides quantifiable data that show your executive team HR's influence in improving productivity and in making the culture changes that allow quality programs to work.

Customized assessments are a useful tool your department can use to provide relatively painless lifestyle changes your company can undergo immediately.

Employees may know how well they are doing, but they don't know how well they could do if they had better skills

---

as hogging the glory and failing to publicize her team's performance because she did not want to share the recognition.

Clearly, until this problem was corrected, it was unrealistic to consider moving her up a level. Feedback from her manager alone would not have turned up the fact that she was often meeting his needs at the expense of her peers and her staff.

Sales and service training

Unfortunately, salespeople work in a feedback vacuum. Because they usually work far from headquarters and their managers, their primary measure of success becomes their sales figures. But sales figures don't provide them with the subtle and more specific feedback they need in order to improve their bottom-line contributions.

They know how well they are doing, but they don't know how well they could do if they had better skills. This is particularly true today, when many sales departments are implementing team selling, which requires team skills and conflict resolution abilities in addition to product knowl-

then conduct follow-up interviews with customers who need special attention.

Customer service can be expanded to include internal customers as well. Research shows that employees tend to treat customers the way their managers treat employees.

Culture audits and employee opinion surveys

A company can also use customized assessments to inquire into employees' overall satisfaction with a company's culture and procedures.

You can group questions according to the company's values or by particular areas of potential concern. Such an assessment can also look at general morale and specific problems employees are having with company practices. Again, there is great value in comparing what employees currently have against what they want. This is more useful than the traditional approach, which uses industry norms as a base of comparison.

It is important that a company measures up against similar companies with which it is competing for qualified candidates. But it is also important

Putting Values Into Evaluation

Learn how corporate values are adopted and converted into action at British Airways.

By ROGER POULET and GERRY MOULT

Managing People First overtly works on the emotions, intellect, values, beliefs, and self-image of the participants

Managing People First (MPF) is a British Airways training program that relies on participants' values for its effectiveness.

MPF is part of a wider culture change process designed to make British Airways sensitive to its customers' needs and responsive to market changes. It also encourages caring, achievement, creativity, innovation, and profit. In addition to MPF, the process includes changing organization structure, communication and consultation policies, appraisal and reward systems, and training.

A one-week residential program for all management personnel—about 1,400 people—MPF uses organizational data on values in use and values to which the airline aspires. It also incorporates data about each manager's performance rating by his or her staff and colleagues. It employs vision-building techniques and creates support groups and networks and also covers, either overtly or by implication, the skills of listening, giving feedback, managing communications, and self-

Based in England, Poulet and Moult are, respectively, managing director of Human Resource Management Ltd. in London and development projects manager at British Airways in Hounslow, Middlesex.

presentation. All of this is directed toward empowering managers to create the new culture by taking control of their tasks and working relationships.

The first problem to confront us was how to quantify the intangibles: value shift, commitment, and empowerment. And we needed to do this without compromising any of the possible outcomes by narrowing our focus to one or two easily measurable criteria.

Measuring the unmeasurable

A traditional response to this problem is to look for the delivery of "ultimate value" to the organization, a pursuit well described in 1974 by A.C. Hamblin in his *Evaluation and Control of Training*. In our case we could have defined "ultimate value" as greater customer and employee satisfaction. But we wanted to get results from the evaluation long before reliable measures of such satisfaction would be available. And this approach, even if successful, would have provided no information on the causal chain linking input with outcome. We wanted to illuminate the process by which values are adopted and transformed into action.

Before working on MPF, we each had concluded that the role of values in training had been neglected and that human resource departments were paying a high price for that neglect. Far too many training interventions are wasted, misused, or somehow fall short of expectations for reasons that remain a mystery to their designers and evaluators. Even training programs that appear to be exclusively technical, skill-based, or conceptual in content are mediated by the value systems, self-image, expectations, and ambitions of their participants. And it is these factors that determine the level of identification, acceptance, and implementation; that is, whether people really will do anything with their learning. This is the missing link in the transfer chain, and trainers disregard it to their own disadvantage.

In MPF we biased the balance between behavior-skill input and value input toward the latter. This gave us a unique opportunity to study the impact of values on the outputs—values both personal to the participants and inherent in the course. We found that implementation of learned skills and behaviors was profoundly affected by whether the individual and corporate values were in or out of sync.

As a result we have been able to measure logically how well the program generates both commitment to the course values and empowerment to take appropriate action. Only after commitment and empowerment occur can we apply testing and tracking mechanisms to the outputs.

Developing an interpretative framework

To interpret the complex data gathered in our course participant interviews, we first had to cluster the responses into a manageable number of categories. The next step was to develop a conceptual model—illustrated in Figure 1—explaining the relationships between them.

We based our model on the notion that, just as the MPF program itself comprised two parallel strands of vision-values and behaviors-techniques, so the participants' responses can be identified under the same two strands. Subsequent use of the model has indicated that this concept applies to all the other courses we have tested so far.

The process

In our study we interviewed a sample of 70 people, selected at random from those who had completed MPF by the end of October 1985. This represented about 15 percent of that population. We briefed the interviewees on the intention of the survey and the process we would follow. We supplemented the interview data with a tailormade data base containing relevant personnel information.

The interviews lasted from one to two hours and were designed to encourage the interviewees to describe, in their own terms and with their own sequence and emphasis, their experience as program participants. We particularly wanted to know their intentions and actions after the program. We captured in our notes as much as possible of the actual words people used.

We prompted discussion of specific topics only when a principal theme or element of the program was not mentioned spontaneously by the interviewee. This ensured that the interviews were comprehensive and comparable and, at the same time, preserved the perspectives, interpretation, and values of the interviewees.

During the interviews we discovered that people were relating two distinct experiences to us:

■ the objective experience of taking the course in their managerial capacity; attending the sessions, workshops, and meetings; and of learning new techniques and ideas;

■ the more subtle impact of the course and its values on the participants as individuals outside the work environment—*who* they were rather than *what* they were, *what it meant* to them rather than simply *what happened*.

Figure 2 illustrates this duality of experiences.

Active values change

Both subjective and objective elements were working in the participants' minds, and they often did not differentiate between the two. When we compared these

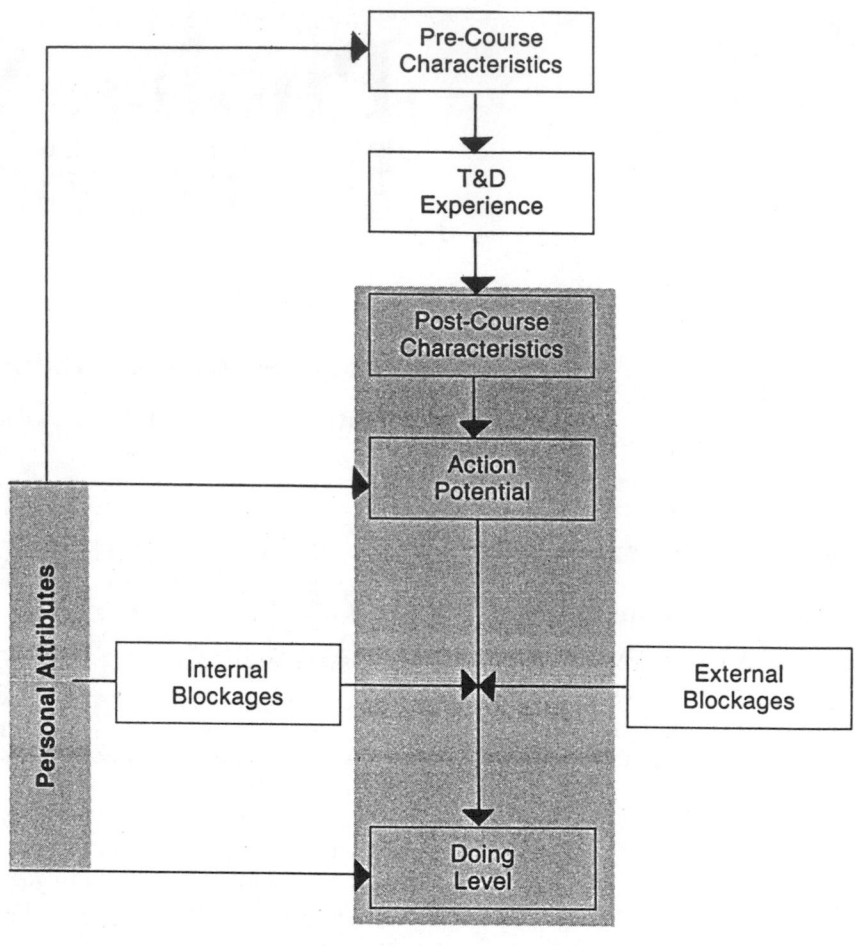

Figure 1—Analysis model

Figure 2—Duality of participant experiences

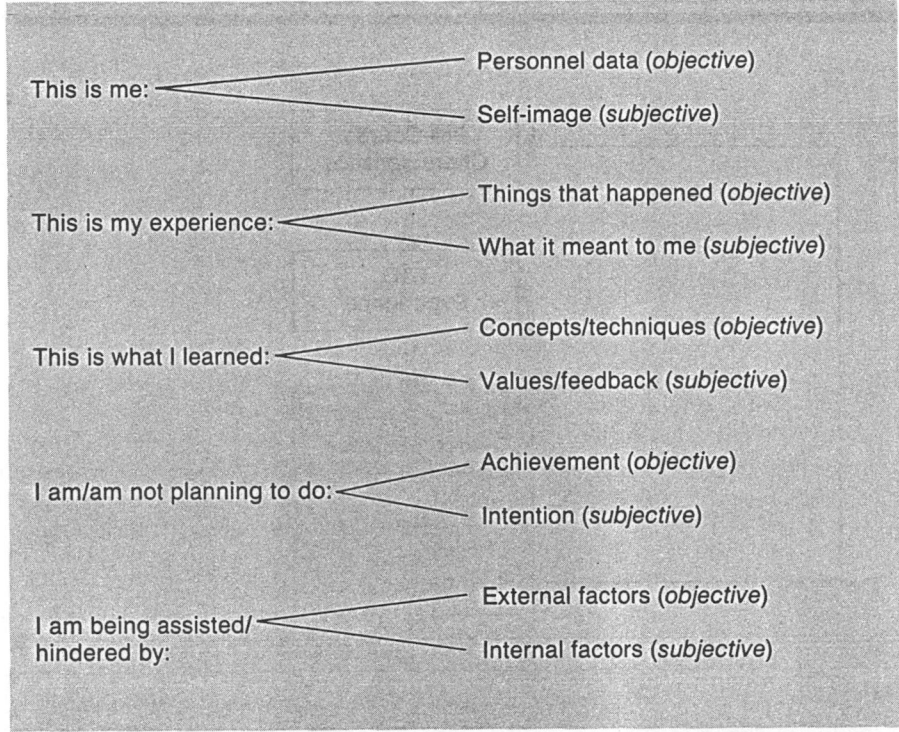

This is me:
- Personnel data (*objective*)
- Self-image (*subjective*)

This is my experience:
- Things that happened (*objective*)
- What it meant to me (*subjective*)

This is what I learned:
- Concepts/techniques (*objective*)
- Values/feedback (*subjective*)

I am/am not planning to do:
- Achievement (*objective*)
- Intention (*subjective*)

I am being assisted/hindered by:
- External factors (*objective*)
- Internal factors (*subjective*)

Figure 3—Want-do scales

Want Level

Level 3—high—is defined as strong statements of *personal* commitment.
Level 2—medium—is defined as impersonal, third-party statements of support or commitment with reservations.
Level 1—low—is defined as no discernible commitment or a negative reaction.

Do Level

Level 3—high—is defined as visible actions, often in excess of what might reasonably have been expected from the course, that have already taken place and require a significant input of time, energy, and will.
Level 2—medium—is defined as action that is limited in scope, is low-risk, and is possibly in the future.
Level 1—low—is defined as no action, or the minimum unavoidable action.

reactions with the course objectives and content, we found that the participants' responses mirrored the course structure. Although we focused a great deal on changing certain behaviors, we discovered we were also affecting people's values. It was this discovery that enabled us to extend our analysis to include the subjective element.

When the values implicit in the course were congruent with those of participants, the course set up a strong resonance in them that came out as a desire to be involved with the program. If, on the other hand, there was no overlap of values, commitment to the program was almost nonexistent.

We call the participants' desire to be involved with the program the *action potential*. We differentiate action potential from Donald Kirkpatrick's "reaction"; the latter is usually measured on "smile charts" immediately after a course. Particularly emotional or exciting training sessions, delivered with more panache than substance, can often "outsmile" a well-constructed and relevant course presented in a lower key. We have found that a more considered view from the same participants a few months later yields a much more balanced, realistic impression of the program *content*.

Action potential is the energy generated in the participants by the course and its trainers in such a way that there is synergy between the participants' expectations and wants and the values expressed in the course. It occurs in all learning experiences and describes the willingness of the participants to implement what they have learned.

What happens to this potential once the participant leaves the training environment? In ideal circumstances people are empowered both by their own commitment and by their organization to convert this potential to true action. It is the results from this *conversion* that traditional evaluation techniques attempt to measure, whether by testing new skills or tracking changes in behavior over time. But because conditions are rarely ideal, trainers may know the output from the training course but may have no idea of the real *potential* of the course.

Impediments to conversion

Several roadblocks stand in the way of empowerment. Like a boulder poised at the top of a slope, action potential is ready to roll down and release its kinetic energy at the bottom. But while rolling down, it will be subject to frictional forces that slow it. It may be pushed off course by other boulders in its way or even halted altogether if the barrier is large enough. We might call these *external blockages* to full conversion. Often we know something about them, and in some cases we can predict their effects.

External blockages include organizational weaknesses and barriers, the absence of adequate systems to enable full use of the skills or behaviors, and senior management resistance to change. These are easy to identify and to report, and they provide invaluable data about what needs to be done to improve the organization before fully implementing the training strategy.

But a boulder might not be completely spherical and, hence, might roll inefficiently. Or it might be riddled with internal cracks that would expand on the slightest impact, shattering the boulder into small stones and effectively stopping conversion. These are *internal blockages*, and they are often hidden.

Internal blockages are at least as powerful as external ones in limiting the conversion of action potential. Internal blockages are in each participant and are manifested as unsureness, inability to accept the value system expressed during the course, or lack of confidence in the message.

Internal blockages limit the action potential present in the course and they drastically reduce the conversion to action—effects that will occur even though the participant knows and understands the skills and behaviors in question and is perfectly capable of applying them.

So unless trainers take into account the action potential generated by a course, simple output measures can't tell them why performance doesn't meet their expectations; nor do these measures tell trainers all they need to know to improve the training next time around.

Levels of potential and action

To help our understanding of these con-cepts, we developed a *want level* for the action potential and a *do level* for the action. Each level is divided in thirds, as illustrated and explained in Figure 3. By using these scales, we were able to achieve complete reliability in assigning "want" and "do" ratings to individuals from interview data. The results are shown in matrix form in Figure 4.

The matrices form the bridge between the value analysis and the external analysis of outputs. For example, a tracking process can show that a substantial change has occurred in the behavior patterns of participants. By relating this result to the want-do matrices, a trainer can tell immediately whether this is more or less than was expected from the action potential. If less, blockages are preventing complete conversion. If more, then it is open to question whether the behavior changes are permanent or merely the reaction of people who feel they ought to comply for a period.

To illustrate these outcomes, we developed result categories—shown in Figure 5—that serve as useful metaphors in our quest to eliminate blockages that prevent conversion.

Key results

As we said earlier, the want-do ratios succinctly describe the success or otherwise of MPF in generating suitable actions; Figure 6 highlights the main results.

Briefly, we found that

■ one quarter of all participants are as committed to MPF values as it was possible to measure and, moreover, they ac-

Figure 4—Want-do matrices

Do				
Numbers	3	2	1	Total
W **a** 3	15	13	4	32
n 2	3	8	6	17
t 1		1	6	7
Total	18	22	16	56

Do				
Percentage	3	2	1	Total
W **a** 3	27	23	7	57
n 2	5	14	10	29
t 1		2	11	13
Total	32	39	28	99

Figure 5—Result categories

Want-Do Level	Category	Definition
3–3 to 3–2	Apostles	People who have seen the light and are doing something positive (potential role models)
3–1 to 2–1	Peters	People who have seen the light but are cautious about action
1–1	Thomases	People who are doubters
2–3 to 1–2	Pork Barrellers	People whose action is greater than their commitment because MPF may be the "flavor of the month"

Figure 6—Results of want-do ratios

Want		Percentage
3	High	57
2	Med	29
1	Low	13

Do		Percentage
3	High	32
2	Med	39
1	Low	28

tively pursue its aims;

■ nearly 60 percent of participants are fully committed to the aims of MPF although they are not all as active as they might become;

■ one third of all participants are actively converting enthusiasm into highly committed action;

■ only 13 percent remain truly uncommitted to the idea, and 28 percent have taken virtually no action at all.

These findings have already proved useful, not only in assessing whether the corporate investment has paid off, but in identifying the location and impact of factors that impede commitment and implementation. The findings, fed back in summary form to corporate management, enable management to make better-informed decisions on what actions would have the most leverage in the change process.

Applicability

So far we have talked mainly about a training program affecting a number of people from one organization. But the process is equally valuable for single-person analysis. Many organizations today use external courses as a way of providing, say, a management development experience that the organization can't set up on its own for reasons of cost or limited relevant training resources.

Individual programs are notoriously difficult to evaluate by traditional methods. The objectives often are unclear, and the course content of commercial programs has to be aimed at a wider audience than would be the case for a similar course developed internally. By using the concepts outlined above, a trainer can identify with a trainee the action potential and conversion in order to discover that trainee's want-do ratio and level of empowerment.

When working with an individual, the trainer should develop the matrix to cover a variety of specific actions that stem logically from the course. Each element forms one part of the action plan for the individual and, because his or her commitment to each one is established, realistic targets can be set and future training needs reconsidered.

What does this mean for effective training? MPF is an unusual program because of its large and overt value content. But the evaluation principles we developed for the program apply equally well to more straightforward training interventions. In a recent basic supervisory management trial program with no overt value element, the results were strikingly similar to those found in MPF.

We consider that all training programs have a value-behavior balance inherent in their design. The only difference between those programs and MPF is where the balance is struck. That most courses, even HRD programs, are biased toward the skill-behavior end of the spectrum does not deny the existence of the value factor. In fact, its effect is out of all proportion to its size.

How Effective Is Outdoor Training?

**DO OUTDOOR
TRAINING PROGRAMS
REALLY WORK?
THIS RESEARCH REPORT
SHEDS LIGHT ON
THE TOPIC, AND OFFERS
SUGGESTIONS ABOUT
HOW TO INTEGRATE
OUTDOOR PROGRAMS
INTO OTHER TRAINING
EFFORTS.**

Executives climb trees, swing through the air, and fall backwards into each other's arms. The proponents of outdoor-based experiential training (OBET) programs say that such activities, as part of a well-thought-out training program, can lead to better teamwork and leadership skills. Critics call the exercises a waste of time, an expensive vacation for burnt-out managers, or harmful indoctrination.

How effective are outdoor-based experiential training programs? That's the question posed in a continuing research project that began in the spring of 1989. The study's goal is to evaluate the effectiveness of various outdoor-based experiential training (OBET) programs.

OBET programs come in two major types: low impact and high impact.
▶ Low-impact programs generally use initiatives with limited physical risk. Activities tend to involve an entire work group.
▶ High-impact programs use initiatives that have a relatively high level of perceived risk. They can involve indi-viduals as the focus of the activity.

The initial evaluation efforts in this study have looked at one type of OBET program that is commonly used by organizations today—the one-day, low-impact, team-focused program. In such programs, the primary training objective is to improve the overall functioning of a group or team of workers.

We originally began our study of OBET programs when we designed a project to evaluate one organiza-tion's program. Our research efforts have grown rapidly since then. The results summarized here are from evaluation studies of six organiza-tions, which have conducted more than 80 OBET programs and trained

BY RICHARD J. WAGNER AND CHRISTOPHER C. ROLAND

more than 1,200 employees in them.

Current research involves another 20 organizations and more than 5,000 participants. These studies are looking at OBET programs that vary in length from one to five days, with objectives focused on team building and such individual traits as leadership and risk taking. They explore both low-impact and high-impact programs, including wilderness programs. (Wilderness programs are high-impact programs in which participants live outdoors and engage in such activities as whitewater rafting, mountain climbing, and sailing.)

Evaluation methodologies used in our research:

▶ participant responses on questionnaires filled out both before and after an OBET program

▶ supervisory reports on the functioning of the work group before and after an OBET program

▶ interviews with managers other than the immediate supervisors to get their reactions to individual and work-group performance after an OBET program.

The initial goal of our evaluation efforts was to determine what behavioral changes, if any, occurred after people participated in OBET programs. Based on previous research in the area of team building, we evaluated two types of behaviors: individual behaviors (including self-esteem, locus of control, and faith and confidence in peers), and group behaviors (including group cohesiveness, clarity, homogeneity, problem solving, and the overall process of the group).

Affect on individual and group behaviors

The participant self-reports we gathered have consistently shown a significant improvement in the overall functioning of a work group after the group attended an OBET program. On the other hand, no significant changes have been reported in *individual* behavior after an outdoor-based program. The results are shown graphically in Figures 1 and 2.

In addition, the improvements in group behaviors have been confirmed by supervisors as long as 15 months after the original OBET program, and by top managers as long as 18 months later. We found it interesting

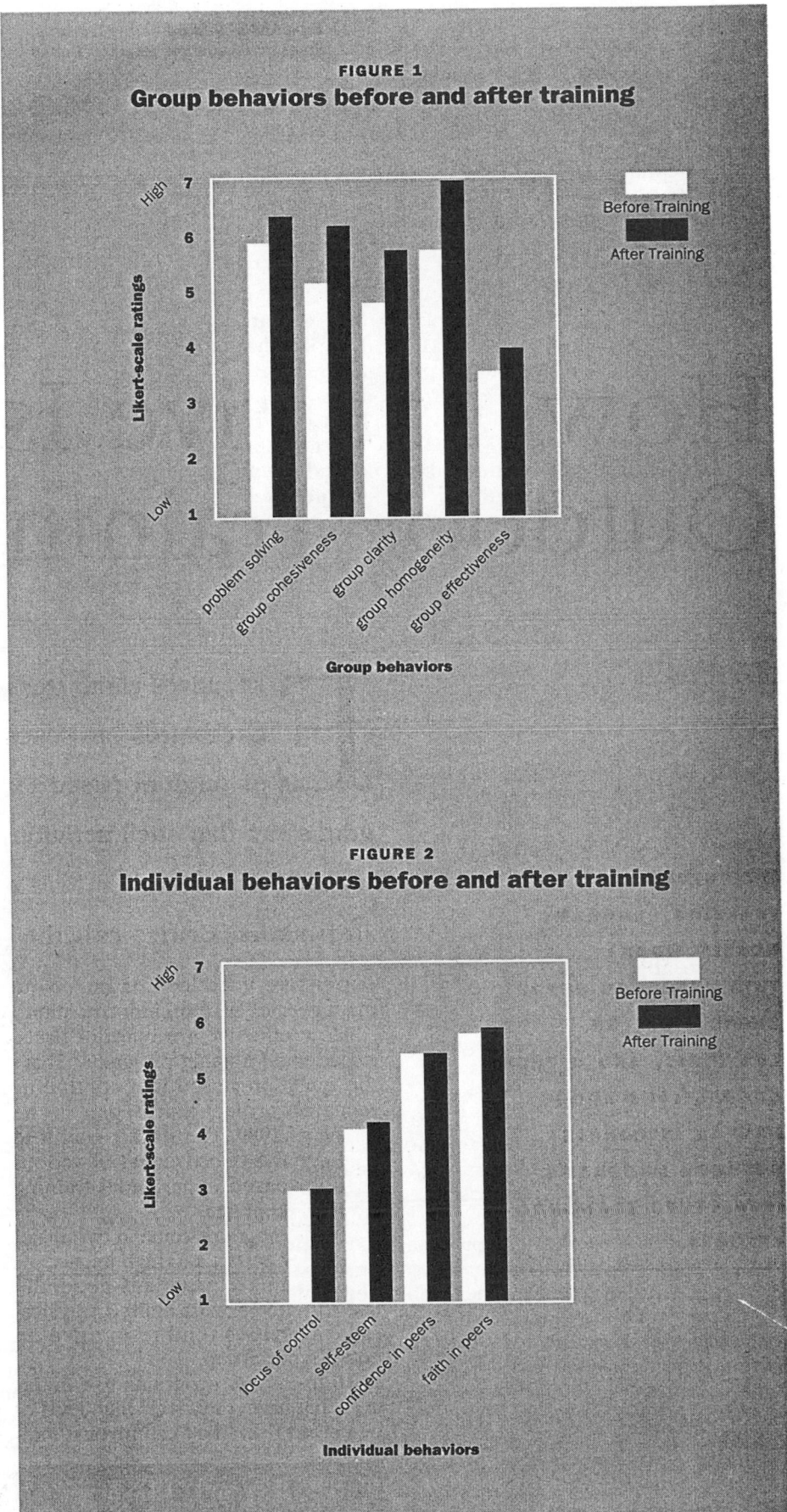

FIGURE 1

Group behaviors before and after training

Before Training

After Training

Group behaviors

FIGURE 2

Individual behaviors before and after training

Before Training

After Training

Individual behaviors

that the programs we evaluated produced no reported significant changes in individual behaviors such as self-esteem and locus of control, though these are some of the very reasons often mentioned for using OBET programs.

It is important to note that the programs studied to date have been almost exclusively low-impact programs with a primary objective of team building. As our research looks at other types of OBET programs (including high-impact programs), it is likely that the evaluations will produce different results.

The participants

The evaluation results described above strongly suggest that low-impact OBET programs are effective in improving group process and interaction skills. But the improvement did not occur uniformly for all participants. In the programs we studied, we looked at several partici-

pant variables that could influence OBET effectiveness:
▶ whether a program was conducted with intact or non-intact work groups
▶ whether participation was voluntary or mandatory
▶ gender composition of the group
▶ whether the group's supervisor was present for the program.

We'll look at each of those four variables separately.

Intact or non-intact work groups. Some team building programs train workers who are strangers to each other at the start of the training. Some organizers of such programs rationalize that strangers often react more honestly with each other than do people who work together every day.

The results of our research on OBET suggest otherwise. Our evaluations show that intact work teams (those who interact at work on a

regular basis) benefit significantly more by attending OBET programs than do people from non-intact work groups.

Volunteers or nonvolunteers: Most training manuals suggest that only volunteers should attend training programs. Our research suggests that this is not true for outdoor-based experiential training.

Almost half of the participants in our OBET programs did not volunteer to participate in the training. Despite having been "forced" to attend OBET, the nonvolunteers' behavioral changes were not significantly different from those of the participants who volunteered to attend the programs.

Gender composition. The groups in the programs we studied ran the gamut from all-female to all-male. But most groups attending OBET include both men and women.

Our research showed a statistically significant relationship between the

balance of men and women in a group, and changes in two key variables: group problem solving and overall group effectiveness. Groups with a balance of men and women showed more improvement in those areas than did groups that were male-dominated or female-dominated. And the groups with a balance of men and women also reported that they enjoyed the OBET program more.

Presence of the supervisor. Slightly more than half of the groups we evaluated attended the program with their supervisors.

We expected the presence or absence of the supervisor to have an influence on the effectiveness of the programs. We were surprised to discover that the only significant effect the supervisor's presence had was that the groups that attended OBET with their supervisors liked the pro-

gram better. As far as changes in group behaviors, we found no significant differences attributable to the presence of the supervisor.

The training program itself

A full evaluation of the variables involved in OBET programs was well beyond the scope of our initial efforts, but we were able to evaluate some aspects of the training programs themselves.

The selection of activities (indoors or outdoors) was made by the individual facilitators and varied from program to program and from group to group. So the possible combinations of activities were simply too great to attempt to analyze in this study. But we were able to evaluate two key variables in the programs:
▶ the amount of training that was held outdoors
▶ the use of follow-up programs.

The first of these OBET variables is the percentage of time spent outdoors. The percentage of the training that was held outdoors varied from 75 percent to less than 20 percent in the programs we studied, depending primarily on the weather. The average amount of time spent outdoors was about 60 percent.

Many people argue that the outdoor setting enhances the success of this type of training. Our research does not support that assumption. The amount of time spent outdoors in the training programs we looked at was unrelated to the success of the programs.

It will be interesting to compare those results with future evaluations of programs that are held entirely indoors or entirely outdoors. Our research strongly suggests that the process, not the setting, facilitates the behavior changes.

We also studied the use of follow-up programs three weeks after an initial OBET session. Transfer of training normally declines after a session is over, so most trainers recommend follow-up sessions to reinforce the original training.

The different programs had three follow-up conditions:
▶ no follow-up
▶ a one-hour general discussion program three weeks after the training
▶ a one-and-a-half-hour session

MANY PEOPLE ARGUE THAT THE OUTDOOR SETTING ENHANCES THE SUCCESS OF THE TRAINING. OUR RESEARCH DOES NOT SUPPORT THAT ASSUMPTION using experiential activities, three weeks after the initial program.

Due to the relative briefness and low intensity of the follow-up sessions used, we were not surprised to find no significant difference in participant behaviors among the three groups. The supervisors' comments confirmed that finding. They reported significantly improved behaviors from OBET participants as long as 18

tating an OBET program.

The facilitators for the second year of programs, on the other hand, participated in an intensive facilitator-training program. In addition to the OBET activities, training emphasized the relationship of OBET to on-the-job activities, as well as debriefing and human behavior skills.

The second year's OBET sessions were significantly more effective than

recreation and counseling degrees but little business experience.

4. Train-the-trainer sequences should pay particular attention to human behavior skills and organizational knowledge. The initial training should focus on activities that require minimal (and low-level) experiential equipment, such as "low ropes" elements. (Ropes courses are specially designed facilities, often using ropes, for conducting outdoor training. "Low ropes" refers to rope-course activities that take place at about eye level or below.)

5. Focus your low-impact OBET programs on intact work groups.

6. Let the supervisor of a group decide whether to attend.

IF POSSIBLE, MIX MEN AND WOMEN EVENLY IN OUTDOOR TRAINING GROUPS

months after training—regardless of whether a follow-up was used.

The OBET programs we studied focused on improving team building skills, so we believe that the effect of the training may increase over time, rather than decrease. That's especially likely when the participants are allowed to practice the skills repeatedly at work.

As we study OBET programs with more intensive follow-up strategies (such as one-on-one coaching, onsite seminars, and additional experiential programs), we hope to shed more light on exactly how this variable influences the effectiveness of OBET programs.

The facilitators

A key element in any training effort involves the trainers or facilitators. Twelve different trainers facilitated the 80 programs that we studied.

We compared the behavioral changes found for each group of participants, based on who facilitated the group. We found significant differences in two key group behaviors: effectiveness and awareness. This suggests that something about the facilitator does make a difference in determining how effective an OBET program will be in improving some behaviors. We looked at two groups of facilitators, each from a different year of OBET programs.

For the first year's facilitators, training was mostly on the job. It emphasized primarily the activity process and safety aspects of facili-

the first year's, especially in the area of group effectiveness. The only key change from the first year to the second year was the training of the facilitators. So we believe that training OBET facilitators in business and human behavior can significantly increase OBET effectiveness.

Ten recommendations

The following 10 recommendations grow out of the results of our research to date. They may be helpful for organizations using—or considering—OBET.

1. Determine the objectives of your program before you select the type of training to use. For example, if your organization needs more dominant leadership, then the OBET programs we studied—which focus on teamwork—would be inappropriate. But such programs could be an excellent fit if a more team-oriented approach is called for.

2. Use OBET only if it appears that this kind of training will meet your objectives. Low-impact OBET programs seem most effective in such areas as problem solving and team building. Other types of OBET programs may be more effective in developing other skills.

3. Select a consulting and training firm (and its consultants) carefully. Decide whether the firm's role will be to design and facilitate the entire program or to train your in-house facilitators. Pay attention to educational background and experience. Many consultants in this new field have

7. If possible, mix men and women evenly in the OBET groups.

8. If an outdoor setting is not feasible, use experiential activities as a part of your indoor training programs. The quintessential ropes course is not an absolute necessity for a successful experiential training program.

9. Focus on the quality of the initial OBET training. But be sure to design appropriate follow-up options.

10. Evaluate the effectiveness of your OBET program. Maximize the results by modifying the program based on the evaluation.

Further study may yield new information and further recommendations. But the preliminary results of our research suggest that OBET is an effective means of training people in certain "soft-skill" behaviors, including problem solving and other team building skills.

Richard Wagner *is an assistant professor of management at the University of Wisconsin's College of Business and Economics, Whitewater, WI 53190.* **Christopher Roland** *is president of Roland and Associates, 67 Emerald Street, Keene, NH 03431.*

This article is a follow-up to "Outdoor Training: Revolution or Fad?" by Richard J. Wagner, Timothy T. Baldwin, and Christopher C. Roland. For that article and more on outdoor programs—including a list of providers and a recommended reading list—see the March 1991 issue of Training & Development Journal.

PART 6.
Results

Who's Afraid of Level 4 Evaluation?

A Practical Approach

BY SANDRA SHELTON AND GEORGE ALLIGER

There is no escaping it: Increasingly, trainers are having to account for training dollars spent. And they are having to do it in terms of business results and return on investment.

In 1959, Donald Kirkpatrick proposed a four-level model of criteria for evaluating training: learner reactions, learning, job application, and observable business results. But many organizations still don't evaluate training. Others base their evaluations only on trainees' reactions. Now, more sophisticated assessments are called for, including Kirkpatrick's Level 4 evaluation—observable business results.

A 1988 poll of about 300 leading organizations, conducted by the American Society for Training and Development, found that only 20 percent evaluated training in terms of its economic effect on the organization. In fact, many training professionals seem to think that measuring training results in terms of dollars and cents takes too much time, is too costly, and is susceptible to extraneous factors that may affect results.

A 1990 IBM study of six large corporations (including IBM itself) and several training consultants, including Kirkpatrick, found that even organizations that say they examine the economic impact of training

MANY ORGANIZATIONS AVOID LEVEL 4 EVALUATIONS BECAUSE THEY CAN BE COSTLY AND TIME-CONSUMING. BUT THOROUGH PLANNING CAN EASE THE PROCESS AND HELP ENSURE VALID RESULTS.

don't do so directly; they rely on people's opinions. Asking trainees whether training has improved their performance or their organization's performance isn't the same as assessing performance directly.

Why avoid Level 4 evaluations?

One reason organizations may shy away from Level 4 evaluations is that collecting and interpreting the data is more difficult and time-consuming than surveying trainees. But a Level 4 evaluation, which should include a cost-benefit analysis, can provide data that are more thorough and more credible than information collected by surveying trainees.

In some instances, conducting a Level 4 evaluation can be fairly easy. When data are routinely collected—such as the number of hours worked, units produced, and defects—a Level 4 evaluation may simply be a matter of obtaining, organizing, and analyzing already available data.

But sometimes organizations try to conduct Level 4 evaluations when they're not appropriate. A Level 4 assessment may not be the proper evaluation method for training that doesn't affect observable outcomes—for example, training that aims to change only attitudes. Such training isn't likely to show changes in organizational output. If a Level 4 evaluation is used in such a case, its inappropriateness is sure to reinforce any negative opinions about the Level 4 approach.

Another example of the inappropriate use of Level 4 evaluations is that of a training manager who promises to conduct a Level 4 study of a particular program and then discovers that objective data aren't available or accurate. Simply put, the manager can't finish what he or she started, and the Level 4 approach again looks impractical.

Last, some people are intimidated by Level 4 evaluations. Level 4 studies often are expensive; if they don't produce positive results, trainers believe they'll be held accountable to management. In other words, if the objective data show that the training failed, the trainer can't avoid culpability.

For example, if a training program aims to increase the number of units produced and reduce the number of defects, it's difficult for a trainer to deny the conclusion of a pretraining/post-training assessment that shows the goals weren't attained. If the data show that more units weren't produced, with fewer defects, then the inescapable interpretation is that the training failed. The credibility associated with objective data is one of the reasons that Level 4 studies should be carefully planned.

Clearly, there are situations in which Level 4 evaluations aren't the appropriate assessments to use. But if you are considering using a Level 4 evaluation, here are some guidelines.

Step 1: Should I conduct a Level 4 evaluation?

First, ascertain whether the training is likely to have a detectable effect on business results.

If there is a business plan, refer to it to examine the results for which the training was designed. For example, the plan might stipulate that the training should improve customer satisfaction by 25 percent, increase sales by 30 percent, and so forth. Unless the training is linked to clear business outcomes, there is no reason for a Level 4 evaluation.

If the training contains material that requires frequent changes, it might be difficult to get meaningful results from a Level 4 evaluation. Changes in the training content during training—and while the Level 4 study is in progress—could affect the validity of the data.

It is best to use a Level 4 evaluation in combination with Level 2 (learning) and Level 3 (job application) evaluations. Levels 2 and 3 are used to evaluate the quality of training, and they provide additional support for conclusions about the training's effects on business results. In particular, a Level 3 evaluation provides evidence of the transfer of knowledge, skills, and attitudes back to the job. If transfer of training didn't occur, a Level 4 evaluation can't show any results.

Level 4 results can help you make decisions about the value of the training, such as whether to modify or eliminate certain programs or develop new ones. If a Level 4 evaluation demonstrates a favorable return on investment, it can help trainers get managerial support for additional training.

Step 2: Is a Level 4 study feasible?

Once you decide that a Level 4 evaluation is appropriate, you still have to determine whether it's doable. The following criteria can help.

Business measures. In a Level 4 study, you will have to specify the business measures that will be collected in the evaluation.

Line managers can help you establish baseline data against which to measure improvement. Baseline data might include such business measures as the pretraining rates of accidents and absenteeism, the numbers of processing errors and units produced, the unit and operating costs, and the frequency of safety violations.

Time. When specifying data for a Level 4 evaluation, you need to consider the amount of time it will take to collect the data and whether such data are available and accessible. It's possible that some of the data you need already exist and are routinely collected in quality measures; in production and repair records; and in sales, activity, and expense reports.

If such data aren't available, you need to find ways to obtain them. For example, if processing errors aren't being measured, you can work with line managers to identify the errors that are critical. Then you can establish and coordinate a method for gathering data on errors.

Data must be available on a timely basis. If it takes too long to evaluate the training, key people may have transferred to other jobs by the time the results are in.

You need time to collect enough data for a reliable measurement. The more reliable your evaluation is, the more confidence you'll have in the conclusions. For example, if you collect data on the number of units produced, make sure the collection period is long enough to reflect temporary fluctuations in output caused by such factors as seasonal work schedules and employee sick leaves. Also, you may need extra time to convert data into monetary terms for calculating return-on-investment.

A Level 4 evaluation requires the time of several people. You need a project leader to coordinate the evaluation and a co-worker to provide backup. And you may need administrative support to help compile data and generate reports.

Extraneous factors. Variables other than the training might influence the data, making it difficult to determine the actual effect of the training. It's important to identify those variables.

Extraneous factors might include trainees' ages and work experience as well as seasonal sales patterns, economic changes, shifts in managerial styles, equipment breakdowns, and customer attitudes.

A good way to measure the effects of extraneous factors is to compare the results of a control group with the results of the trainee group. If you can't use a control group, you should establish some baselines with which to compare post-training results.

Costs. The costs of a Level 4 approach vary from evaluation to evaluation. Some Level 4 studies are inexpensive; some produce benefits that far exceed the costs. But if every aspect of a Level 4 evaluation requires a lot of time and effort, expenses can be high. In such cases, you should weigh the potential costs against the potential value of the results.

An evaluation design that uses pre- and post-training measures and no control group is likely to be less expensive than a control-group design. An evaluation that involves a control group usually costs more because it requires the collection and analysis of more data.

If relevant data are already being routinely collected and recorded, data collection shouldn't be expensive. But creating a system for collecting data can be costly. Talk with the appropriate people to make sure money is available for new data-collection systems. To keep expenses down, try getting people who have access to the data you need to save the information for you. If you need more than one source or type of data, there may be added costs for coordinating and matching information. If the data are coming from a single site, collection is less expensive and easier. Also, it costs less when the site or sites are nearby and people don't have to travel.

Obviously, contracting with training suppliers raises costs. Using internal consultants and experts usually saves money.

Step 3: Which design should I use?

Once you have determined that you will conduct a Level 4 evaluation and you have verified its feasibility, you need to select a design and write a measurement plan.

The main criterion for selecting the evaluation design is that it should enable you to conclude whether the training affected business outcomes. There are several designs to choose from.

You can use a design that compares pretraining and post-training results and that uses no control group. With this design, data are more easily collected because the sample size is small. But this approach doesn't eliminate the possibility that pre- and post-training changes are due to variables other than training.

Another design measures only the post-training results achieved by trainees and a control group. This design requires no pretraining measure. But it doesn't take into account possible pretraining differences between trainees and members of the control group. It also requires a larger sample size than the first design.

A third design measures both the pretraining and post-training results achieved by trainees and a control group. It's the most effective design for eliminating factors other than training as the cause of positive or negative results. But it requires more data collection and analysis than the first two designs.

Evidence of your training's effectiveness is strongest when you can rule out the possibility that something other than the training is the reason for any improvement. Taking pretraining measurements can help eliminate the possibility that trainees were operating at post-training levels prior to the training. Also, using a control group can help determine whether the trainees would have performed just as well without training.

At this stage, you should write a measurement plan to distribute to stakeholders. Your plan should document evaluation work that has already been done and define work that is going to be done. The plan can include such elements as a timetable for the evaluation, a list of necessary resources, a statement of the responsibilities of stakeholders, explanations of your data-collection and data-analysis methods, and rationales for the design selection and for conducting a Level 4 evaluation, including the net value to the organization. You also can include any project-management documentation required by your department.

You can submit your plan for review to peers or to measurement and evaluation experts. The feedback can help you avoid pitfalls.

Step 4: What will the training cost?

In addition to data on business measures, a Level 4 evaluation requires data on the total cost of training. You will need to assemble some information in the following categories.

Personnel. Trainers, designers, developers, subject matter experts, training suppliers, and trainees represent various training expenses.

To calculate the cost of personnel, subtract the number of vacation days and holidays from the number of working days per year to get the number of productive days per year. Then divide a person's annual salary by the number of productive days to get the daily cost per person.

Multiply the daily cost or hourly rate—of consultants, for example—by the number of days or hours worked or attending training, to get the total cost of each person's time.

Facilities. To calculate the cost of classrooms, labs, and learning centers for external courses, multiply the hourly, daily, or weekly rental fee by the number of hours, days, or weeks the facility is used for each training session. Multiply that cost by the number of sessions.

Equipment. To determine the annual cost of such equipment as computers and VCRs, divide the purchase, rental, or maintenance price of each piece of equipment by how long the equipment is expected to last. Distribute the annual cost equally among all training programs or sessions.

For example, a $2,000 personal computer with a two-year life span has an annual cost of $1,000. If it is used for 10 training programs or sessions, the cost per program or session is $100.

Course materials. To calculate the cost of course materials such as manuals and software, multiply the price of the materials by the number of trainees.

Travel. To get your total travel costs, multiply the average travel expenses per person per day by the number of travelers.

For example, 35 people travel to take a five-day course. Housing costs $40 per day, per person. The per diem rate is $25; incidental expenses are $5 per day for each person. Airfare is $450 each, which averages out to $90 per day. Shared rental cars cost $4 a day per person, which brings the transportation cost per day to $94 each. So the total travel cost for the 35 travelers is $28,700.

You can use the data on training cost to help you calculate ROI.

Step 5. How do I analyze the data?

Has the organization benefited from the training? Are organizational units in which employees have completed training doing better than those in which employees haven't completed training?

You can find the answers to those questions by comparing pretraining data to post-training data, or by comparing trainees to a control group.

The data may be analyzed to obtain a return-on-investment ratio. There are other ways to analyze data, but ROI is easy to compute and almost universally understood in the business arena.

Calculate ROI by dividing the dollar value of the training by the total training cost. If you calculate ROI on training gains that are statistically significant and that aren't due to random errors in the trainee or control group, then the gains reflect the actual value of the training.

For example, in a training program for customer-service representatives, the anticipated business result was a gain in productivity; in other words, an increase in the number of monthly service calls or visits, without an increase in the number of reps. The training goal was to provide reps with a standard troubleshooting procedure that would reduce the time required to complete a service call.

Before calculating ROI, it's necessary to calculate the cost and the dollar value of the training. The cost of the training is the total expenses of personnel, facilities, equipment, course materials, and travel. In our example, the total cost of training the customer-service reps was $58,090.

The "dollar value" of the training is the difference between the value of calls made before and after the training. The Level 4 evaluation confirmed an increase in the number of calls per day made by service reps after they were trained. The productivity gain enabled reps to make more customer calls without increasing staff.

To figure ROI, subtract from the post-training dollar value of training (number of calls per day × cost per call) the pretraining dollar value. The post-training value is 27 reps × 6.1 calls per day per rep × $24.50 per call × 228 work days per year = $920,014 in costs per year. Subtract the pretraining dollar value, 27 reps × 5.1 calls per day per rep × $24.50 per call × 228 work days per year = $769,192 per year.

The difference is 27 reps × 1.0 calls per day per rep × $24.50 dollars per call × 228 work days per year = $150,822 yearly cost. Divide $150,822 by $58,090, the training cost, and

you get $2.59. Multiply that number by 100 to get ROI.

ROI is 259 percent, or $2.59 saved for every dollar spent on training. The rule of thumb is that any value greater than 100 percent represents a return on investment that is beneficial.

As a precaution, we recommend testing for significant differences before drawing conclusions about the value of the training. To do that, compare the mean or average results of trainees with the mean results of the control group. Using statistical methods, test to see whether the difference is significant; in other words, whether the difference is likely to be due to chance, or to the training. If you didn't use a control group, you can test for significant differences between pre- and post-training measures.

Testing the statistical significance of the difference in means between results achieved by the trainee and control groups can help determine whether the calculation of ROI is based on actual gains due to the training. You can use one of the many software programs on the market to conduct the statistical analysis.

Step 6: How do I report the results?

Step 6 involves writing and distributing a report on the evaluation's findings. The report should include an executive summary, a statement of business needs, an evaluation and

measurement of business results, actual findings, conclusions, implications, and recommendations.

You can add an appendix that describes the measurement instruments and shows a tabulation of the data. Once the report has been distributed, be sure to follow up. The feedback from people receiving the report can tell you which ones are using the report, whether they understand and believe the information, whether the report matches their interests and needs, how they are using the information in decision making, and how the report could be improved.

Though the concept of evaluating training on the basis of business results is hardly new, it has yet to be widely used. That may be changing. Organizations are recognizing the need to measure training in the same ways that they measure other areas of business. Level 4 evaluations can help companies demonstrate that training dollars are a wise business investment.

George Alliger *is an associate professor of industrial and organizational psychology in the department of psychology, State University of New York, Albany, New York.* **Sandra Shelton** *is in the department of learner-directed education, Skill Dynamics, Box 2150, WF04A, Atlanta, GA 30301-2150.*

The Bottom Line

Here's how one company proved the business impact of management training.

By BASIL PAQUET, ELIZABETH KASL, LAURENCE WEINSTEIN, and WILLIAM WAITE

Most HRD practitioners believe that management training makes a real difference in the workplace, but many of us avoid proving it. However at CIGNA Corporation's corporate management development and training (CMD&T), which provides training for employees of CIGNA Corporation's operating subsaiaries, we set out to do just that—prove that our management training makes a real business contribu-

Paquet is assistant director of corporate management development and training of CIGNA Corporation in Hartford, Connecticut. Kasl is an adjunct professor of adult education at Teachers College, Columbia University, New York City, and a partner in Training Measures Associates in Stamford, Connecticut. Weinstein is an associate professor of management and department chairperson at Sacred Heart University, Bridgeport, Connecticut, and a partner in Training Measures Associates. Waite is vice president of corporate management development and training, CIGNA Corporation.

tion. We wagered that an investment of creativity, hard work, and some budget dollars would pay handsome dividends for our organization.

For the sake of brevity, we will describe just three of the 14 in-depth cases from our pilot study showing how managers used their training to improve productivity. These three cases reflect savings and income totaling $280,000, or two-thirds of the full cost of training the entire population who were involved in that program for the full year—including participants' salaries, overhead, and program development costs.

Since our companies operate in a profit-oriented environment where results are the measure of ultimate worth, our CMD&T evaluation team reasoned that if we were going to prove our training's worth, then we should use management's own language—business results and return-on-investment. That meant we needed to show that management training could be linked directly to improved productivity in the workplace. If we could create a bottom-line impact evaluation, we believed we would benefit in our efforts to

■ justify the worth and budget of the training group to top management;

■ market our training products and services to our internal clients;

■ redesign our programs and develop new ones.

Targeting productivity results

Perhaps the primary reason proving the bottom-line impact of management training is not easy is that, like the managers you train, you achieve results through others—in this case, your trainees. We began by setting up some guidelines for our work. First we took a hard look at what an impact evaluation is supposed to accomplish. Generally speaking impact evaluation is the study of whether an educational intervention brings about intended results. Reformulated to fit our business environment, the question became, "Does management training result in improved productivity in the manager's workplace?"

The question was straightforward and suggested our first guideline: Our evaluation would have to follow the manager back to the workplace in order to find out if the work unit's productivity improved after the manager attended training.

Second we took a hard look at what it means to conduct evaluation research in a profit-oriented business environment. We realized that we couldn't expect people to put much time into helping with the research unless they were also benefitting in some way. We also recognized that any viable strategy for ongoing impact evaluation needed to be cost-effective for CMD&T budget and staff time.

These two considerations led us to establish our second guideline: Impact evaluation data collection needed to be built into the training program itself. This strategy addressed both of our concerns. If managers could use the evaluation data for their own benefit as part of their training, they would be more likely to cooperate. If data were collected as part of an on-going training assignment, CMD&T staff would be spared the time-consuming effort of adding a new activity—evaluation data collection—to already busy schedules.

As our impact project was getting under way, it fortuitously happened that another CMD&T project team was redesigning the training program that CIGNA offers to all new first-level to mid-level subsidiary managers, Basic Management Skills (BMS). Our two teams joined forces, working together to embed the framework for impact evaluation into the new program design.

The BMS program was an ideal choice for the impact evaluation project. Its curriculum included a broad range of management skills that, if practiced in the workplace, should have an impact on work-unit productivity. Content areas included planning, problem solving, motivating direct reports, communication, leadership, delegation, performance appraisal, plus a host of corporate human resource management policies and procedures ranging from compensation administration through employment practices.

CMD&T already had positive feedback on this program. Since its inception, BMS had received superior ratings on the "smile sheets" collected from the trainees. Every staffperson who had provided training during the program had been stopped in a hallway or the company dining room by happy graduates with success stories to share. All we needed to do was transform intuitive beliefs and anecdotal evidence into bottom-line language.

The chain of impact

The linkage between training and workplace results takes place through a chain reaction along the levels of impact. Training changes the participant's attitudes

Table 1—CIGNA CMD&T impact model

Chain of Impact		Research Tool		Time Period
Opinions	D i f f i c u l t y	Trainee self-report	P o w e r	Throughout training and at three-month follow-up
Learning		Trainee self-report		At end of training
Behavior		Survey of trainee's subordinates		Before training and at three-month follow-up
Results		Trainee's work unit records, action plan, and BMS workbook		Tracked from three months preceding training to three months following training

and knowledge so that the participant is then able to change his or her management behavior back on the job. If the training has targeted appropriate behaviors, then a change in those behaviors produces the results intended—improved work-unit productivity in the case of a management program like BMS.

Each link in the chain of impact can provide evaluative feedback about the training program. The further along the chain the information is, the more removed it gets from the training experience and the more difficult it is to obtain. But along with that difficulty comes the power to show bottom-line results. Each step on the chain of impact moves us closer to proof that the training is making a real impact on the corporation.

The decision of when and how to measure these types of change is the task of any impact evaluation design team. Table 1 summarizes the design our team produced.

The first level of information—participant opinions about the job-relatedness and effectiveness of the training—we obtain by asking trainees to rate the content areas. We do this throughout the BMS program.

The next level is knowledge, skill, or attitude acquisition. Trainees complete an overall program evaluation that collects information on what they have learned.

The next link in the chain of impact is behavior back in the workplace. Our evaluation uses a survey of the trainee's direct reports. This scale measures employee observations of manager behavior in the BMS content areas of planning, leadership, motivation, performance

management discussions, setting clear performance standards, and the work unit's communications environment. We survey the direct reports before management training begins and three months after the training has been completed and the manager is back on the job.

Finally the last critical step: We base our results on repeated measures of work-unit performance *before* and *after* training. We directly relate these measures to productivity action plans created during the management training and base them on actual business records.

Let us be clear that CMD&T did not set out to prove scientifically that the chain of impact exists. And though we did observe strict research methods, CMD&T is not a research unit and CIGNA is not a laboratory. We laid out a model on the types and possible timing of evaluation data and collection to clarify for ourselves what might be possible to accomplish. The business results that trainees achieved exceeded our expectations. Without a model and methodology we never could have documented these results.

Designing evaluation in

In its original form, the program included 12 training days. The project team working on the new training design had been charged with lowering costs by reducing the program to six days: one overview session and a five-day training week. To facilitate this reduction in time without shortchanging the skills training that had made this program so successful, all the corporate policy material was moved into a self-study workbook for trainees to complete before attending the group sessions.

Building our impact evaluation model into BMS required major design changes in addition to those already planned. We radically redesigned the planning and productivity modules, added a follow-up day, created a pre-posttraining survey of management behavior, and made other changes that made an outstanding program stronger still.

Results: changes in behavior

Our BMS training makes a difference in management behaviors. Our pre-posttest data from the new management skills survey backs us up.

The survey consists of 36 Likert-type five-point scale items that are used to create seven different indices. Six of these indices measure the manager's behavior, and the seventh measures general organizational climate for which the manager is not necessarily responsible. The six behavior indices are related to separate content units in the BMS training design and are reported to the trainee as separate scores in order to help that trainee set personal learning objectives.

Because the intercorrelation among the six scales is relatively high, for the purpose of impact evaluation they are treated as one global measure. The six indices are weighted equally and combined into one grand mean, referred to in this report as "Manager Behavior." The seventh scale is called "General Climate." The average intercorrelation among the six scales is $r = .75$. The correlation between Manager Behavior and General Climate is $r = .57$.

Statistical properties are based on $N = 427$. Trainees in the management program the year before the new design was launched were used for the developmental work.

Our impact evaluation uses these scales for three different comparisons. Time 1 is before training and Time 2 is after training. Comparison 1 equals Time 1 minus Time 2 on Manager Behavior, Comparison 2 equals Time 1 minus Time 2 on General Climate, and Comparison 3 equals Comparison 1 minus Comparison 2.

The five-point scales assign *1* to the most positive opinion and *5* to the most negative. A positive difference for Comparisons 1 and 2 indicates a positive change, respectively, in Manager Behavior and General Climate. Comparison 3 is important because we need to know that if there is a change in General Climate, this change alone is not the sole cause of a change in productivity. Thus it is important for Comparison 3 to be positive and significant.

We are not arguing here that climate is immaterial to performance. To the contrary we trust that long-term positive changes in behavior will affect climate, which ought to in turn affect performance. Our immediate objective is, however, to change behavior, and it is those shifts we measure in our pre-posttesting by using Comparison 3 to factor out general climatic factors that might obscure behavior change.

Table 2 demonstrates that when the first three sessions of BMS were analyzed as a group, the three comparisons support the assertion that training makes a positive difference in the manager's behavior. Further the data indicates that if these managers' work units improve their productivity, that improvement is likely to be linked to the manager's behavior and not to other climate factors in the corporation.

Individualizing productivity measures

When challenged to measure results in management's productivity-oriented language, we took a bold and startlingly simple step: we decided to measure productivity in the work unit.

As noted in the start-up phase of the project, the productivity that counts is what takes place in the trainee's individual work unit. Since it could never be cost effective to send a team of data-gathering evaluators to each graduate's work unit, we set out to design into BMS the mechanism by which each trainee would provide us with the needed data.

We call our strategy "individualizing productivity measures" and we believe that it achieves a type of validity in impact evaluation that most strategies fail to reach. As set forth in our start-up phase, our efforts are predicated on the assumption that the intended outcome of management training is improved productivity in the workplace. From this assumption it follows that the most valid measure of management training's impact is a measure of workplace productivity.

Our impact evaluation methodology demands that a management program set out to affect productivity in both method and purpose. Linkage to training occurs in the specific case of our BMS program, for example, because

■ productivity is a central focus of this management training program;

■ participants are taught how to create productivity measures as part of their training;

■ participants are also taught how to use productivity data as performance feedback and as support for performance goal setting;

■ participants write a productivity action plan as part of their training, and are contracted to bring back measurable results to a follow-up session;

■ individualized productivity measures (IPMs) are put in place as part of the action plan. IPMs are tailored to measure plan results specifically, accurately, and objectively.

These steps ensure core validity; the strongest proof of the true impact of training will be found by measuring the produc-

Table 2—Comparisons of manager behavior and general climate for first three sessions of basic management skills

	Comparison #1 Manager Behavior Time 1-Time 2	Comparison #2 General Climate Time 1-Time 2	Comparison #3 Manager Change— Climate Change
N*	22	22	22
Ex	3.10	0.51	0.91
Ex²	3.3512	6.9503	3.0565
x	0.155	-0.023	0.178
	$t = 1.98$ $p < .05$	$t = 0.19$ $p = $ n.s.	$t = 2.49$ $p < .025$

*BMS participants are included in this analysis only when the Time 1-Time 2 comparison of matched surveys is based on two or more employee opinions. Five participants who have only one employee matched at Time 1 and Time 2 are deleted from this comparison.

tivity of the trainee's work unit. In a large corporation, where trainees in a management skills course come from a variety of production and staff environments, a measure of impact that has core validity by definition *must vary from manager to manager.*

Thus we have gone in the opposite direction of most HRD practitioners who have sought to prove the results of training. We believe that those who track absenteeism, turnover, morale, and promotion patterns limit their measures to surrogate results. True bottom-line results are found in the work records of the individual participants and their units.

Training people to demonstrate results

As part of their training, BMS participants are required to follow step-by-step workbook procedures that lead them through a thorough analysis of their unit from both a technical or work management perspective and from a human resource management perspective. It also initiates a systematic approach to planned changes that will positively affect their units' business results.

Previously we believed it was enough to teach good planning and problem solving, with evaluation embedded in the planning model. We discovered, however, that the majority of new managers did not know the rudimentary techniques of work measurement and writing performance standards for both their units and individual workers. Management by objectives, performance management, and productivity were abstractions for our first-level management population. We discovered—and it was a revelation despite how obvious it seems—that we could not prove results if our trainees did not know how to document them, and that they were far less likely to produce them if they did not know how to use productivity measurement as a management tool.

So we devised training materials that teach managers how to write performance standards for both their units and individual performers; how to use simplified work measurement techniques; how essential performance management is used as a tool to complement management by objectives, positive reinforcement, and constructive feedback on performance; and other essentials of good management. Measuring the results of their action plans becomes as important to our trainees as obtaining these data is to us, the training department.

The principle is simple. Many managers never go beyond measuring performance in terms of production. We push our trainees to establish standards and to create true measures of productivity based on this familiar formula: Divide your output by a partial factor of input consumed to produce that product or service.

These measures and the action plans they will track are reviewed and critiqued by fellow trainees and trainers during the BMS action planning module, which lasts a full day. Participants are challenged to produce the best plans and measures possible. Then they are required to go back to their work records and establish a base period of performance for at least three months prior to their training and action plan. When they leave training they

implement their plans, track progress, and deliver their results 16 weeks later at a follow-up session. Before, during, and after the training we will have cycled the trainees through our impact model at all levels.

Business results

Table 3 is a set of case studies representing two BMS pilot groups whose results CMD&T studied. In some cases the business impact can be converted into dollars. In others it is reported as productivity gains. In all cases these managers were asked to produce verifiable results based on actual work-unit records to support pre-posttraining and measures of performance.

Table 3—BMS case studies on bottom-line results

Case No.	Manager Function	Plan Results
1	Reinsurance coding unit	18 K in savings, 730% ROI
2	Premium collections	80 K increased investment income, 3,100% ROI
3	Premium collections	150 K increased investment income, 5,900% ROI
4	Exchange department	95 K increased revenue, 4,475% ROI
5	Benefits administration	60% improvement on timing standard
6	Benefits administration	80% improvement on accuracy standard
7	Benefits administration	80% improvement on accuracy standard
8	Systems processing unit	20% productivity increase, 110 K in budget savings, 4,500% ROI, plus increased tape mounts, improved turnaround on service requests
9	Financial analysis unit	35% improvement on on-time service, plus 25% productivity increase on job set-ups
10	Systems testing unit	165% efficiency increase against production standard
11	Actuarial unit	15% improvement on timing standard, 70% improvement in accuracy while production volume increased by 20%, no staff increases
12	Actuarial unit	15% improvement in on-time reporting, 30% increase in quality review standard
13	Actuarial unit	30% increase on service timing and 10% improvement in accuracy standard
14	Actuarial unit	5% increase on service timing and 10% improvement in accuracy standard

In a fast-paced business environment, a lot of changes can occur in a quarter of a year. Many CIGNA company trainees change their jobs and sometimes their jobs change. About two-thirds of our trainees return for the follow-up; about one-half the original class gets the opportunity to see their action plans through to completion. For many of our trainees their business reality changes so significantly within a quarter of a year that tracing their IPM results loses its validity.

We have compared all relevant demographics on managers who provide follow-up data with those who do not. We are unable to detect any differences and do not think self-selection is biasing our results.

Because our IPM methodology creates different results for each participant, it does not lend itself easily to summary analysis or inferential statistics. Right now our cases rest on their own merits as narratives of actual results. There is considerable power in realizing that a handful of trainees contribute enough savings and income from the success of their BMS action plans to fund the training of an entire managerial population for a full year.

But we do think that inferential analysis is possible. By randomly selecting a subset of trainees, results could be classified according to degree of success, links could be established between productivity data and behavior-change measures, and in-

ferences could be drawn to the entire BMS population.

Such an inferential analysis might reveal powerful data, but it is also considerably more research-based than a functional management training unit like CMD&T can probably afford to pursue, and it is unlikely we will decide to fund such an extension of our study. For now, we will continue to collect case study results for the illustrations they draw of the business results that management training can generate. We also are eager to examine larger sets of data on behavior change as

measured by our survey instrument.

Details from three of our case studies can illustrate how trainees use a synergistic application of BMS-learned skills along with their action plans and IPMs to create and document productivity changes.

■ *Case One*—In this case, the manager of a reinsurance coding unit confronted problems with processing backlogs, work-flow bottlenecks, work distribution issues, and morale. This manager tackled them head-on in her BMS productivity action plan. She created new standards, increased in-

dividual worker responsibility, and streamlined the work flow and processing sequence for her unit.

To measure her action plan results in true productivity terms, she divided a weighted production output—number of cases produced multiplied by a weight representing the degree of case difficulty—by the total available worker days in a given month, thus factoring out vacation time as well as time spent by staff on other kinds of unit production. Bottom-line results in six months showed that every productivity goal for each worker

Our efforts are predicated on the assumption that the intended outcome of management training is improved productivity in the workplace

and the entire unit had been achieved or surpassed.

Figure 1 illustrates both productivity and production charts for a six-month period. Notice in this manager's case the difference between measuring productivity and production. Had she been monitoring only production, as was her practice before BMS training, she would not have understood that her unit was performing well in spite of the severe staffing problems in October.

Based on calculations using salary and

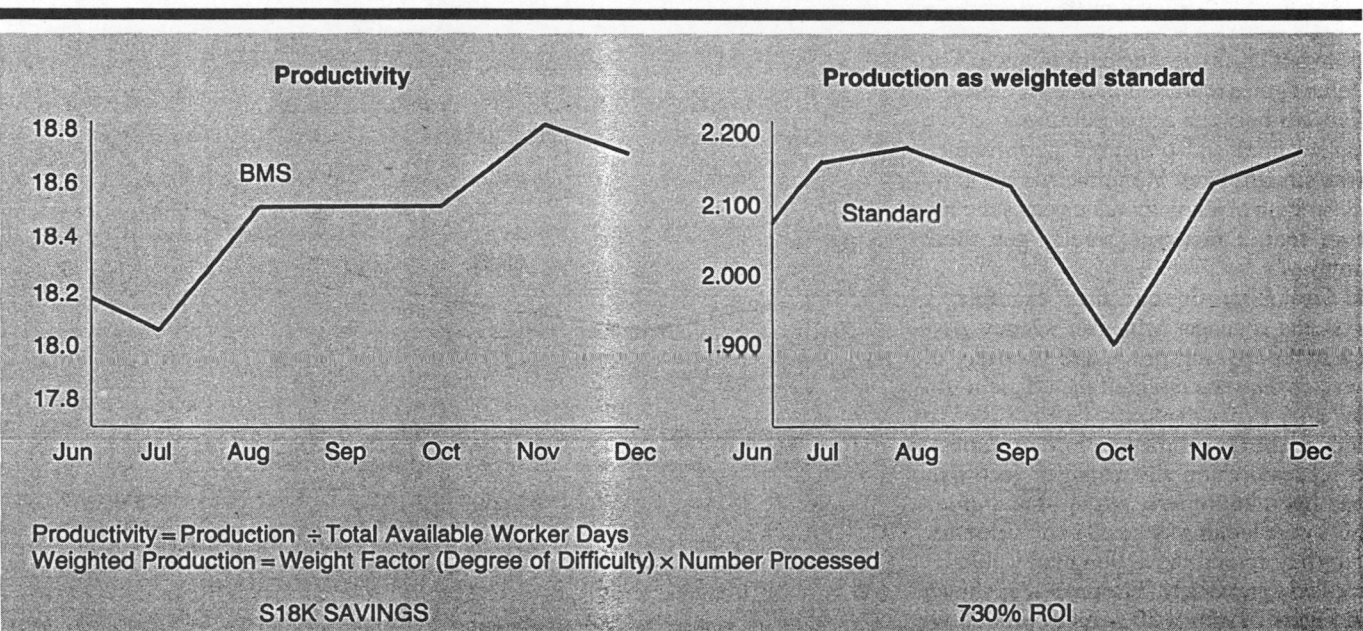

Figure 1—Productivity and production in the reinsurance unit

Productivity = Production ÷ Total Available Worker Days
Weighted Production = Weight Factor (Degree of Difficulty) × Number Processed

S18K SAVINGS   730% ROI

fringe for the unit, net improvements equaled $18,000 in savings, for an ROI of 730 percent on the cost of training this manager. In most cases an ROI estimate based on the value of partial input instead of output is an underestimate of actual benefits to the company. Even with the underestimate this manager was pleased with the estimate of her action plan's bottom-line value. She noted that $18,000 was approximately the cost of hiring a new worker and realized, with satisfaction, that hiring a new worker had been her boss' original idea for solving the backlog problem.

■ *Case Three*—This case features a premium collections manager. Collecting premiums on time is important in our business because late premiums represent a lost investment opportunity. Through survey feedback from her subordinates this manager learned her problems were largely due to poor human resource management skills and failure to set clear performance goals. In-depth posttraining interviews with her workers by CMD&T revealed a manager who dramatically altered numerous management behaviors.

Her unit's collection rate also changed dramatically—from an average of 75 percent of premium on time to an impressive 96 percent. This improvement yielded extra investment income of $150,000 per annum, for an ROI on training dollars of 5,900 percent.

The premium collection manager's productivity graph in Figure 2 is very interesting. She asked us to project where she would have been *without* her training. We plotted a "line of best fit" projected from her unit's baseline performance. The dollar figures represent premium value differences between actual performance (the jagged line) and projected performance (the straight line). We think this is a good illustration of training's value versus the notion that a manager would "get there anyway."

■ *Case Eight*—In our final example, a systems manager who ran agency pay-cycle reports and processed a variety of service requests targeted the efficient use of analyst-programmer (A-P) time to reduce the downtime involved in respecifying, researching, and testing projects that had been improperly set up. He and his boss were looking for small improvements, but they agreed the action plan would be tackled aggressively. The results, as shown in Figure 3, were a 20 percent productivity increase that bottom-lined at $112,000 in savings on the budget, for a 4,500 percent ROI of training dollars.

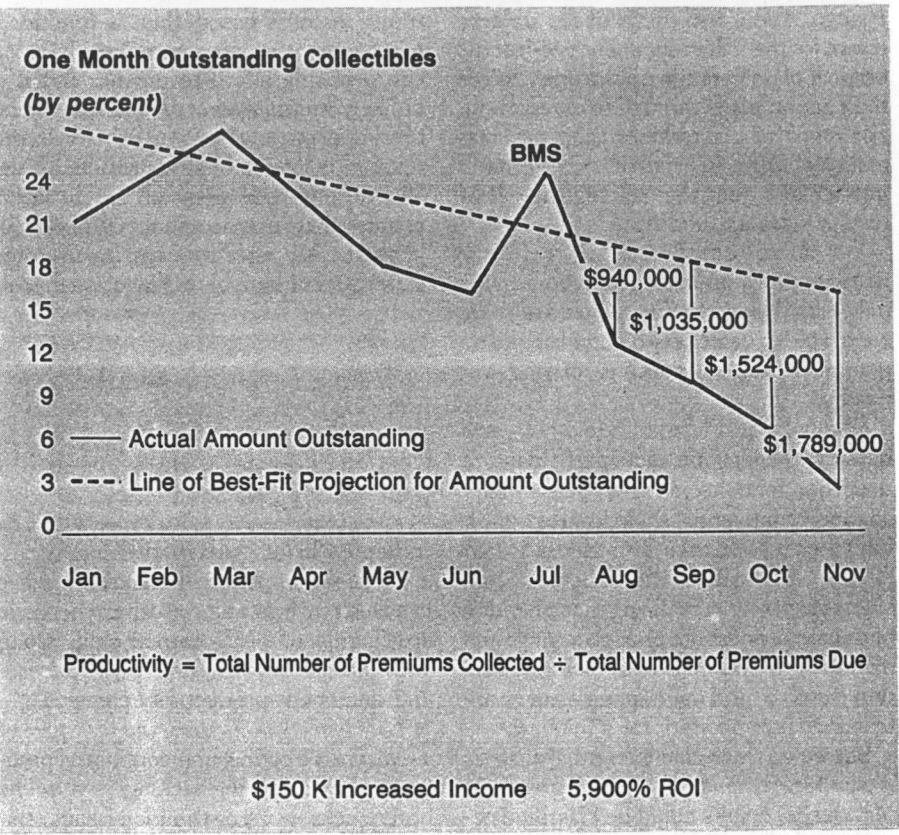

Figure 2—Improvement in premium collections

One Month Outstanding Collectibles (by percent)

BMS

$940,000
$1,035,000
$1,524,000
$1,789,000

—— Actual Amount Outstanding

- - - Line of Best-Fit Projection for Amount Outstanding

Jan Feb Mar Apr May Jun Jul Aug Sep Oct Nov

Productivity = Total Number of Premiums Collected ÷ Total Number of Premiums Due

$150 K Increased Income   5,900% ROI

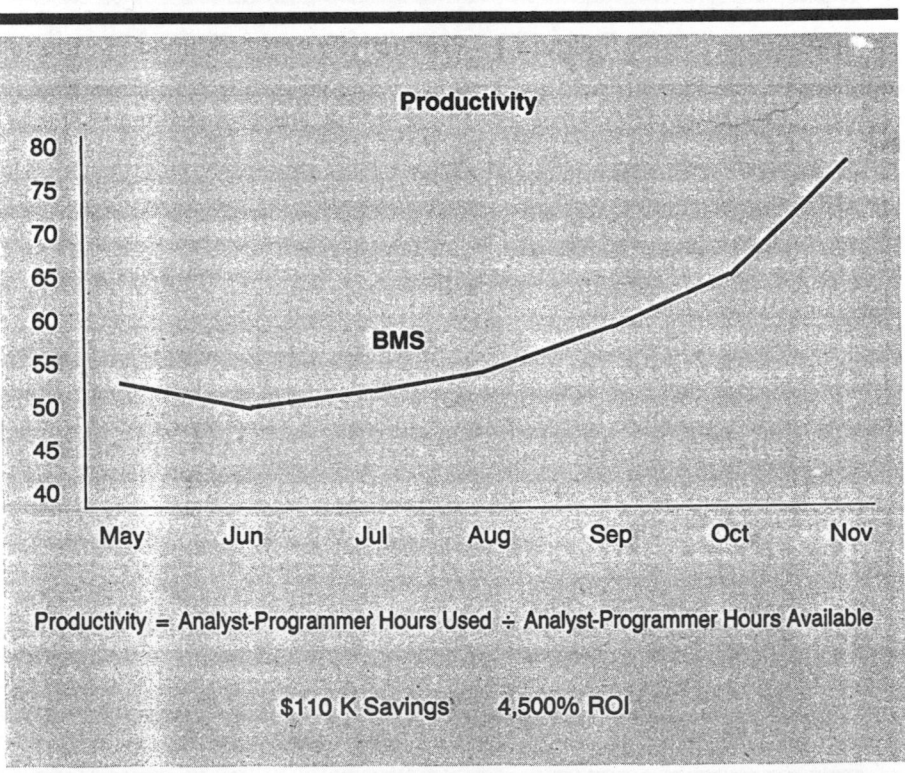

Figure 3—Analyst–programmer utilization index

Productivity

BMS

May Jun Jul Aug Sep Oct Nov

Productivity = Analyst-Programmer Hours Used ÷ Analyst-Programmer Hours Available

$110 K Savings   4,500% ROI

The IPM used to track plan results was a calculation of A-P hours used divided by A-P hours available; this generated an A-P utilization-efficiency index. Productivity spin-offs included an increase in the average percent of tape mounts per month and a decrease in the average number of days it took to turn around client service requests. It's worth noting that management reinforcement of training was present in this case.

We have built into BMS numerous opportunities for joint discussion, planning, and actions by the trainee and his or her manager. Such reinforcement contact increases the likelihood of both changed behavior and actual plan results.

Calculating return-on-investment

When we were able to price the value of an action plan, we then could calculate ROI for the corporation.

The two halves of the ROI equation are benefits and costs; ROI equals benefits minus costs divided by costs. We used a model for calculating training costs that has been described by two of the authors in previous issues of the *Training & Development Journal*.[1, 2] This cost estimate is much more complete than those usually accounted for in a typical training department budget. We include classroom, program development, (amortized, in this case, over a projected 25 programs), trainer preparation time, general administration, and corporate overhead.

Altogether one BMS seven-day program costs approximately $11,850. Figuring an average of 20 trainees per session, this is about $590 per person.

For some purposes the time spent by trainees away from their regular jobs is a cost of training, so we also estimated this. Using the mid-point salary range for a typical participant job grade, and figuring

in 24.4 percent fringe, we compute a worth of $145 per day for each BMS participant. This generates a lost-time cost of $1,015 per participant and $20,300 per program.

When costs of $1,600 per participant are compared to some of the dollar values we were able to attach to action plans, the ROI leaves little doubt that the program produces amply.

References

1. Weinstein, L.M. "Collecting Training Cost Data." *Training & Development Journal* (August 1982): 30-34.
2. Weinstein, L.M., and Kasl, E.S. "How the Training Dollar Is Spent." *Training & Development Journal* (October 1982): 90-96.

By James Heideman and Bruce Sanderson

Zooming in

How do you know if your training program will produce meaningful results? Here's how trainers at an automobile manufacturer used "focused evaluation" to tailor their training program to their company's bottom-line concerns.

In This Story

▼ evaluation
▼ technician training

Evaluation usually signals the end point of a training program, but even a last step shouldn't be given short shrift. Effective evaluations are planned carefully from the beginning: A well-crafted end-of-training evaluation will assess only those things that are important and measurable, without wasting time on insignificant details.

We call this preplanned, posttraining analysis "focused evaluation." Trainers who use this method can better determine who should be trained and what the training content should be. The result is a training program that improves the company's overall performance by

▼ producing the greatest possible performance results from the time and money spent
▼ indicating when to look beyond training for answers to performance problems
▼ providing a baseline for employee job performance that can be used to define future training needs.

Customer Satisfaction

We first used focused evaluation while developing a training program for dealership technicians at a major automobile manufacturer. The program covered all the right things according to the instructional systems design model: It assessed training needs, prescribed the technical training needed to improve job perfor-

mance, provided for necessary training, and evaluated technician performance afterward. Since the needs assessment was thorough, the correct needs had been identified; and since training was individualized and competency based, the training was sure to improve the job performance defined in the training objectives.

When we found performance didn't improve, we took another look at our training evaluation methods. We found that management and the training department were evaluating different things in determining the training program's worth. As trainers, we measured success by the technicians' mastery of the skills taught during training, while management looked instead for improvements in the company's Service Quality Index (SQI) scores, tallied from customer satisfaction surveys. Although the relationship between appropriately skilled technicians and SQI was a logical one, we found our training did not produce those changes.

We knew our training objectives were sound, so we set about finding how to produce measurable improvements in technician skill performance and customer satisfaction. Thus, the term "focused evaluation" was born.

It was no easy task. We had to create an evaluation process that would allow us to identify whether performance prob-

on Training Goals

lems had to do with the learner's skill level or the job environment, then to identify what technical issues needed to be resolved.

Five Steps of Focused Evaluation

We used these five steps to accomplish the focused evaluation:

Step 1: Assess the job environment in which poor performance occurs. SQI scores reflect percentages: An SQI score of 100, for example, indicates that every customer surveyed reported that his or her vehicle was repaired correctly the first time it was brought in for service. Since

returned surveys. By focusing our training efforts in those dealerships, we had the best opportunity to improve the manufacturer's overall performance.

Step 2: Identify elements with the greatest impact on the organization's performance. Targeting dealerships with SQI scores below 70 enabled the trainers to identify low performing dealers. These became the dealers with the greatest potential for improving overall SQI performance. At this point we did not know whether training would provide the solution to the performance issue, only that they were our target group.

For example, in one dealer we considered SQI performance in each technical area of vehicle repair. Some dealers had approved repair procedures but didn't have enough technicians trained to perform them. A dealership with 10 technicians is required to have at least three who have mastered the objectives for automatic transmission service and repair, and at least five who have mastered the objectives for electrical and engine control. When this standard was not met or maintained, technicians were scheduled for training. At other dealerships, technicians had received training more than five years ago, and thus their skills lagged behind technological advances. These technicians were selected for skill updates.

Step 4: Develop a training program to meet the needs of specified employees. Not every technician in the identified dealerships needed training. We examined the SQI score of the dealership and training history of each technician to determine whether other conditions in the dealership hampered the desired performance. For example, if a technician had mastered certain skills after training but did not perform on the job, the cause might be the dealership's reward system or an environmental factor. In those cases, we needed to explore nontraining strategies that could affect SQI performance.

If only some of the company's technicians need training, broad-based training strategies offer questionable hope of achieving measurable results

every dealer's SQI score is weighted based on the number of surveys returned, response rate is the most significant factor in determining SQI.

Many training programs target dealerships that have the most technicians or sell the most vehicles. Our strategy was based on identifying the dealerships with the greatest impact on total SQI. Thus, we looked at all dealerships in seven areas nationally, and targeted those with SQI scores below 70 and those with 45 or more

Step 3: Identify job performance deficiencies of individual technicians. After the dealers were sorted by SQI score, we looked at each dealer individually to determine the potential for technical training to improve these scores. Training was recommended for those dealers with an inadequate number of technicians trained in a specific area of repair, or when individual technician SQI scores were low in any specific area of repair.

Step 5: Compare job performance changes with evaluation criteria. Once all the selected technicians were trained, the trainers' task was to measure results based on the criteria for success: mastery of the training objectives and improvement in SQI scores. Now we had a solid foundation for measuring the impact of training on dealership performance. Only if training was the solution for a dealer's low SQI score would improved dealership performance result.

We found that by focusing on each technician who completed training, we were able to determine whether changes in performance occurred or did not occur.

In assessing performance, we found that performance improvements were greater in dealerships with fewer technicians. In larger dealerships, training did not produce results that were as significant. While training might have made a difference, there were clearly other factors that also played a significant role in technician performance.

Applying Focused Evaluation

There are three important factors to consider in applying and assessing the results of focused evaluation:

1. Accurate record keeping makes focused evaluation effective. You must have accurate dealer training records, individual technician training records, and job performance records.

2. You need realistic expectations for improvement. Do not promise more than you can expect to deliver. A one- or two-point increase in SQI score for a manufacturer is significant.

3. Effects of focused evaluation are not immediate. Not all technicians apply the skills they mastered in training on their first day back on the job. In the automobile industry, it usually takes six to nine months to complete at least one customer survey period. Allow time for tools such as SQI surveys to "catch up" with the effects of training.

Focused evaluation is a way to maximize training effectiveness and accurately measure the results of training efforts—in this case, in terms of automobile dealership performance. Because not all dealerships or technicians need training, broad-based training strategies offer questionable hope of producing measurable improvements in a company's customer satisfaction rating. Focused evaluation is a useful tool for replacing broad-based training approaches, in which everyone participates, with targeted training tied to the manufacturer's goal of improved quality of customer service.

James Heideman is a professor in the Department of Technology at California State University, Los Angeles (5151 State University Drive, Los Angeles, CA 90032-8154; 213/343-4573; fax: 213/343-4571; e-mail: jheidem@calstatela.edu). **Bruce Sanderson** is with Nissan Motor Corporation, and teaches at the University of Phoenix.

Cost-Benefit Analysis Techniques for Training Investments

By Werner J. Schmidt

Learning a few simple financial techniques can go a long way in helping design, evaluate, and implement training programs. And much of the work needed to show costs and benefits for training investments can be done with the skills that most technical trainers already possess.

In This Story

▼ return-on-investment
▼ training investments

With limited resources to support our technical training endeavors, it has become imperative for us to focus on the quantitative impact of our training interventions. Unfortunately, for many of us in the training field, quantitative analysis is a daunting task. In the past, we have not always needed to prove ourselves financially, but times are changing, requiring us to justify not only how our training programs improve performance, but also how those improvements contribute to the bottom line. The days of automatic annual training budget increases are over.

One reason for our concern about financial analysis is the dread of having to learn new skills that we have distaste for and that we feel are too difficult to learn. Another perceived concern is that even if we take the time to acquire these skills, they would be of little value because we cannot get legitimate results due to the nature of "soft" data. A third concern is the fear that the financial tools that are intended to support our goals may actually reveal disappointing numbers. All these fears are groundless; it's really the fear of strange territory that is the problem. Learning a few simple financial techniques can go a long way in helping us design, evaluate, and implement training programs.

The fact is, a few simple financial analysis techniques can yield tremendous insight into the impact of training and performance interventions. The simplicity lies in using everyday math and logic, while the power lies in helping us make decisions that accurately show why the approach we are taking gives the best "bang for the buck." What little financial analysis we have done in the past has too often been a reactive defense of our budgets; these tools can allow us to become proactive and inform management what the anticipated return-on-investment of our planned training programs will be.

Figure 1: Comparative Cost Analysis Matrix

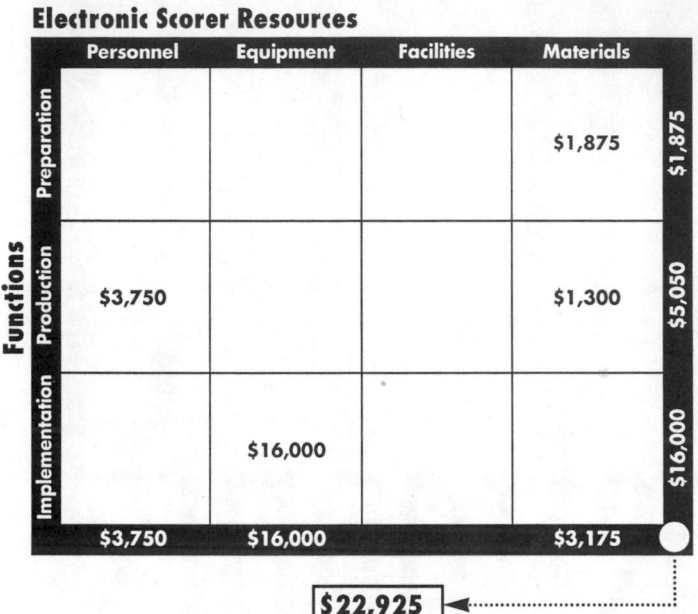

Electronic Scorer Resources

Functions	Personnel	Equipment	Facilities	Materials	
Preparation				$1,875	$1,875
Production	$3,750			$1,300	$5,050
Implementation		$16,000			$16,000
	$3,750	$16,000		$3,175	

$22,925

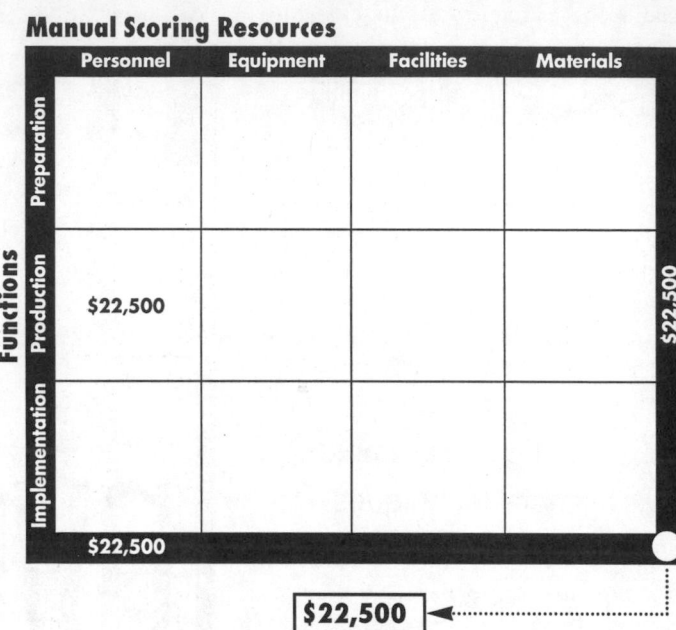

Manual Scoring Resources

Functions	Personnel	Equipment	Facilities	Materials	
Preparation					
Production	$22,500				$22,500
Implementation					
	$22,500				

$22,500

And in the author's experience, at least 90 percent of the work needed to show costs and benefits for training programs can be done with the skills that the average technical trainer already possesses.

Three-Step Plan

To compare the costs and benefits of competing approaches, there are three basic steps to follow:

1. Determine what you want to accomplish, clearly state your objectives, and review methods to accomplish your objectives. This is much like conducting a training needs assessment—you need to ensure that the program is focused on the best method to accomplish your objectives.

2. Set up a matrix laying out the specific costs for each option that meets your stated objectives. One of the biggest challenges trainers face in performing a financial analysis is in organizing costs. The easiest and most foolproof method is to set up a

matrix comparing functions with resources. Although the categories may differ, you will not go astray if you break down resources into generally accepted categories: personnel, equipment, facilities, materials, etc. Functional categories

A matrix helps you organize your thoughts on how to better design, produce, and implement your program

can include preparation, production, and implementation (See Figure 1).

In addition to providing a good financial breakdown, a matrix also helps you organize your thoughts on how to better design, produce, and implement your program. Completing this process of organizing numbers is 90 percent of the battle. When finished, run it by your colleagues

for feedback, to get consensus on your figures, and to make adjustments.

3. Finally, you must select an option based on cost and value. The matrix analysis will reveal which option is best based on cost alone. However, you need to balance the cost against both the short-term and long-term value to your organization. For instance, if one option is very costly during the first year, how does it stack up as a long-term investment compared to other options? A payback analysis is needed to provide that information.

A Utility's Experience

An example of the impact that basic financial analysis can have—one that forever changed the author's view of the value of such techniques—occurred while he was serving as training and performance improvement financial analyst with a large utility company in California. The financial analysis not only simplified a seemingly complex problem, it also served as a teaching tool to convince the company's training staff to apply the techniques in all their major activities.

While performing a financial analysis of a major curriculum project for the utility, the training manager and her staff attempted to assess whether it was better to hand-score test papers for an upcoming companywide training program—or to purchase an electronic scoring machine. At issue was how to justify an initial outlay of $16,000 for the electronic scorer when old-fashioned hand-scoring had sufficed in the past. The more the training department tried to resolve the issue, the more they got tangled up in numbers. They turned to the author and asked if there was a way to justify the purchase in financial terms.

The training department had always been able to grade test papers manually with help from available clerical staff. However, they were now preparing to launch a major training effort expected to handle as many as 4,000 trainees during the first year.

First, the training staff conducted an assessment to clarify their objectives, based on the need to regularly score tests for 4,000 trainees during the first 12 months. The scoring volume would taper off to 500 or less in following years.

The second step, where the training managers had earlier been stumbling, became more clear when organized in the matrix. The numbers for the electronic scoring device, which was an inclusive cost, was $16,000 and placed in the implementation box since it was to be used during the final phase of training. Manual scoring does not involve any equipment costs.

The personnel or labor issue required more careful analysis. From interviews with the vendor and training staff at another utility using the same electronic scorer, the program staff learned that 15 minutes or less were needed for a clerk to set up, scan, and record a set of 20 tests. This is equivalent to 12.5 hours per 1,000 forms. For 4,000 participants with a total of 20,000 forms (based on five tests each), the time is 12.5 x 20 for a total of 250 hours. The labor rate is $15 per hour and, therefore, the cost is $15 x 250 or $3,750. This figure is placed in the personnel/production box because the activity begins prior to implementing training.

Figure 2: Monthly Cash Savings

The utility's monthly cash savings are calculated below:

Manual Scoring

	75	Hours
x	1.5	
	112.50	Hours
x	$15	Per Hour
	$1,687.50	
-	281.25	
	$1,406.25	Monthly Cash Savings

Electronic Scoring

	12.5	Hours
x	1.5	
	18.75	
x	15	Per Hour
	$281.25	

$$\text{Payback} = \frac{\$16,000}{1,406.25/\text{Mo.}} = 11.38 \text{ Months}$$

For manual scoring, the trainer calculated that it takes 1.5 hours for a clerk to score 20 forms. This is equivalent to 75 hours per 1,000 forms. For 20,000 forms the time is 75 hours x 20 for a total of 1,500 hours. At $15 per hour, the labor cost is $22,500. We place this figure in personnel/production, also. Already we see a comparison of the labor costs, which are continuous, with the initial high outlay of equipment costs. The labor savings calculated in this analysis is 83 percent ($3,750 versus $22,500).

Looking at materials, we are only concerned with the cost of electronic test forms. The forms cost approximately $32.50 per 500. When 4,000 participants complete five training programs and tests each, the total is $1,300. Additionally, the vendor's one-time charge to customize a form is $625. Since three different assessment forms need to be used for the various training programs, the figure multiplies up to $1,875. The $1,300 is placed in the materials/production box, and the $1,875 is placed in the materials/preparation box. There is no dollar amount placed in any of the manual scoring boxes because in-house forms are considered part of the overhead.

When the figures for each matrix are compiled and compared, the sum total for a 12 month period to cover 4,000 participants in the training program is almost equal for both the electronic scoring and manual scoring methods ($22,925 versus $22,500).

Payback Calculations

If a decision between the two options were made at this juncture of the analysis, there would appear to be little difference in terms of costs and benefits. This is deceiving, however, because thus far, the comparison weighs only fixed costs using a set volume of test scoring—specifically, the requirement of scoring 4,000 tests in the first year. As the number of participants increases, saving increases considerably when using the electronic scorer over manual scoring. The converse is equally true.

For example, when 5,000 participants proceed through the training program, a savings of 15 percent can be realized when choosing the electronic scorer over manual scoring. Now, if the electronic scorer were to be used on an annual basis, the increase in participants over time would also show a substantial savings over using the manual method.

At this point in the analysis, the training staff were convinced that the electronic scorer provided better long-term savings. This realization was based not on intuition, but on an objective, justifiable cost analysis. While cost analysis is important in itself for budgeting purposes, it is the investment part of the analysis that catches the eye of the CFO and upper management and ensures approval of the budget.

The third step entails looking at the value of the options in terms of payback. Payback is defined as the initial investment divided by the periodic cash flows resulting from the investment. In the case of the electronic scorer, the period is monthly and the cash flows are the labor savings obtained by using the electronic scorer over the manual method. We generally think of cash flows as income from sales; as trainers, we need to be aware that savings can also be considered a form of cash flow.

The payback calculation for the electronic scorer is the initial investment divided by the monthly cash savings, which worked out to 11.38 months for the utility. In other words, the electronic scorer would pay for itself in less than one year. After the first year, there is only a nominal figure for maintenance and upkeep. This is based on the figure of 4,000 participants going through the training program per year or about 300 per month. At 300 per month, we have 5 instruments per participant or 1,500 forms a month.

Since it takes 75 hours of manual labor to score 1,000 forms, it will take 112.5 hours to score 1,500 forms. It takes 12.5 labor hours to score 1,000 forms with the electronic scorer or 18.75 hours to do 1,500 forms (See Figure 2).

"Soft" Variables

The final ingredient of our cost-benefit analysis requires a look at nonfinancial values—the subjective benefits that need not and should not be quantified. Examples include improved customer service in terms of faster feedback on scores, freeing up clerks to perform other services, reduced space requirements for an electronic scorer, less chance for error, and use of equipment to assist with additional grading requirements for other training programs. This last item is particularly beneficial if it can be quantified because it can be added into the figures, making for an even shorter payback period.

Admittedly, not every cost was included in the analysis. The purpose was to develop a reasonable analysis providing a quantifiable means of comparing two options—and to communicate those findings to those holding the purse strings. It's amazing how much more agreeable management can be when training investment proposals are presented based on such quantifiable methods.

While the case study dealt with a comparison of a product used by the training department, it serves equally well when looking at the costs and benefits of training programs. An example would be the comparison of one format against another, such as individualized versus class instruction or centralized versus decentralized training. If we take the time to detail our figures and secure buy-in from all parties involved, we can come up with reasonable cost estimates. From these, we can perform a payback analysis that will steer us toward the best option.

Werner J. Schmidt is a faculty member at San Francisco State University and the University of California at Berkeley Extension School, and is also administrator of technical training programs for the East Bay Municipal Utility District (375 Eleventh Street, Oakland, CA; 510/287-1349; e-mail: wschmidt@ebmud.com).

PART 7.
Return-on-Investment

By Tom Barron

Is There an ROI in ROI?

While many Americans were tuned to a high-profile series of debates last fall—namely, the presidential debates—a lesser known but equally dynamic debate was unfolding in Cincinnati. There, technical trainers gathered for the 10th annual ASTD Technical & Skills Training Conference and Exposition witnessed a lively discussion between advocates of a return-on-investment approach to evaluating training efforts and those who find that approach too difficult and unreliable to justify its own ROI.

The debate, a spirited exchange between four training professionals, centered on whether calculating ROI makes sense as a way for training departments to demonstrate their value to a company's bottom line. And though it failed to yield a hands-down victor, the discussion certainly raised some thoughtful considerations on both sides of the issue.

Advocating ROI as a worthy enterprise were Kim Ruyle, president of LaCrosse, Wisconsin-based Plus Delta Performance; and Pat Wrobel, training manager for Universal Instruments in Binghamton, New York. Arguing that better approaches to evaluating training exist were Cynthia Baerman, director of training and development for Kraft Foods Incorporated in Northfield, Illinois; and Barry Cahill, manager of human resources development for Clayton, Missouri-based The Earth Grains Company. Moderating the debate was Mark Spear, technical training coordinator for Miller Brewing Company at its Milwaukee, Wisconsin, headquarters. (See sidebar, next page.)

"Whether you want to call it training, performance, operating efficiencies, wise utilization of our human assets, or

In This Story

▼ evaluation
▼ return-on-investment
▼ transfer of training

Does calculating return-on-investment for training programs provide a payoff for training managers—or merely divert resources better spent on improving performance? Four leaders in the training field debate the pros and cons of calculating ROI; judge for yourself which argument is more sound.

whatever, what we do has to show up on the bottom line," said Mark Spear in opening the discussion. "Over the past five years, we've been deluged with books and formulas showing us ways to formulate ROI. But is calculating ROI *essential* for demonstrating the value we bring to our organization? That's what we're here to find out."

Following is a summary of key points made by each side in the ensuing battle.

The Case for ROI

Likening ROI to a tetanus shot that must be endured to inoculate against shrinking training budgets, Kim Ruyle and Pat Wrobel raised the following points in championing the methodology.

Training contributes to a company's profit. And because training contributes to profits, we can measure the return on training investments. In doing so, we can show how we successfully use training dollars to contribute to the profitability of the organization, and can thereby justify future training expenditures.

The basic question is this: Is ROI a valid method to evaluate business functions? If it is, and most businesspeople would agree with that proposition, the next question is whether or not training should be considered a business function. If it is, as is the case with a growing number of companies, shouldn't ROI be applied to training just as it is with other business functions?

Every time you make decisions about whether or not you are going to train, how you will train, how much you will invest in training, whether to contract it out or stay in-

Should ROI be applied to training just as it is with other business functions?

house, you are making implied ROI calculations. The point is, whether you realize it or not, you're already applying ROI thinking, but by adopting a formal ROI methodology, the training function becomes leaner, more relevant, more behavioral in nature, and more bottom-line oriented.

ROI ensures needs-driven training. In pursuing ROI for training, department

managers and employees become partners in the effort. The ROI approach generates needs-driven training that ensures a focus on the proper skills. ROI creates an increased commitment from managers in terms of the types of training; and an increased commitment from employees because, in order to compute ROI, we would be measuring the transfer of training—the effectiveness of the training on workplace performance.

One of the criticisms trainers frequently get is that we're not attuned to business—that we're not real businesspeople. ROI is one way to put that criticism to rest. ROI encourages careful planning for the expenditure of training dollars. Couple that with the fact that ROI creates a commitment with other business components and forces us to partner with line managers, and the planning process takes on new meaning.

ROI encourages objective rather than subjective decisions. All too often, training decisions are made based on subjective

views on how training dollars should be spent. As training managers, we need to provide some objectivity to that process. We need to be able to help our clients see objective reasons for training, what realistically can be done, and what the payoffs are. When we begin to use more systematic and formal methods for doing that, we bring discipline to the process—and as our decisions become more objective, they become sounder.

ROI doesn't have to be calculated for every training program. The important thing is to calculate ROI for major programs and initiatives—those that create a significant change. Building credibility by measuring ROI for the major programs will produce a halo effect—adding credibility to training programs for which ROI is not measured.

We can't afford not to measure ROI. In the past, many companies have forged ahead with accustomed business practices in blissful ignorance of how well or how badly their training programs worked. There are horror stories aplenty of training

Isn't Training Evaluation the Real Point?

The great ROI debate generated more than friction between opposing sides—it also yielded some interesting observations from a decidedly split audience. Perhaps most profound was this comment, from a training manager who said he had come "all the way from St. Louis" to hear the discussion.

"Both sides are missing a point. The arguments are valid, but what's missing is that we're using the term ROI, but forgetting about trainee feedback. For any training program to be successful, we need to know how well it applies to the field. It's not important to me as a trainer that someone gets 100 percent on the training test, if they go out in the field and what I taught them is not what they need. I don't think our reputations can be upheld if we don't gather the feedback to tell whether we're doing a good job.

"It's not important whether you use ROI or any other calculation to determine results, you need to get feedback, and we need a system that's easy for one and two-person training operations to use in the field to determine whether we're doing our job well."

---

budgets being downsized largely because the value of training programs could not be ascertained. If we can demonstrate the effectiveness of a training program by its contribution to the bottom line, by the dollar amount of improvement, or by the amount of improvement evident to productivity, then we have done a lot to advance the role of training and its importance.

Trainers should step up to the challenge. We don't avoid getting a tetanus shot just because it's painful, and there is some expense and discipline involved in calculating ROI. There are some new skills we need to learn as trainers to be able to do it, but we shouldn't shrink away from the challenge. Rather, we should see it as an opportunity.

Why ROI Doesn't Work

In making their case against ROI, Cindy Baerman and Barry Cahill argued that several misperceptions about organizations and the role of training are driving the ROI bandwagon. In examining those 'myths,' they say, it becomes clear that ROI endeavors distract from training's mission and fail to live up to their promises. Here's a summary of their key points.

Myth #1: Training alone can yield performance improvement. The truth is that training is only part of the solution to improved individual and business performance. Training must be tightly linked to business objectives and learning interventions in an integrated, systematic approach, so you can't measure training individually and conclude that it alone is going to improve performance.

Myth #2: Training's customers are clamoring for ROI. Customers include everyone from senior leaders, managers, other training suppliers, trainees and their supervisors, operations managers, and even training professionals themselves. They all rely on the results of training and provide key inputs into the training process. Do these customers really want ROI, or do they want improved performance? Would senior leaders choose a training program that gave them an ROI of 25 percent but didn't create linkages to goals and objectives of the company? Wouldn't they prefer a training program that might have a 15 or 20 percent ROI but is really linked to where the organization is going?

Myth #3: We need to calculate ROI to prove our worth. There's no avoiding the

> **You can't measure training and conclude that it alone is going to improve performance**

perception that in the era of downsizing and cost-cutting, we need to calculate ROI to show that we're holding up our end. However, it's everyone's job to manage and use learning, and when you start calculating ROI, you're just measuring that slice of ROI that relates to the training department. What we're really striving to achieve is a sense of ownership throughout the organization. If we're really creating a learning organization, the entire organization should be held accountable for learning.

Myth #4: ROI can be calculated without overburdening the training function. A great example to contradict that view can be found in a "best practice" identified by a training consultancy last year involving a formula to calculate training ROI. At the company where it was developed, the formula required $500,000 to develop and implement. It took three years to implement and required one-and-a-half full-time equivalents to oversee. In a time of tightening belts and downsizing, do we really have the time and resources to be calculating these types of formulas? Or would we be better off focusing on tying our training programs to business performance objectives?

Myth #5: ROI assumes that transferring of learning is certain to occur back at the workplace. That assumption is often false, for several reasons: It assumes that transference of the training is independent of the trainee, the trainee's supervisor, and the organization as a whole. In the real world, however, the ball typically gets dropped by one or more of these players. Trainees may not be rewarded for adopting their learning, and may not see the payoff in doing a task the "new way." Supervisors sometimes prevent transfer of training because they might not have been involved in training content, or might not be aware of the training need that's being addressed.

Pro-ROI Rebuttal

We heard the argument that you can't measure training ROI individually because of all the other variables involved. However, can you think of a business process that doesn't have similar constraints? Take the design engineering group, for example. If the product is manufactured poorly or marketed poorly, how does the design group know they've succeeded?

The fact that there are mitigating factors that affect the impact of training shouldn't deter us from measuring training ROI. Indeed, those factors actually make a good reason for measuring ROI, because they foster partnering with other functions in the organization. And if we're not looking to see whether our training is transferred back in the workplace, we're not doing our jobs.

It's interesting that our opponents chose a $500,000 example of an ROI model, when you can use a common spreadsheet application with an embedded help system to achieve a reasonable ROI formula. All you really need to know is your initial investment, and be able to quantify what your payoff is going to be over a given period of time. Certainly, you can't get it down to the gnat's eyebrow, but you can derive numbers with some credibility. These calculations are simply not that difficult.

Anti-ROI Rebuttal

It's interesting indeed to hear that our opponents are not talking about calculating ROI for all training programs, and are even saying that the formulas do not have to be exact and precise. We're clearly seeing some backpedaling on their part over those points.

How many other components of business offer ROI on their activities? Do our corporate attorneys provide us with their ROI? Do our accounting folks present ROI data on their activities? Perhaps the reason the training function is being pushed to do ROI is that some training departments aren't as tied into business goals as well as they should be—and are being asked about ROI as a result.

Costing Out the Value of Training

Calculating the DIF value (difficulty, importance, and frequency) of tasks helps to demonstrate the dollar value of training.

By Paul E. Brauchle

T rainers are often plagued by the lack of simple and effective methods for showing the benefits of training. So they resort to "smile sheets" or assume that if the training is based on a needs analysis, it probably is effective. But these methods cannot tie training activities to the dollar values that most organizations consider important. This puts trainers at a disadvantage when dealing with their more financially literate colleagues. The accountants are likely to know exactly how much training costs, but they may have little idea of its value.

When translated into dollars, the costs of training are a powerful measuring scale with enormous emotional impact on managers. Next to a dollar measure of cost, smile sheets or assumptions based on a needs analysis seem like weak arguments. What we need are methods that show the dollar value of training in terms that managers can understand.

Several methods can estimate the monetary benefits of training. Looking at the consequences of *not* training is one way. Others involve analyzing performance records or costing out the training curve under training and nontraining conditions. But the best known method is cost/benefit analysis.

Cost/Benefit Analysis

Usually a lengthy and tiresome process, cost/benefit analysis calculates training costs, assesses some performance value that occurred as a result of training, and computes an index of benefits. Performance values are normally derived from productivity measures such as the number of items produced per shift. If more items are produced after training, the "gain" calculated by subtracting the training cost shows the benefit. But these measures do not normally account for factors such as the difficulty of a task or the value of that task to the organization.

This article shows how to calculate training benefits in a way that considers four factors:
▼ the difficulty of the task for which a person is trained
▼ the importance of that task to the job or the overall organization
▼ the frequency with which that task is performed
▼ the improvement in job performance that occurs as a result of training.

The results are expressed in dollars and can form powerful arguments for the effectiveness of training. This method can be called "DIF Analysis," for its examination of task difficulty, importance, and frequency. The DIF approach is based on a 1989 book by Terence Jackson and on research, published in 1986, conducted by Wayne F. Cascio of the University of Colorado and Robert A. Ramos at Bell Communications Research. It uses three phases to determine the value of training: a job analysis, a performance assessment, and a calculation of training's value and benefits. Each phase contributes important information to the analysis.

Phase 1: The Job Analysis

1 **Conduct a Job Analysis**
First, conduct a job analysis by listing all important tasks for the person on the job. For simplicity, consider only tasks or activities that comprise at least 10 percent of overall work time over a one-year period. Let's assume there are

In This Story

▼ cost/benefit analysis
▼ evaluation
▼ performance-based training

Criteria for DIF Analysis

Figure 1

Scale	Difficulty	Importance
1	Easy to learn. Little concentration needed. No knowledge of basic principles.	Of little importance to performance of unit. Errors do not matter.
2	Some practice required to learn and maintain proficiency. Needs concentration. Some grasp of basic principles desirable.	Has some importance to performance of the unit. Errors may cause inconvenience.
3	Constant practice required. Knowledge of basic principles essential. Decision making required.	Has major importance to performance of unit. Errors and failure to perform adequately may give rise to business or financial loss.
4	Difficult to learn. Experience increases performance. High level of decision making and concentration required. Many factors and concurrent activities.	Unit cannot function without this key activity being competently performed.

Excerpted from: *Evaluation: Relating Training to Business Performance,* Terence Jackson. San Diego: University Associates, 1989. Reprinted with permission.

Job Performance Criteria

Figure 2

Rating	Criteria
0.0	Has produced no results in this function.
0.5	Has produced results in this function which are sometimes (about 50 percent) consistent with standards of quality and quantity.
1.0	Has consistently produced results in this function consistent with standards of quality and quantity.
1.5	Has sometimes (about 50 percent of work) produced results in this function which are well in excess of standards in both quality and quantity.
2.0	Has consistently produced results in this function well in excess of standards in both quality and quantity.

Excerpted from: *Evaluation: Relating Training to Business Performance,* Terence Jackson. San Diego: University Associates, 1989. Reprinted with permission.

five main tasks for a job. Those tasks should be listed in column 1 of Table 1 under "Activity."

2 Set Difficulty Ratings

Always consider the difficulty of a task when evaluating training. Difficult tasks should be weighted more heavily than easy tasks. In Step 2, you establish "difficulty" ratings for each task. The criteria for DIF analysis in Figure 1 establishes the appropriate difficulty level for each task. The difficulty rating goes in column 2 of Table 1.

3 Establish Task Importance

The importance of each task to the unit must also be considered. This procedure allows us to weigh the relative significance of the task that a person performs. Use the Figure 1 criteria to establish the importance of each task. The ratings are listed in column 3 of Table 1.

4 Establish Frequency/Time Values

Frequency/time is the percentage of time that each task requires to complete. Since we are dealing with the five most important tasks, we will assume the other tasks are of minor importance. So let the total of the frequency/time column equal 100 percent. Then suppose Task 1 occurs 20 percent of the time; Task 2, 40 percent; Task 3, 10 percent; Task 4, 10 percent; and Task 5, 20 percent. The frequency/time values are listed in column 4 of Table 1.

5 Obtain DIF Products

We now have DIF values in place for each task. Next, we need to obtain the DIF products as follows:
a.) Multiply the first three columns for each task.
b.) Record the results in the DIF column.
c.) Total this column.

The total can be any number. In column 5 of Table 1, the total is 630.

6 Get DIF Weightings

Next, we need to express the relative weight of each task. This involves calculating what percentage of the total DIF product is contributed by each task's individual DIF product. For example, Task 1's DIF product of 120 is 19 percent of 630, so 0.19 is entered in column 6 of Table 1.

For the other tasks:
a.) Divide the total for the DIF column into each DIF product.

b.) Enter these values in the "Weighting" column.

c.) Total all weightings. They should add up to 1.00.

Now, take a break. You have completed the Job Analysis phase.

Phase 2: Determine Value of Performance

In this phase, we look at three factors: worker cost, task value, and performance rating.

7 **Start With Weightings** For this step, record the DIF weightings from Table 1 in column 2 of Table 2.

8 **Enter Worker Cost** Worker cost is the worker's annual salary (include fringe benefits if you want to use a more complete figure). The worker's annual salary for each task is recorded in column 3 of Table 2.

9 **Determine Task Value** The weightings are multiplied by the personnel cost for each task, which will give you the task's annual value to the unit. The total of the task value column should equal the worker's annual salary. These task values are entered in column 4 of Table 2.

10 **Establish Job Performance Ratings** Now let's look at the individual's on-the-job performance. The value of a person's work should be related to how well the work is performed. In most jobs, performance is distributed around what we call "nominal" performance—i.e., the level of performance that a well-qualified, experienced worker is ex-

Job Analysis Worksheet

Table 1

Activity	Difficulty	Importance	Frequency/ Time	Product (D x I x F)	DIF Weighting
1._____	2	3	20	120	0.19
2._____	1	3	40	120	0.19
3._____	3	1	10	30	0.05
4._____	4	1	10	40	0.06
5._____	4	4	20	320	0.51
Total	—	—	100	630	1.00

Adapted from: *Evaluation: Relating Training to Business Performance,* Terence Jackson. San Diego: University Associates, 1989. Reprinted with permission.

Performance Value Worksheet

Table 2

Task	DIF Weighting	Worker Cost	Task Value	Performance Rating	Performance Value
1._____	0.19	$24,000	$4,560	1.25	$5,700
2._____	0.19	24,000	4,560	1.00	4,560
3._____	0.05	24,000	1,200	0.50	600
4._____	0.06	24,000	1,440	1.75	2,520
5._____	0.51	24,000	12,240	1.50	18,360
Total	1.00	—	$24,000	—	$31,740

Adapted from: *Evaluation: Relating Training to Business Performance,* Terence Jackson. San Diego: University Associates, 1989. Reprinted with permission.

Performance Gain Worksheet
(Comparing Performance Values Before and After Training)

Table 3

Task	Task Value	Performance Rating	Performance Value	New Performance Rating (After Training)	New Performance Value	Performance Gain
1.____	$4,560	1.25	$5,700	1.50	$6,840	$1,140
2.____	4,560	1.00	4,560	1.25	5,700	1,140
3.____	1,200	0.50	600	1.00	1,200	600
4.____	1,440	1.75	2,520	1.75	2,520	—
5.____	12,240	1.50	18,360	1.75	21,420	3,060
Total	$24,000	—	$31,740	—	$37,680	$5,940

Adapted from: *Evaluation: Relating Training to Business Performance,* Terence Jackson. San Diego: University Associates, 1989. Reprinted with permission.

DIF Analysis Worksheet

Table 4

Now it is your turn. Assume you have a worker with the following data. What is that worker's performance value? How much performance value was added by the training?

Job/Task Analysis:

Task	Difficulty	Importance	Frequency/Time	Product (D x I x F)	Weight
1.	1	2	25	_____	_____
2.	3	2	25	_____	_____
3.	2	1	30	_____	_____
4.	4	3	10	_____	_____
5.	3	4	10	_____	_____
Total	—	—	100	_____	100

Performance Value:

Task	DIF Weight	Worker Cost	Task Value	Performance Rating B/T*	Performance Value B/T	Performance Rating A/T*	Performance Value A/T
1.	_____	$25,000	_____	1.5	_____	1.50	_____
2.	_____	25,000	_____	0.5	_____	1.00	_____
3.	_____	25,000	_____	1.5	_____	1.50	_____
4.	_____	25,000	_____	1.0	_____	1.25	_____
5.	_____	25,000	_____	1.0	_____	1.25	_____
Total	1.0	($25,000)	_____				

*B/T stands for "before training"; A/T stands for "after training."

pected to produce. Most workers fall into this central category; a few are above or below this standard. Figure 2 lists the criteria for employee job performance along a continuum from low to high, in which 1.0 is nominal. The rating for job performance may be assigned by the individual's supervisor, a committee, or the individual and the supervisor together. The job performance ratings are recorded in the "performance rating" column of Table 2.

11 Find Performance Values

Next, calculate the value of each task's performance at the performance level determined in Step 10. To get the performance value, multiply each task's value by the employee's performance rating. For example, column 6 of Table 2 shows that Task 1 has a task value of $4,560 and a performance rating of 1.25. That is, the worker performs Task 1 a little better than

the level expected of a competent worker. Therefore, the value of the worker's performance is a little higher than average. The performance value for Task 1 is $5,700, or 1.25 x $4,560.

When all the performance values are calculated, add them up. The total may be any number. It may even add up to more than the task value.

Treat this information as confidentially as you would other salary-related data.

Phase 3: Determine the Benefits of Training

This phase determines the benefits of training by subtracting the worker's pretraining performance value from his or her post-training performance value. Let's go to the final step.

 **Compare Performance Values for Each Task Before and After Training**

At this point you have the performance value for each task before training and the individual's total performance value for all job tasks. The only step left is to assess the worker's performance after training. To do this, get a rating of the worker's performance after training and calculate his or her performance value.

Suppose that after training the performance value for Task 1 improved from $5,700 to $6,840. The actual performance gain for Task 1 would be $1,140, or $6,840 - $5,700. You can add all the gains in performance value, as in column 7 of Table 3, and get a total gain in performance value for that individual, which in this case is $5,940. Let's assume that the training cost was fairly high, about $1,000. The actual gain posted by the training is then $4,940. Also suppose that this individual was average and was part of a class of 20 similarly average workers. This adds up to a total group performance increase of more than $98,000.

DIF Applications

Although this method of cost/benefit analysis works for trainers and training managers with some background in statistics and accounting, their experience need not be extensive. Moreover, DIF works for both individuals or groups.

Data for individuals can be accumulated and averaged, or else a spreadsheet can be used to accumulate the data for an entire group. The results can then be calculated by a computer.

This method of DIF analysis has two key applications:

▼ *monitoring the results of training*. Results are monitored by giving meaningful weights to different factors that influence the training results. Trainers will appreciate that their efforts at delivering complex and difficult courses are given more weight than simple ones. Such data may provide discussion points for planning training that meets organizational needs and better utilizes resources.

▼ *planning training*. Information from the DIF weighting column or the performance value column can form the basis for a Pareto display of rank order importance of tasks to be trained. After consulting such a rank order listing, a trainer can recommend allocating resources to tasks that will make the most difference. By using the analysis on a spreadsheet and asking "what if" questions, the training manager can project what levels of performance increase will yield a performance gain to make training cost-effective.

Finally, DIF cost/benefit analysis is quite adaptable to different situations. Trainers can use scales of difficulty to calculate weightings and different units for the ratings. No one is required to use the four-point scales discussed here.

Now that you understand DIF analysis, try working through a problem. Use the Job/Task Analysis and Performance Value worksheets in Table 4 to calculate the training performance value for an individual. Then subtract the performance value before training from the performance value after training to get the performance gain. You may be surprised at the numbers you see.

References

Cascio, W., *Costing Human Resources: The Financial Impact of Behavior in Organizations* (2d ed.). Boston: PWS-Kent Publishing Company, 1987.

Cascio, W., and R. Ramos, "Development and Application of a New Method of Assessing Job Performance in Behavioral/Economic Terms." *Journal of Applied Psychology*. Volume 71, Number 1, February 1986. Flamholtz, E., *Human Resource Accounting* (2d ed.). San Francisco: Jossey-Bass Inc., Publishers, 1985.

"How to Conduct a Cost/Benefit Analysis." *Info-Line*. Alexandria, Virginia: American Society for Training and Development, July 1990.

Jackson, T., *Evaluation: Relating Training to Business Performance*. San Diego: University Associates, 1989.

Swanson, R., and D. Gradous, *Forecasting Financial Benefits of Human Resource Development*. San Francisco: Jossey-Bass Inc., Publishers, 1988.

Paul E. Brauchle is the coordinator of graduate studies for the Industrial Technology Training & Development department of Illinois State University (Turner Hall 210 E, Normal, IL 61761-6901; 309/438-2696).

Measuring Training's ROI

ARE YOU A TRAINER WHO STILL BELIEVES THERE'S NO NEED TO CALCULATE ROI? YOU MIGHT DECIDE TO RE-LENT—AT LEAST SOME OF THE TIME—WITH THIS LOOK AT THE PROS AND CONS, PLUS FOUR MEASUREMENT METHODS AND CASE REPORTS.

BY SCOTT B. PARRY

"**T**RAINING DOESN'T COST...it pays! HRD is an investment, not an expense."

Rare is the trainer who doesn't believe this. Far more common is the trainer who doesn't believe that return-on-the-training investment can (or even should) be calculated.

Should all training programs be required to show a return-on-investment (ROI)? Not at all. However, courses of three days or more that are offered many times to reach a large number of trainees (say 100 or more) represent a significant expense. The professional trainer should justify this expense by calculating the return on this investment.

We're talking about Level Four: Results on Donald Kirkpatrick's evaluation model, and it's the most difficult to measure. Level One: Reaction and Level Two: Learning can be measured with relative ease in class, using paper-and-pencil instruments and simulations. Level Three: Application at work is more difficult, because it means measuring performances on the job where many variables are affecting the performance of our graduates. Level Four: Results is usually shown as a return-on-investment...the dollar value of the benefits of training over and above the cost of the training itself.

And there's the rub. Many factors make this level of measurement the most difficult by far. Here are some of the more common difficulties that are cited as reasons for not doing a level-four evaluation:

▶ The costs of training are known and expressed in dollars, but the benefits are often soft, subjective, and difficult to quantify and convert to dollars.

▶ We have enough trouble getting managers to send people to training without imposing additional requirements to collect data to document the impact.

▶ Costs are known up front, before training, but benefits may accrue slowly over time. At what point after training do you attempt to measure impact?

▶ As trainers, we lack the time and the accounting skills to do a cost/benefit analysis. Besides, our requests for data disrupt productivity.

▶ We probably will continue to run most of our popular training programs even if costs exceed benefits. So why bother? We're not a profit center.

▶ The outcomes could be damaging to the HRD staff and to budget support from top management. We may be better off not knowing.

▶ People at work perform the way they do for many reasons, only one of which relates to training. How can we take credit or blame for their performance?

▶ The very act of collecting data on the dollar value of performance will tend to bias the information we get, making it hard for us to present a true picture.

If you've been looking for some reasons for not evaluating the ROI of your training efforts, read no further. This list should enable you to persuade the most insistent believer that any attempt to prove that training pays for itself is sheer folly! Let sleeping dogs lie— what we don't know can't hurt us. Right?

Wrong! Lest we be accused of favoritism, let's give equal time to a list of reasons why we should take the time and effort to calculate the costs and the benefits of our major training programs. Here are some supporting reasons:

▶ HRD budgets can be justified and even expanded when training can contribute to profit and is not seen as an act of faith or a cost of doing business.

▶ Course objectives and content will become more lean, relevant, and behavioral with focus on monetary results rather than on the acquisition of information.

▶ Better commitment of trainees and their managers, who become responsible for follow-up and ROI, and not just for filling seats.

▶ Action plans, individual development plans, and managers' briefings will be taken seriously, thus strengthening the trainee-manager partnership.

▶ Better performance by HRD staff in containing costs and maximizing benefits. They become performance managers and not just instructors.

▶ HRD staff has solid data about where training is effective and where it is weak, so that courses can be revised and fine-tuned to produce the best returns.

▶ The curriculum of courses offered can be determined on a financial basis and not just on popularity, rank of the manager requesting it, and so forth.

▶ Course enrollments will be serious, with trainees aware of the expectations that follow graduation. We'll get the right faces in the right places at the right times.

▶ By calculating ROI on the courses where it is possible, we are more apt to be trusted on the ones we can't evaluate at level four.

Four ways to measure ROI on training

Now that we've examined the pros and cons of calculating the ROI of a training program, let's look at four ways of doing so. The nature of the training and the course objectives will determine which method is most appropriate.

1. When hard data exists. Performance data is routinely collected on many jobs for which we provide training. Examples include driver safety (monetary value of reduced accidents, lower insurance); machine maintenance (fewer repairs, less downtime); sales training (increased volume, fewer returns); bank tellers (fewer "overs and shorts," more services and customers handled per hour).

Many technical training programs have data on existing performance before the course is launched. By comparing the costs of inadequate performance prior to training with the reduced costs of better performance after training, we can see the return-on-investment.

Even courses that teach "soft skills" can have a "hard data" side to performance. Examples include writing skills (time saved via shorter letters, understood without subsequent clarification); meeting leadership (shorter meetings, better follow-up); EEO and diversity (fewer grievances and lawsuits).

Notice that our examples focus on the quantitative aspects of performance—things that can be counted in minutes, dollars saved or gained, and so forth. To be sure, these courses also have qualitative aspects. But these are more difficult to measure (such as courteous driving, more professional selling, clearer writing, more participative meeting leadership). Hard data probably doesn't exist to evaluate these qualities, so we have no way of comparing pre-training and post-training performance.

Conclusion: If we want to take credit for the impact of training on workplace performance, we must establish a "bench level" of what the performance was before we launched the training program.

2. Estimates by trainees and their managers. This method is the easiest way to estimate ROI, but also the most subjective. Several months after completing each cycle of a training program, send a memo to each graduate and manager (sponsor). State the actual cost to the organization of the trainee's participation in the course. Ask the two to get together, discuss the actual improved performance that has taken place since the course, agree on a dollar value of this improvement, and project the total value over the coming year (or whatever period is appropriate to the application of the concepts and skills that were learned).

The two then send this projected dollar value in, along with a one- to two-paragraph explanation of how the estimate was made. By comparing the costs of those who responded with their dollar estimates of value added to workplace performance, we can arrive at a crude estimate of the cost/benefit ratio.

In situations where bench levels were not established before the course was launched, this method of estimating ROI has appeal. What it lacks in accuracy it makes up for in getting trainees and their managers to recognize that the responsibility for making training effective is primarily theirs and not the trainer's.

3. Action plans, managers' briefing. During a training program, each participant prepares an action plan that spells out how the concepts and skills learned will be applied back at work. If the course involves teaching the entire job to a new employee, then the action plan will resemble a job description. If the course is for present employees (such as supervisors, team leaders, project managers), then the action plan spells out those actions the participant will take back to the job, which will differ from other participants whose needs are different.

After the training program, participants share their action plans with their managers and anyone else who is a stakeholder in their growth and development. This helps to build the participants' managers into their development—as coaches, mentors, and overseers of the implementations of the action plan. (A pre-training meeting with the participants' managers is important: to cover course objectives, how the action plans work, and how managers can help their enrollees in the post-training follow-through.)

Several months after the training, participants and their managers come together for a two- to three-hour meeting at which each participant reports on the results since implementing the action plan, along with the cost of doing so and the value of the benefits. Managers work with their participants prior to this meeting to arrive at the dollar value of the costs and the benefits. By tallying the numbers reported by the participants and adding the cost of the course, the return-on-investment is obtained.

4. Cost/benefit analysis via accounting. This method is the most demanding way to calculate ROI, but also the most accurate. Costs can be listed under seven categories, as noted below:
▶ course development (time) or purchase (price, license fees)
▶ instructional materials: per participant (expendables) and instructor (durables)
▶ equipment and hardware: projectors, computers, video ("fair share" use)
▶ facilities: rental of conference cen-

■ *Benefits accrue long after training, and can be projected typically one to five years* ■

ter and "fair share" use of classroom overhead
▶ travel, lodging, meals, breaks, shipping of materials, and so forth
▶ salary: of instructor and support staff (prorated), consultants' fees, and so forth
▶ lost productivity (if applicable) or cost of temporary replacements for participants.

These costs are of three types: one-time (such as needs analysis and design), cost per offering (such as facility rental, instructor's salary), and cost per participant (such as meals, notebooks, coffee breaks). Costs must therefore be calculated over the life of the training program.

Benefits fall into four major categories as shown below:
▶ time savings (less time needed to reach proficiency, less supervision needed, and so forth)
▶ better quantity (faster work rate, less down time, not having to wait for help, and so forth)
▶ better quality (fewer rejects, lost sales, reduced accidents, lower legal costs, and so forth)
▶ personnel data (less absenteeism, fewer medical claims, reduced grievances, and so forth).

Benefits accrue long after training, and can be projected over the life of the trainees in the job for which they were trained (typically one to five years). While costs can be calculated by HRD managers, the benefits should be calculated by the trainees and their managers after they have had enough experience in the workplace to collect enough data to project the benefits over the payback period. A comparison of the total costs to the total benefits yields our return-on-investment.

Eight observations on conducting a cost-benefit analysis

▶ Some courses should be offered without expectation of a measurable return on the investment (such as orientation of new employees and retirement planning). Because the benefits of conducting such programs are difficult if not impossible to measure, and because organizations offer them without expectation of any tangible return on the investment, it is foolish to attempt a cost-benefit analysis.
▶ Training programs for employees whose jobs have well-defined and quantified expectations (standards, goals, quotas) are the most appropriate ones for measuring return-on-the-training investment because performance measurement systems already exist.
▶ By contrast, training for supervisors, managers, technical experts, project coordinators, and others for whom performance measurement systems do not exist are much more difficult to evaluate via level four (cost-benefit analysis). The responsibility rests with each participant to generate pretraining data and posttraining data on performance, and to assign dollar values to these two sets of data.
▶ Most cost-benefit analyses are comparative studies that show how the performance levels obtained by

COST-BENEFIT WORKSHEET

Costs

	One-Time Costs	Cost per Offering	Cost per Participant
1. Course development (time) or selection (price, fees)			
▶ needs analysis and research..........................		NA	NA
▶ design and creation of blueprint......................		NA	NA
▶ writing, validating, and revising......................		NA	NA
▶ producing (typesetting, illustrating, reproducing)		NA	NA
2. Instructional materials			
▶ per participant (expendables: notebooks, handouts, tests, and so forth)	NA	NA	
▶ per instructor (durables: videotape, film, software, overheads)......................			NA
3. Equipment (hardware)			
▶ projectors, VHS, computers, flipcharts, training aids.........			NA
4. Facilities			
▶ rental or allocated "fair share" use of classrooms, and so forth	NA		NA
5. Off-site expenses (if applicable)			
▶ travel, hotel accommodations, meals, breaks...........	NA	NA	
▶ shipping of materials, rental of A/V equipment, and so forth	NA		NA
6. Salary			
▶ participants (number of instruction hours X average hourly rate)	NA	NA	
▶ instructor, course administrator, program manager, and so forth	NA		NA
▶ fees to consultants or outside instructors			NA
▶ support staff (audiovisual, administrative, and so forth)	NA		NA
7. Lost productivity (if applicable)			
▶ production rate losses or material losses...........	NA	NA	
A. Total of all one-time "up front" costs		NA	NA
B. Total of all costs incurred each time course is offered	NA		NA
C. This sum (box B) X number of times course is run (_____).	NA		NA
D. Total of all costs incurred for each participant.............	NA	NA	
E. This sum (box D) X number of participants (____) over life of course.............	NA	NA	
F. Total costs (sum of boxes A, C, and E)			

Benefits

	One-Time Costs	Cost per Offering	Cost per Participant
1. Time savings			
▶ shorter lead time to reach proficiency (hours saved X $)	NA		NA
▶ less time required to perform an operation (hours saved X $).............	NA	NA	
▶ less supervision needed (supervisory hours saved X supervisory $).............	NA		
▶ better time management (hours freed up X $)	NA	NA	
2. Better productivity (quantity)			
▶ faster work rate ($ value of additional units, sales, and so forth).............	NA	NA	
▶ time saved by not having to wait for help (hours saved X $)	NA	NA	
▶ less down time ($ value of reduced nonproductive time).............	NA	NA	
3. Improved quality of output			
▶ fewer rejects to (scrap, lost sales, returns, and so forth....$ value)	NA	NA	
▶ value added to output (bigger sales, smoother castings ...$)	NA	NA	
▶ reduced accidents ($ value of savings on claims, lost work).............		NA	
▶ reduced legal costs (EEO, OSHA, WC settlements)......$)		NA	
▶ improved competitiveness (change in market share...$).............		NA	NA
4. Better personnel performance			
▶ less absenteeism/tardiness (self or subordinates...$ saved)	NA	NA	
▶ improved health ($ saved on medical and lost time).............		NA	
▶ reduced grievances, claims, job actions ($ saved)		NA	
▶ same output with fewer employees ($ on jobs eliminated)		NA	NA
A. Total of all one-time benefits		NA	NA
B. Total of all benefits occurring once per participant	NA		NA
C. Total value of all improvements per participant per month	NA	NA	
D. Length of payback period in months	NA	NA	
E. Number of employees affected during this period (D)	NA		NA
F. Total of B X E	NA		NA
G. Total of C X D X E	NA	NA	
H. Total benefits (sum of A + F + G)			

installing a new training program compare with the performance levels obtained by no training (such as safety, drugs, stress reduction) or by some alternative form of training (such as on-the-job training versus classroom, individualized versus group, centralized versus regional, and so forth). As in the third item, pretraining data on performance prior to installation of the new program may not have been collected. This must be done prior to carrying out a cost-benefit analysis.

▶ When training is conducted to accompany the installation of new equipment (procedures, products, policies, technology) and no prior training of a similar nature existed, a cost-benefit analysis is inappropriate for two reasons: There are no prior performance measures with which to compare the results of the new training; and the impact of installing the new changes makes it impossible to separate "performance attributable to training" from "performance attributable to innovation." (Examples: moving from manual to PC operations and learning to use e-mail).

▶ The costs of training are known upfront and should be calculated by HRD managers and others whose budgets are funding the program. The major unknown is based on the shelf life of the course—how many times (cycles) it will be run before it is no longer needed (such as when all eligible trainees have received it or when changes in technology have rendered it obsolete). Costs should be calculated over the shelf-life of the program.

▶ Similarly, the benefits of training should extend well beyond the final offering (cycle) of the program. Different behaviors that were "shaped" by training have different life cycles. The payback period on skills that are practiced regularly (such as time management) might be projected over the employment life of the trainee, whereas skills that are called on less frequently (such as selection interviewing in a downsized econo-

my) may have a much shorter payback period.

▶ Although training costs are best calculated by HRD managers, the benefits should be identified, quantified, and converted to dollar values by management (the trainees' supervisors, department heads, and so forth). There are two reasons for this: They are in the best position to observe changes in performance attributable to training; and their data is more objective and less suspect than if HRD specialists attempted to collect it.

Four examples of applications of cost-benefit analysis

▶ A rapidly growing fast-food chain had a three-week apprentice training program that prepared employees for promotion as assistant managers. The corporate HRD manager felt that training time could be reduced to one week with a formal training program at headquarters. The one-week formal program required travel and hotel costs not associated with the three-week local apprentice training program. However, the company's ability to place assistant managers in outlets two weeks earlier resulted in savings that more than offset the cost of developing the program and bringing the trainees to a central location. It also assured uniform quality of instruction which was lacking in the decentralized apprentice training that had taken place in each outlet.

▶ A major corporation had relied on two professors from the state university to conduct their supervisory training program, using their own handouts, visuals, and hands-on exercises. Some 93 supervisors went through the five-day program in classes of 15 to 16 participants each. Three years later when the company offered supervisory training again, they purchased a packaged course with videos, workbooks, and instructor guidelines for their own internal instructors. Although the package cost $27,000, they ended up saving

$16,000 (the professors had charged $36,000 for labor and $7,000 for materials). Moreover, post-workshop evaluations showed that transfer of training from workshop to workplace had improved significantly.

▶ A government agency ran a three-day workshop on project management with six offerings for 20 participants each. During the year following each workshop, the trainers surveyed the graduates to see how their posttraining performance on projects compared with their pretraining behavior (as assessed during the needs analysis prior to training). Factors evaluated included: percentage of projects completed on time and within budget, level of client satisfaction, and estimate of time/money saved as a result of improved project management. The agency concluded that a $95,000 training investment had saved an estimated $670,000. This figure did not include one reported savings of $2 million projected over five years.

▶ An automotive manufacturer installed a management development program as part of the company's TQM/empowerment efforts and put 220 managers at an assembly plant through the program. The average length was six days. After the first day of assessment, each manager attended only those workshops that dealt with the competencies and skills that received lower scores. Six months after the training, participants were assessed again. Benefits were evaluated on three factors: the degree to which each manager's individual development plan had been implemented, the change in productivity of the manager's work group, and the improvement in scores (percentiles against nationwide norms) by each manager on the two assessments. All three measures showed that the benefits far outweighed the costs.

Scott B. Parry *is chairman of Training House, P.O. Box 3090, Princeton, NJ 08543-3090. Phone: 609/452-1505; fax: 609/243-9368.*

Demonstrating ROI of Training

TRACKING THE RETURN ON A TRAINING INVESTMENT CAN BE TIME-CONSUMING, BUT IT'S MORE IMPORTANT NOW THAN EVER.

BY ERIC A. DAVIDOVE AND PEGGY A. SCHROEDER

Too many training professionals have no idea how training investments relate to their companies' business objectives.

When asked to report a training investment, trainers usually measure the payroll and resource expenditures necessary to produce, deliver, and maintain training. Most trainers determine whether training improves employee and business performance by looking at opinions given on self-report surveys distributed after training. These usually show whether employees enjoy the training, whether instructors think the course materials are easy to use, and whether experts believe the training content is accurate, relevant, and complete.

Such information gives senior managers no basis for making strategic business decisions, allocating resources, or controlling internal operations. Line managers can only guess at whether performance discrepancies are caused by a lack of knowledge and skills, low motivation levels, environmental constraints, or procedural issues. Also, such evaluations provide trainers with no clue about how to improve instructional materials.

In short, decision makers have no way to assess the connection between the investment in training and the accomplishment of specific business objectives.

Line managers who send employees to training expect results. Scores on tests completed immediately after training can show how much the employees have learned, but test scores alone do not guarantee that employees will transfer the skills to the job. To answer questions asked by managers, trainers must track and report employee performance well after initial training.

Tracking training investments

Business leaders have no generally accepted definition or accounting procedure for tracking training investments. Generally accepted accounting principles do not allow for human asset accounting, or accounting for the value of human capital. But the success of today's companies relies heavily on intangible assets that don't appear on balance sheets—items such as a trained workforce.

For our purposes, the terms "training expenditures," "training costs," and "training investments" are used interchangeably.

The largest direct training cost is payroll for people to plan, develop, deliver, and attend training. Other direct costs include travel, food, lodging, training room rental, and the purchase of training materials. Indirect training costs such as overhead cannot be readily identified with a particular project, but are also necessary.

After trainers understand how their training investments are allocated, they can compare old training to revised or updated training, or compare existing training to proposed training. When there are no plans to revise old training or develop new training, trainers can benchmark training investments against industry standards.

Tracking performance

A lower training investment is not automatically better for an overall return on investment. The training department must also determine how training affects individual and business performance.

To begin tracking performance, trainers should first review the results of a needs analysis conducted before training development begins. The purpose of the analysis is to identify business problems (expressed in terms of organizational results) that are important to decision makers and that are caused by employees' lack of knowledge and skills.

The analysis will point out the differences between what employees should be able to say or do, and what they actually can say or do. These differences are then expressed as learning objectives.

A survey is a good tool for identifying learning objectives that relate to business objectives. Confirm survey results by observing or interviewing successful employees. Once you have an idea of which skills affect business results, confirm the findings with key decision makers. Asking decision makers for input can ensure that the evaluation collects and reports evidence they view as helpful and important.

After training, trainers can use the results of the needs analysis to see whether employees have mastered pre-determined learning objectives, have transferred the learning to the job, and are helping to accomplish business objectives. Surveys can help determine whether line managers believe employees are applying the newly learned skills.

Once trainers have established the relationship between learning objectives and business objectives, they can measure training effectiveness. Written tests designed to measure a particular set of learning objectives provide information about training's effect on performance. Trainees may take tests before training, immediately after training, and a few months after training.

Tests must be valid and reliable. A test is valid when content experts agree that the test questions reflect the learning objectives they were written to measure—objectives that represent the knowledge and skills necessary to perform a job.

A test is more valid if it simulates real-life conditions. For example, a true-or-false question is less valid than a role play for measuring how well someone can put out a chemical fire at night. In general, true-or-false, multiple-choice, and matching items are less valid than essays items. They are popular because they require less time to score, can be given to many people simultaneously, and are unaffected by the subjectivity of graders.

A test or measurement instrument is reliable if it provides consistent and accurate information about the knowledge or performance being evaluated. Generally, reliability is higher for surveys and tests that have more questions, and for questions that require respondents to select an answer from two or more options.

Simulations are useful when performance must be observed in controlled settings; for example, in situations in which working conditions vary greatly due to environmental distractions or assistance from others.

In some situations, trainers may want to measure performance by observing employees' work and documenting certain behaviors. Another measurement approach is to ask experts to judge the quality of employees' results. In such situations, increasing the number of raters and ensuring that they use the same performance criteria will enhance the credibility of the evaluation.

Because employees who come to training differ, measurement controls are necessary to ensure that all training classes whose results are being compared are made up of similar trainees. In most cases, there should be no significant differences between training classes with respect to reading ability, experience levels, motivation, age, and gender makeup. Such differences can make it impossible to tell whether training alone has affected performance.

Demonstrating return on investment

To show a credible return on investment for training, describe results in the context of the financial and performance models that the company's decision makers already use to measure business results. Three key business objectives—quality, timeliness, and operational costs—are often important to senior and line managers, are usually achievable with good training, and are generally possible to monitor.

One way to calculate a return on the training investment is to divide operational savings or revenue increases resulting from training by the training program costs. Then multiply the result by 100.

For example, suppose half the salesforce is randomly selected for training that costs $100,000 to develop and $100,000 to deliver. Six months after training, if the trained salespeople sell $50,000 more than the people who received no training, the ROI is 25 percent. If the trained salesforce sells $50,000 more in the next six months as well, the ROI is 50 percent.

This ROI formula has serious limitations. There are no standard guidelines for defining and tracking training program costs, revenue benefits, and operational savings, so evaluators often do a poor job of measuring performance indicators.

Several problems can result in evaluation reports that are meaningless to key decision makers. They include an inability to isolate the effect of training on the accomplishment of business objectives, inconsistent measures of business objectives, and a lack of needs analysis information at the beginning of training projects.

Training may have a low ROI for several reasons. Employees who attend training long before they need to use the new skills on the job may forget what they learn before they can apply it. Instruction may not be available on demand. It might take too long to complete. Or the training may neglect slow and fast learners.

Sometimes the effect of a training program on business objectives is difficult to determine. Employee skill and knowledge deficiencies may have little to do with increasing revenue or decreasing operational costs. Failure to accomplish business objectives might be due to worker motivation levels, working conditions, or forces outside the organization.

Critical to measuring the ROI of training is understanding ways to reduce training costs and increase training effectiveness through instructional design. The easiest ways to keep training costs low are to reduce trainee and instructor hours and to conduct training in employees' work areas. Another way to decrease training costs is to use self-paced media such as computers and books. Job aids and desk references can reduce instructional time and the time supervisors spend helping employees.

Training departments can minimize the time employees spend in the learning environment by providing integrated performance support tools. Such tools give immediate, on-the-job access to integrated data bases as well as productivity software, practice opportunities, and expert consultation. Employees learn what they need, when they need it.

Another effective way to reduce costs is to simplify, automate, re-engineer, or eliminate job tasks. It may be more cost-effective to eliminate resource and procedural problems than to train employees to work around them.

Recommendations and conclusions

Training departments are under continuous pressure to demonstrate how training investments relate to business objectives. But standard approaches fall short of giving meaningful evidence for reducing training investments and increasing training effectiveness.

Many business leaders still view training as an overhead expense. With thorough ROI evaluations, training departments can convince businesses to view them instead as partners in creating the assets that are crucial to organizational success. ∎

Eric Davidove *and* **Peggy Schroeder** *are with Andersen Consulting's Change Management Services, 1 Financial Plaza, Hartford, CT 06103.*

Selling Technical Training to Top Management

Before asking top management to fund training, calculate the potential return on investment.

The value of any activity is measured by its impact on long-term success. As the president of two corporations, I will not spend money unless I see the potential return on investment (ROI). Usually, the person who calculates that ROI is the one who has an idea that needs funding. This "idea" person must present the idea, along with the investment rationale, so that an executive can make a decision. I know—I once stood on your side of the desk, trying to gain top management's support for my ideas. Here's the process I used.

Enter the Executive's Mind

Profitability is first in the minds of most executives. The executive is accountable to a board of directors and stockholders. These people want to know if the company is getting more profitable with each passing year. They tend to be unfriendly when business declines or when return on capital decreases. The executive doesn't think in "micro" training terms, such as business writing, spot welding, etc. He or she thinks in "macro" business terms:

▼ How can we reduce overall business costs?
▼ In what ways can we increase market penetration?
▼ In what ways can we capture the market, gain visibility, and meet our customers' needs?

Learn the Executive's Language

Money and time are in short supply today. To maintain or capture a competitive advantage, top management must stay in touch with a multitude of dynamic conditions and options. So trainers must

By Mary Coeli Meyer

understand how to communicate with these executives, to know their concerns, priorities, and ways of thinking.

And remember: When the economy is in disarray, many executives instinctively resort to "tried and true" methods rather than take risks on new approaches. So your presentation must be relevant to the executives' perspective, keeping in mind that risk taking is not their long suit.

Communicating with top management means talking and listening with an executive ear, not a training ear. Training is a means to an end—the organization's goals. Hence, you must understand the mission of your organization and its business.

Generically, we can say that all companies strive to attract and keep customers. You must know how your company goes about doing that. Ask yourself what are the organization's concerns and plans for the future. Learn what succeeded in the past and why.

If you understand these issues, you're in a better position to get an executive's attention.

For instance, when one of the Big Three auto companies tried to reduce its costs by five cents per car, top management apparently did not consider that fewer recalls could lower their costs by even more than a nickel per car. Instead, managers looked at the front end of the equation while the answer may have been at the production process end. What does training have to do with that situation? Therein lies the answer to funding for training.

Understand Your Training's ROI

The first rule is this: Know your bottom line. How much does a training program cost? Line items such as program designers, trainers, media, room, travel, lost productivity time, etc., need to be included.

In This Story

▼ cost/benefit analysis
▼ evaluation
▼ productivity improvement

These are the actual numbers to which an executive pays attention. All too often, an executive says, "You know that training room down the hall—let's expand the information systems department over there." That's because the executive sees a greater value from using that space for some purpose other than training.

Why? Because some training is seen as a waste of time and money. The result is that ROI is not captured to justify training. To that degree, training is not seen as a value-added activity.

Indeed, people may come out of a training course smiling because it was the best training course they have attended in the last 10 years. But then they ask, "What did we *get* for our money?"

That is the only question an executive asks when considering the expense of a training program. In other words, can your program reduce the cost of unit car manufacturing by five cents?

I have made speeches about calculating ROI to some two thousand trainers in the past 18 months. Unfortunately, many trainers do not want to do such calculations. Determining ROI means conducting pretests and post-tests, in addition to knowing the cost of a trainer, a designer, the space occupied, the lost productivity time, etc. You must account for the company's investment in the training process.

Never forget that top management commitment to training is predicated on a desire that every dollar spent will result in a return of multiple dollars—if not directly, then indirectly. And even if they change from profit-driven to customer-driven, *you* still must speak in terms of ROI.

Determining ROI—Part 1

The first step to calculating ROI is prequalification, which consists of three items:

▼ *Skill testing*, which ascertains how well a person knows how to execute a particular skill prior to training. This must be measured so you can calculate the ROI at the completion of skill training.

▼ *Cost assessment*, which involves working with the client to determine what the present skill delivery costs. In other words, how many units of "x" does a worker produce per hour? This is a cost when translated into dollars. Or another example: What is the waste produced by the average worker on a particular task? This too translates into dollars.

▼ *System nonsupport*, which requires determining whether a deficiency is a knowledge problem or process/system problem. Process/system problems need to be ruled out before you make a request for

Recouping the Cost

Consider this example: It costs $145 to train an employee on a drill press operation. The drill press operator produces five units per hour. The unit cost for the item drilled is $25, which includes raw material cost, defects for raw materials, and waste as a result of a miss-drill.

The drill press operator goes through training. You measure the productivity afterward and you find that he is now producing six units per hour and the cost per unit is reduced to $22.34. An executive is interested in the unit cost of $22.34—a reduction of 10.6 percent. A plant manager is interested in the increase of one unit per hour. These measurements show an ROI over a period of time.

In other words, the end result of a training program on drill press operations should produce the following results:

▼ recouping the cost of $145 per person
▼ reducing the $25 unit cost of the item drilled
▼ increasing the quality of the units per hour.

There are several other potential returns on this $145 investment, but you won't know them unless you conduct post-testing. It should show the following improvements:

▼ The drill press operator can increase the number of units completed successfully per hour.
▼ The operator can reduce the amount of waste generated through error.
▼ The operator can spot defective raw materials before beginning work and reject them.

training funding. In fact, tracking the number of incidents of process/system problems identified through training is one way of calculating ROI. Although training cannot solve process/system problems, advising top management of the problems that training uncovers will increase your positive visibility.

These three prequalifying activities provide you with several benefits:

▼ baseline data for measurements of later success
▼ "executive" information that makes sense to a profit-oriented top manager
▼ task analysis data for the design and delivery of the training program.

Determining ROI—Part 2

To speak relevantly to top management, you must know the results your training activities produced. Again, it is not a matter of warm fuzzies and good times. It is a measurement. In fact, it is three measurements:

▼ Measure the immediate sentiments of the trainees. Was the course relevant? Did it hit the mark concerning job-relatedness? Was the delivery responsive to their specific needs? Was the trainer prepared? Were the materials, the room, etc., up to standard?

▼ Measure the skills. Using a parallel test to measure the same criteria as the pretest, find out what a person has learned and to what degree his or her skills improved.

▼ Measure the actual on-line delivery of the skill. Are more units produced per hour? Are there fewer defects? Has waste been reduced? What does this mean in dollars and cents? Overall, how does this affect the company's bottom line?

Now remember this: Top management doesn't care about the first two measurements. Those measurements assist you in refining the course. Top management only cares about the last one. Will these results improve their image to the board? Will they improve return on equity to stockholders? How do they impact the long-term or short-term financial success of the company?

Measuring Costs

Three measurements help you calculate ROI. The first two are pretraining measurements:
1. Unit cost of training
2. Cost per unit produced.
The third is a post-training measurement:
3. New cost of unit produced.

To determine measurement 1, you must first determine the overall cost of training. Then, divide that cost by the number of participants in the training program. This amount gives you a unit cost of training, which is one figure used to determine ROI. In fact, it is a target figure for recouping the dollars spent.

To determine measurement 2, use the cost assessment measurements mentioned earlier to determine the cost of a particular unit produced. In most instances a plant manager, information systems manager, or senior department manager can provide you with this kind of data. If not, it is up to you to get the information.

To determine measurement 3, you must examine the productivity of the person who underwent training. The newly trained worker needs to return the unit cost of training in some fashion (see "Recouping the Cost"). This should show up as a productivity improvement, such as more units per hour, fewer mistakes, less waste, etc. But you don't know this without the cost assessment from measurement 2.

Presenting Your Case

Having accumulated data for all the points addressed above, you are now ready to speak relevantly to top management when asking for additional funding or support because you can speak in numbers and facts.

You can talk about the bottom-line impact from the reduction of waste.

You can talk about the increased quality the customer will receive.

Or, you can talk about the enhanced productivity that helps your organization meet increased customer demand. In other words, the company increased capacity without increasing costs.

Start the presentation with a statement of your understanding of corporate goals, objectives, and concerns. Then explain, in numerical terms, how you plan to meet those goals, objectives, and concerns through additional training.

Here is a sample presentation:

"I know we are making a significant effort to increase the market for our G-100 gaskets. But presently, 36 percent of our gaskets contain flaws. This quality deficiency has probably turned potential customers away.

"When the plant manager and I reviewed this problem, we discovered that a maintenance deficiency was the cause: Maintenance personnel don't know how to anticipate problems with the cutting equipment.

"Next month, I would like to provide an eight-hour program for maintenance personnel on anticipating and resolving problems with the 396 stamping machine. This program will cost the company $7,000, but will provide a potential return by reducing defects 21 percent.

"Do you think that would increase the customer base?"

Of course, this presentation could not be made unless you had the data for assessing ROI. It does take time to work the calculations, but that time is essential. Few executives will invest in an activity unless you can demonstrate that it adds value.

You now have the tools. Go forth and figure!

Mary Coeli Meyer is president of Cheshire Ltd., an international management consulting/training firm. She also heads a rehabilitative pet products manufacturing and distribution company. She may be reached at Cheshire Ltd., 1601 Shadowbrook Dr., N.W., Acworth, GA 30101; 404/928-2700.

ROI: The Search

FOR MANY YEARS, MEASURING THE RETURN ON IN-VESTMENT for training and development has been a critical issue—on meeting agendas, in the literature, and on the minds of top executives. Although interest has heightened and some progress has been made, the topic still challenges even the most sophisticated and progressive HR departments. Some HR professionals argue that measuring ROI for training isn't possible; others quietly and deliberately develop ROI measures. But overall, most practitioners acknowledge that they must show a return on the investment in training so that they can maintain training funds and enhance HR's status.

Currently, it's difficult to pinpoint the state of ROI within the field. Many HR managers are reluctant to disclose internal practices. And many say that little progress has been made, even in the most progressive organizations. It's also difficult to find case studies that show specifically what organizations have done in measuring training ROI.

Recognizing the void, the American Society for Training and Development began collecting case studies with real-life examples of measuring the return on investment in training. ASTD contacted more than 2,000 people in HR directly, including senior practitioners, authors, researchers, consultants, and confer-ence presenters. ASTD also contacted organizations that were perceived as profitable, respected, and admired. In addition, announcements appeared in the *National Report* and other ASTD publications inviting case submissions. To meet requirements, organizations had to be willing to describe the specific steps, issues, and concerns involved in their efforts to measure training ROI.

More than 150 respondents requested specific guidelines for developing a case; 40 were willing to submit cases. ASTD stopped collecting cases once 30 were delivered. Of those, 18 were selected for publication in *Measuring Return on Investment* (ASTD, 1994). A second volume is planned for publication in late 1996.

Building onto level 4

The ROI process adds a fifth level to the Level-4 evaluation model developed by Donald Kirkpatrick. See figure 1. At level 1, participants' satisfaction with the training program is measured, and a list of their plans for implementing the training is included.

At level 2, measurements focus on what participants learned during training. At level 3, the measures assess how participants applied learning on the job. At level 4, the measures focus on the business results achieved by participants when the training

BY
JACK J. PHILLIPS

for Best Practices

objectives are met. The fifth, and ultimate, level of evaluation is the return on investment. It compares the training's monetary benefits with the costs.

Most organizations conduct evaluations to measure satisfaction; few conduct evaluations at the ROI level. Both are desirable. Evidence shows that if measurements aren't taken at each level, it's difficult to show that any improvement can be attributed to the training.

The model in figure 2 is a framework for developing ROI. Many of the companies in the case studies followed this model. It tracks the steps in measuring ROI—from collecting post-program data to calculating the actual return. The model assumes that training costs will be compared with monetary benefits and that all training programs will also have intangible, but reportable, benefits.

The process begins with the collection methods of post-program data. Such methods are at the heart of any evaluation. Which methods to use depends on the evaluation's purposes, instruments, measurement levels, design, and cost of data collection.

Two common formulas for calculating return on investment are a benefit/cost ratio (BCR) and ROI. To find the BCR, you divide the total benefits by the cost. In the ROI formula, the costs are subtracted from the total benefits to produce the net benefits, which are then divided by the costs.

For example, a literacy-skills training program at Magnavox produced benefits of $321,600 with a cost of $38,233. The BCR is 8.4. For every $1 invested, $8.4 in benefits were returned. The net benefits are $321,600 − $38,233 = $283,367. ROI is $283,367 ÷ $38,233 x 100 = 741 percent. Using the ROI formula, for every $1 invested in the program, there was a return of $7.4 in net benefits.

Typically, the benefits are annual, the amount saved or gained in the year after training is completed. The benefits may continue after the first year, but the effect begins to diminish. In a conservative approach, long-term benefits are omitted from calculations. In the total cost of a program,

This is the first in a series of three articles about measuring the return on investment in training. Real-world case studies provide a look at how the search is going.

Figure 1:
Level-5 Evaluation

Here are some questions for conducting a level-5 evaluation.

Level	Questions
1 reaction and planned action	▶ What are participants' reaction to the program? ▶ What do they plan to do with what they learned?
2 learning	▶ What skills, knowledge, or attitudes have changed? By how much?
3 applied learning on the job	▶ Did participants apply what they learned on the job?
4 business results	▶ Did the on-the-job application produce measurable results?
5 return on investment	▶ Did the monetary value of the results exceed the cost of the program?

the development cost is usually front-loaded and prorated over the first year of implementation. Or, you can pro-rate development costs over the projected life of a program.

Some recommendations

The case studies in the ASTD project represent a wide range of settings, strategies, and approaches in manufacturing, service, and government organizations. The training audiences varied from all employees to managers only to specialists only. Though most of the programs focused on training and development, others included such areas as total quality, performance management, and employee selection.

The cases provide a rich source of information on the strategies and thought processes of some of the best practitioners, consultants, and researchers in the field. The companies' returns on investment ranged from 150 to 2,000 percent.

Several common approaches have emerged. They could be considered best practices or just recommendations. Whichever, they seem to have worked well for the companies in the case studies.

Set targets for each evaluation level. Some organizations set a target for each level of evaluation, a target being the percentage of HR programs that will be measured at that level.

For example, many organizations require 100 percent of their training programs to be evaluated at level 1 because it's fairly easy to measure participants' reactions. Level 2 (learning) is also relatively easy to measure. Typically, the target range is 40 to 70 percent, depending on the type of program.

Level-3 evaluation (on-the-job application) involves more time and expense to conduct follow-up evaluations so targets tend to be lower at 30 to 50 percent. Level-4 (business results) and level-5 (ROI) evaluations require significant resources and budgets so their targets tend to be small: 10 percent for level 4 and five percent for level 5.

Establishing evaluation targets has several advantages. One, it provides measurable goals for assessing the progress of all training or a particular

■ Use a variety of approaches to collect evaluation data ■

segment. It also focuses attention on accountability and communicates a strong message to the HR staff about the need for measurement and evaluation.

Evaluate at the micro level. Measurement and evaluation usually focus on a single program or a few tightly integrated programs. ROI measurement is more effective when applied to one program that can be linked to a direct payoff. When all of the courses in a series must be completed before their common objectives are met, it may be appropriate to wait to evaluate the series as a whole. The decision to evaluate several courses in a series should take into account the training goals, timing of the courses, and cohesiveness of the series.

It can be difficult to evaluate a series conducted over a long period of time. A cause-and-effect relationship becomes more confusing and complex. Also, it is hard to evaluate an entire function, such as management development, career development, executive education, or technical training.

Use a variety of methods. The companies in the case studies use a variety of approaches to collect evaluation data. They don't latch onto one or two practices and stay with them regardless. They recognize that every program, setting, and situation is different. They know that techniques such as interviews, focus groups, and questionnaires work well in some situations and that action plans, contracts, and performance monitoring are needed in others.

These companies use internally developed criteria to match a particular data-collection method with the training program.

Isolate the effects of training. A critical aspect of the evaluation process is

trying to isolate the effect of the training from other factors occurring during the same period that could affect business results. Most of the time, training can take only partial credit for improvements in on-the-job performance. When planning to measure ROI, the case-study organizations go beyond a standard control-group analysis to use one or more techniques for isolating extraneous factors.

Use sampling wisely. It's rare for organizations to use statistical sampling in selecting a sample of training programs in which to measure ROI. For most, the result would be too many calculations. For the sake of practicality, many organizations decide to evaluate just one or two sessions of their most popular training programs. Others select one program from each major training segment. It's recommended that organizations calculating ROI for the first time select only one course to measure, as a learning experience.

If sampling is used, it's important to be statistically sound. But it's more important to consider the tradeoff between the available resources and what kind of ROI calculations management is willing to accept. Remember: The primary goals of an ROI calculation are to convince the HR staff that the process works and show senior-level managers that training can make a difference. With that in mind, it's best to get the input and approval of top management in developing your sampling approach.

The sample number depends on the following variables:
▶ the HR staff's expertise on evaluation
▶ the type of training programs being evaluated
▶ the resources allocated for evaluation
▶ the degree of support from management for training and development
▶ the organization's commitment to measurement and evaluation
▶ the amount of pressure from others to show ROI calculations.

Other variables endemic to the particular organization may apply.

Convert program results to monetary values. The organizations in the case studies seek a specific return on investment. Consequently, data on business results must be converted to

A LOOK AT THE CASE STUDIES

Setting	Target Group	Program Description	Evaluation Process	Results
Bottling company (Coca Cola)	First-level supervisors	Eight half-day workshops covering supervisory roles, setting goals, developing the team, and so forth	▶ Action planning ▶ Follow-up session ▶ Performance monitoring	▶ 1,447% ROI ▶ Benefit/cost ratio 15:1 ▶ Variety of measures
Paper products company	Managers, supervisors, hourly employees study teams, skill	Organization development program (Workshops, action ▶ Performance building programs)	▶ Follow-up with interviews ▶ Survey ▶ Performance monitoring	▶ Variance from standard +$106,000 ▶ Efficiency 4% improvement ▶ Waste 36% improvement ▶ Absenteeism 35% improvement ▶ Safety 25% improvement ▶ Housekeeping 27% improvement
Health Maintenance Organization	All managers and all employees	Organization development program (team building, group meetings, customer service training	▶ Performance monitoring ▶ Management estimation	▶ 20,700 New HMO members ▶ 1,270% RO ▶ Benefit/cost ratio 13.7:1
Large commercial bank	Consumer loan officers	Two-day sales training program—focus on increase in consumer loans	▶ Follow-up ▶ Performance monitoring	▶ 30% increase in consumer loans ▶ 2,000% ROI ▶ Benefit/cost ratio 21:1
Information services company	Supervisors	Twelve two-and-one-half hour sessions on behavioral modeling	▶ Follow-up with surveys	▶ 336% ROI
Electric & gas utility	Managers & supervisors	Applied behavior management which focused on achieving employee involvement to increase quality, productivity, and profits	▶ Action planning (variety of projects) ▶ Performance monitoring	▶ 400% ROI ▶ Benefit/cost ratio 5:1
Oil company	Dispatchers	Skills training program including customer customer interaction skills, problem solving, and teamwork	▶ Follow-up observations ▶ Performance monitoring	▶ Customer complaints reduced by 85% ▶ Absenteeism reduced by 77% ▶ Reduction in pull-outs saved $283,800 ▶ 383% ROI ▶ Benefit/cost ratio 4.8:1
Bakery (Multi-Marques, Inc.)	Supervisors/ administration services	Fifteen-hour supervisory skills training including the role of training	▶ Action planning (work process analyses) ▶ Performance monitoring	▶ 215% ROI ▶ Benefit/cost ratio 3.3:1
Avionics (Litton Industries)	All employees	Self-directed work teams	▶ Action planning ▶ Performance monitoring	▶ Productivity increased 30% ▶ Scrap rate reduction 50% ▶ 700% ROI
Truck leasing (Penske Truck Leasing)	All supervisors	20-hour program on supervisory skills using behavioral modeling	▶ Performance monitoring	▶ Turnover reduction of 6% ▶ Abseeteeism reduction of 16.7%

These cases appear in Measuring Return on Investment, *published by the American Society for Training and Development, Alexandria, VA. 1994. Jack J. Phillips, editor.*

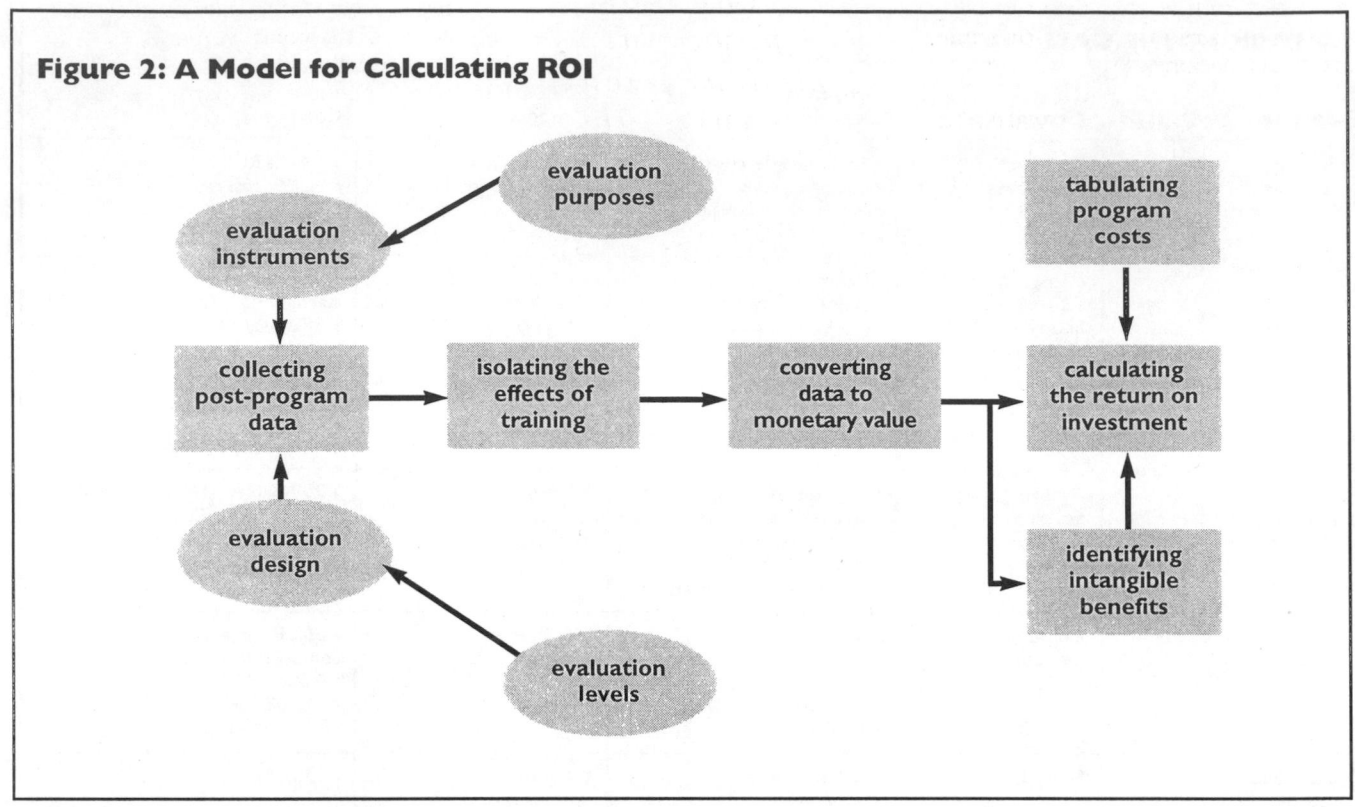

Figure 2: A Model for Calculating ROI

monetary benefits. These companies aren't content to show just that training can result in such improvements as increased productivity and decreased employee turnover. They take the process a step further by converting such improvements to monetary units so that the improvements can be compared to costs and further developed into an ROI calculation.

For such hard-data items as productivity, quality, and time, the conversion to monetary units is relatively easy; soft-data items such as customer satisfaction, employee turnover, and job satisfaction aren't so easy to convert. Still, there are techniques for making the conversions with a reasonable amount of accuracy, and several strategies are used in these case studies. (Note: this important topic is covered in the third article of this series.)

The search continues

Because the variables that can affect performance are numerous and complex, it can be difficult to determine how much of a change is due to training. Most ROI figures aren't precise,

though they tend to be as accurate as many other estimates that organizations routinely make.

After garnering the cases for the ASTD project, there are still unanswered questions about measuring ROI.

Cost standards. The methods used to monitor costs vary widely. What one organization considers a cost of training, another does not. The HR field needs standard cost data. It's becoming increasingly difficult to compare costs from one program to another. Most efforts to solve the problem have failed. In the interim, it's necessary to describe the cost components that make up the total cost of any effort to measure ROI.

Evaluation design. Many organizations don't design their evaluations to isolate the effects of training. Control groups are rarely used even though they can be used effectively without the disruption, problems, and inconvenience usually feared by practitioners. Though a control-group approach is preferable, other evaluation designs such as trend-line analysis, forecasting, and estimations can be useful.

Standard methodology. Evaluation

techniques vary, though there are only so many ways that data can be collected and analyzed. Often, data-collection methods are used without regard to their advantages or disadvantages. The different labels and terminology adds to the confusion. As professionals, we need to standardize and publicize evaluation methods.

Statistics. Many HR practitioners avoid statistics. But statistical analysis can provide a sound basis for conclusions about training results. And though many top managers don't understand statistical analysis, they need to feel confident that any conclusions about training results are supported by appropriate methodology.

In several of the case studies and other evaluation projects, the power of statistics is largely ignored. Sample sizes are so small that the results can't be considered supportable, at least statistically.

Converting data to monetary values. Because of the subjective nature of this process, the results of many HR programs aren't converted to monetary units. Yet, this conversion is an essential step in ROI calculations in which monetary benefits must be compared with costs. It should be a fundamental requirement for some level-4 evaluations.

Evaluate, evaluate, evaluate

In the cases cited in the ASTD project, a variety of evaluation methods were used to determine the success of training.

Follow-up assignments. Perhaps the easiest approach to post-training data collection is to ask participants to complete a task or project, to serve as evidence of their successful application of the training content. Typically, assignment results are reported to participants' supervisors. This approach is especially helpful for level-3 evaluations and when management support isn't strong.

Surveys and questionnaires. These can capture participants' accomplishments and behavioral changes after training. You can collect responses from all participants or a sampling. Surveys and questionnaires are inexpensive, as well as easy to implement and tabulate. They're most appropriate for on-the-job application (level 3) and business results (level 4).

One-on-one interviews. Interviewing participants individually is an excellent way to capture changes in job-related behaviors and to garner specific details. More versatile than questionnaires, interviews can probe issues, concerns, and actions related to the training. They're suited to level-3 evaluations.

Focus groups. An extension of interviewing, focus groups involve collecting post-training information from eight to 12 participants in a structured setting. Focus-group members are asked specific questions about what they have changed as a result of training. The exchange of information often triggers creative thinking among

■ *The most powerful measurement is through the use of action plans* ■

participants, which provides high-quality data. A focus-group follow-up is appropriate for level-3 evaluations.

Observation. Direct on-the-job observation of participants after training can show whether they are applying new skills. This level-3 approach works best when participants are unaware that they're being observed. For example, to measure changes in customer service, an organization can hire shoppers to observe salespeople.

Action planning. The most powerful way to measure training's effect is through the use of action plans. Participants apply their new skills or knowledge in a task or project and then document their progress in achieving measurable objectives outlined in an action plan. Their supervisors may or may not be involved. In some organizations, progress is audited by a training coordinator. Or, participants submit their action plans to the training department to substantiate whether the desired results are attained.

The training should provide a module on how to develop an action plan. Both level-3 and level-4 evaluations can benefit from this approach.

Performance contracts. These contain a pretraining agreement between participants, their supervisors, and sometimes the facilitator. The parties meet prior to training in order to develop measurable goals related to the train-

ing content. Later, they can determine whether the goals were met.

Special follow-ups. It can be effective to reconvene participants one to three months after the initial training segment was conducted so they can report on their success. Follow-up sessions also provide opportunities for additional training, such as refining new skills. This approach is appropriate for level-3 and level-4 evaluations.

Performance tracking. The most credible post-evaluation approach is to track department, work-unit, or individual performance after training is completed in such areas as productivity, quality, cost, and time—and in soft-data areas such as customer satisfaction.

This approach requires examining the organization's overall performance data to obtain before-and-after comparisons of each data item. Because it can provide the most convincing evidence, it's often the preferred approach of senior managers.

The search for the best practices has revealed some important concerns. It's almost universally agreed that more attention regarding ROI is needed. But only a few successful examples of ROI calculation exist. The process isn't as difficult as it may seem. The approaches and techniques can be useful in a variety of settings. Practitioners and researchers must continue to refine the techniques and show successful applications.

In the next issue of *T&D*, the second article in this series will focus on approaches for isolating the effects of training.

Jack J. Phillips *is principal consultant with Performance Resources Organization. He can be reached at Box 1969, Murfreesboro, TN 37111-1969. Phone 615/896-7694, fax 615/896-7181.*

Was it the Training?

● THIS IS THE SECOND ARTICLE (SEE T&D, FEBRUARY 1996) IN A THREE-PART SERIES ON MEASURING TRAINING ROI, BASED ON CASE STUDIES COLLECTED BY THE AMERICAN SOCIETY FOR TRAINING AND DEVELOPMENT. ● EVALUATION EXPERT JACK PHILLIPS DESCRIBES 10 WAYS TO ISOLATE THE EFFECT OF TRAINING SO THAT IT'S CREDITED OVER OTHER VARIABLES AS THE REASON FOR PERFORMANCE IMPROVEMENTS.

BY JACK J. PHILLIPS

IT'S A COMMON SCENARIO. After a major training program, there's a boost in trainees' work performance. Clearly, the two events are linked. But then a manager asks the dreaded question: "How much of the improvement was caused by the training?"

This familiar inquiry is rarely answered with much accuracy or credibility. Performance improvements may be linked to training, but usually nontraining factors have also contributed. As most HR practitioners know, it can be difficult to show a cause-and-effect relationship between training and performance. Up-front planning is essential. This article recommends several approaches for isolating the effect of training, a crucial step in calculating training's return on investment in terms of dollars and cents.

But first, it's important to explain the "chain of effect" implied in the five-level evaluation model shown in Figure 1. To start, it's essential to derive the measurable results of training from participants' application of new skills or knowledge on the job over a specific period of time after training is completed, a level 3 evaluation. Logically, successful on-the-job application of training content should stem from participants having learned new skills or acquired new knowledge, a level 2 evaluation. Consequently, for a business-results improvement (a level 4 evaluation), the chain of effect implies that measurable on-the-job applications (level 3) and improvement in learning (level 2) are achieved. Without this preliminary evidence, it's difficult to isolate the effect of training or to conclude that training is responsible for any performance improvements. Practically speaking, if data is collected on business results (level 4), data should also be collected at the other three levels of evaluation.

Specific approaches

Here are several ways to isolate training's effect on performance.

Use of control groups. A highly credible approach for isolating the effect of training is the use of control groups in an experimental training design. The experimental group receives training; the control group does not. Participants in both groups should be similar demographically, selected at random, and subjected to the same environmental influences.

It isn't necessary to take pre-program measurements of the two groups. Rather, measurements taken after training show the difference in performance between the two groups that can be attributed directly to training.

For example, Federal Express gave 20 new employees training in driving company vans. Their post-training performance was compared with a control group of 20 new employees who hadn't received the special training. The two groups' performance was tracked for 90 days in 10 performance categories, including accidents, injuries, and errors. Experts from engineering, finance, and other groups assigned dollar values to the performance categories. The ultimate outcome was that the training showed a 24 percent return on investment.

A disadvantage of the use of control groups is a misperception that the training staff is turning the workplace into a lab. To avoid this negative image, some organizations conduct a pilot of the training program using pilot participants as the experimental group and nonparticipants as a control group. In fact, the nonparticipants aren't even informed of their status as "the control group."

Sometimes, management may not want to take the time to experiment; it may just want to make sure employees get the training. But using control groups is worthwhile when training programs are costly and linked with organizational objectives.

Trend-line analysis. In this approach, a line is drawn from current performance to future performance, assuming that the current trend will continue even without training. After employees receive training, their post-training performance is compared to their performance predicted on the trend line. It's reasonable to attribute any improvement over the trend-line prediction to training. It's not an exact process, but it does provide a reasonable estimation of training's effect. (See Figure 2.)

In Figure 2, the reject rate for defective components at an electronics manufacturing firm is shown before and after training, with a pre-training downward trend. The training apparently reduced the number of rejects dramatically, though the trend line shows that reject-rate reduction would have continued anyway.

It's tempting to measure the improvement by comparing the average six-month reject rate prior to training (1.85 percent) to the average six-month rate after training (0.7 percent). But a more accurate comparison is to compare the six-month average after training with the trend-line value at the same point (1.45 percent). In this instance, the difference is 0.75 percent. Sometimes, it behooves the training department to use more modest measures to demonstrate the effect of training rather than to make claims that can't be proved.

The main disadvantage of this approach is its potential inaccuracy. A trend-line analysis assumes that the events that influenced performance prior to training still exist after training. It also assumes that no new influences entered the situation. On the positive side, the approach is relatively simple, inexpensive, and effortless.

Forecasting. This approach is more analytical and mathematical than the trend line. Instead of drawing a straight line, a linear equation is used to calculate a value of the anticipated performance improvement. A linear model (such as $y = ax + b$) is appropriate when only one variable influences the results. When several variables intervene, it's necessary to use sophisticated statistical models. Without them, forecasting is difficult to implement. Still, it can be an accurate predictor of performance variables without training, if the appropriate data and models are available.

Participant estimation. This approach involves asking participants to determine how much performance improvement is due to training. Their actions have produced the improvements, so they should have some idea of how much improvement is because they applied what they learned in training. Management tends to find such reports credible because participants are at the center of the improvement.

Participants' input can be obtained by asking the following questions:
‣ What percent of the improvement can be attributed to the application of skills, techniques, or knowledge gained in the training?
‣ What is the basis for your estimation?
‣ What degree of confidence do you have in your estimation?
‣ What other individuals or groups could make an estimation?
‣ What other factors do you think contributed to the improvement?

To be conservative, it's recommended to factor in a confidence level. For example, if a participant estimates that 50 percent of an improvement is due to training but is only 70 percent confident about that estimate, multiply the confidence percentage by the improvement percentage and divide by 100, for a confidence level of 35 percent. Then multiply that figure by the amount of the improvement in order to isolate the portion attributable to training. To calculate ROI, convert that portion to a monetary value.

To enhance this approach, management can approve participants' estimations. For example, in performance-management training at Yellow Freight Systems, participants estimated the amount of savings that could be attributed to the program. Managers reviewed and approved the estimates, confirming participants' estimations.

One disadvantage to this approach is obvious: It's just an estimate. The input data may be unreliable. Some participants aren't comfortable providing estimates; some may not be able to estimate improvements because they don't know which factors contributed. The advantages are that it's inexpensive, time-saving, and easily understood by most participants and others who review evaluation data. And the estimates do originate from a credible source—the people who actually produced the improvement.

Supervisor estimation. Participants' supervisors may provide input in lieu

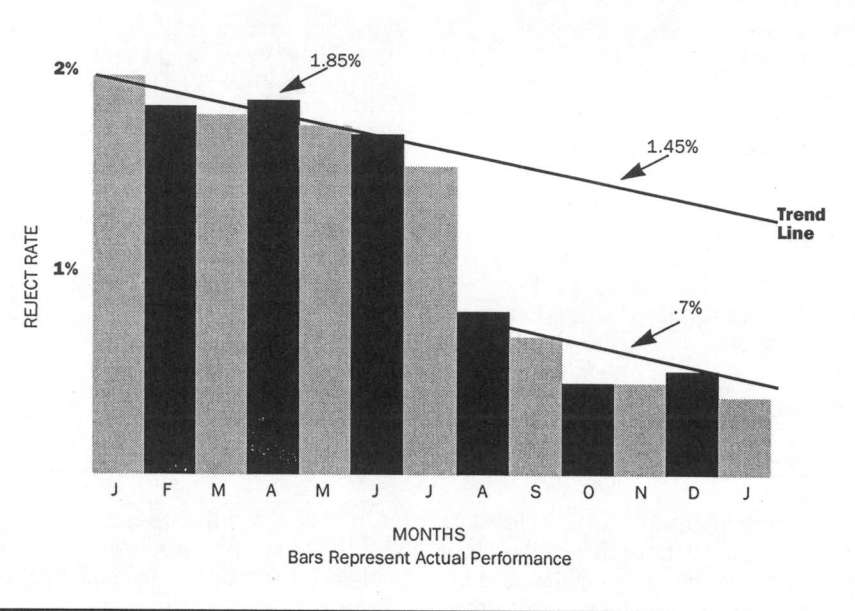

Figure 2: A Trend-Line Analysis

Here's an example of a trend-line analysis, conducted by an electronics manufacturing firm on the rate of rejects for defective parts.

of, or in addition to, participants' estimations. In some settings, participants' supervisors may be more familiar with other factors that could have produced the improvements. It's recommended to ask supervisors the same questions asked of participants.

Supervisor estimation should be treated the same way as participant estimation in summarizing and analyzing the data. The evaluator may not know which estimates to use. A conservative approach is to use the lowest value and include an appropriate explanation. Or, the evaluator can recognize that each source has its own perspective and average the two, placing equal weight on each group's input.

This approach has the same disadvantages as participant estimation. Because it's subjective, it may be viewed skeptically by management. Supervisors may be reluctant to participate. Or, they may be incapable of providing accurate estimates. The advantages are also the same: It's simple, inexpensive, and fairly credible because the information comes from the "horse's mouth"—in this case, the supervisors of people who received

the training. Credibility rises when supervisors' estimates are combined with participants' estimates and when a confidence level is factored in.

A restaurant chain implemented a training program on performance management for manager-trainees. Trainees learned how to establish measurable goals for staff, how to provide performance feedback, how to measure progress toward goals, and how to take action for ensuring that goals are met. Trainees developed action plans for improvement, using the skills taught in the training. The top managers learned how to convert measurable improvements into economic values. They decided employees could focus on any improvement areas (such as inventory, food spoilage, or employee turnover) on the conditions that they use the new skills taught in training and that improvements be converted to either cost savings or profits.

As part of a follow-up evaluation, trainees' action plans were documented to show results in quantitative terms converted to monetary values. Trainees were asked to estimate (conservatively) the percent of improvement that resulted from the application of skills either acquired or enhanced in training. Each improvement was calculated using an annual monetary value. To implement the improvements, trainees worked closely with the restaurant managers, who estimated for the trainees the percent of improvement (outlined on their action plans) that could be attributed to training.

Management estimation. Top-level managers can provide estimates on the percent of improvement they attribute to training. At Litton Guidance and Control Systems, management applied a subjective figure (60 percent) to improvements due to training, after considering other contributing factors such as changes in processes, procedures, and technology. The upshot was that training received credit for 60 percent of the improvements in quality and productivity.

Clearly, this approach can be highly subjective. But then, the input is from people who provide the training funds.

Customer input. Why not elicit input directly from customers? Ask them why they chose a particular product or service. Ask them to explain how their reactions to a product or service were influenced by employees who were using the knowledge and skills taught in training. This approach focuses directly on what training programs are designed to improve.

For example, following a bank's teller-training program, customers indicated in market-research data a 5 percent reduction in customer dissatisfaction with teller knowledge.

Expert estimation. Experts—such as independent consultants and industry sources—must be carefully selected regarding their knowledge of a particular process, program, or situation. For example, an expert in quality can provide fairly reliable estimates of how much quality improvement can be attributed to training—and how much can be attributed to other fac-

tors associated with a TQM effort.

This approach can be inaccurate, unless the new training program and setting are similar to the current program and setting, and the approach may lack credibility because the estimates come from external sources. Still, it's a quick source of input from a reputable source. Right or wrong, management can sometimes place more confidence in external experts than internal staff.

Subordinate input. In some situations, participants' subordinates can provide input on training given to supervisors and other managers on implementing work-unit changes or developing new skills in dealing with employees. Subordinates usually can't estimate how much of an improvement is attributable to training, but they can provide input about specific changes that have occurred since the supervisor received training. And they also can help determine the extent to which other factors have changed.

Subordinate input is usually obtained through surveys or interviews. When the survey responses show significant changes in supervisors' behavior after training and no significant change in the general work climate, improvements in work performance can be attributed to changes in supervisors' behavior.

Typically, subordinates are aware of the factors that have caused changes at work, and they can provide reliable input about the magnitude of such changes. When combined with other approaches, subordinate input is even more credible.

Other factors. In some situations, it's feasible to calculate the effect of factors other than training that may have contributed to some improvement and then to conclude that training accounts for the rest.

For example, a consumer-lending program for a large bank experienced a significant increase in the number of loans after training was provided to loan officers. In addition to the effect of training, other factors included

falling interest rates and loan officers' growing confidence in their knowledge and expertise.

This approach is appropriate when other factors are easily identified and when the necessary mechanisms for calculating their effect are in place. In some cases, it's just as difficult to estimate the effect of factors other than training. This approach is highly credible when the methods used to isolate the effect of other factors is credible.

But which one?

With 10 approaches available, it can be difficult to select the most appropriate one. It's important to consider the following criteria:

- feasibility
- accuracy
- credibility
- costs
- time—including that of participants, managers, and others.

Generally, two approaches are better than one. In using multiple sources, it's recommended to combine the inputs. This conservative approach builds acceptance. The target audience should receive explanations of the approach and the subjective factors.

It's not unusual for the ROI in training to be an extremely large figure. Even when a portion of the improvement is attributed to other factors, the numbers can still be impressive. But it should be understood that ROI figures aren't precise, though every effort is made to isolate training's effect. An ROI figure represents the best estimate given the conditions, time, and resources the organization was willing to commit. Chances are, it's more accurate than other types of data in the organization.

....
Jack J. Phillips *is a principal of Performance Research Organization, Box 1969, Murfreesboro, TN 37111-1969. Phone: 615/896-7694; fax: 615/896-7181. For more information on specific case studies, see* Measuring Return on Investment *(volumes 1 and 2: ASTD, 1994 and 1996).*

How Much Is the Training Worth?

MANY HR PRActitioners consider a training evaluation complete when they can link business results to the program. But for the ultimate level of evaluation—return-on-investment—the process isn't complete until the results have been converted to monetary values and compared with the cost of the program. This shows the true contribution of training.

Here's a basic formula for calculating ROI:

‣ Collect level-4 evaluation data. Ask: Did on-the-job application produce measurable results?
‣ Isolate the effects of training from other factors that may have contributed to the results.
‣ Convert the results to monetary benefits.
‣ Total the costs of training.
‣ Compare the monetary benefits with the costs.

The nonmonetary benefits can be presented as additional—though intangible—evidence of the program's success.

It's useful to divide training results into hard data and soft data. Hard data are the traditional measures of organizational performance. They're objective, easy to measure, and easy to convert to monetary values. Management tends to find hard data highly credible. Hard data is available in most types of organizations, including manufacturing, service, not-for-profit, government, and educational.

Hard data represent the following areas of a work process:
‣ output
‣ quality
‣ time
‣ cost.

For example, a government office that approves applications for visas typically collects data in all four areas to measure overall performance: output (the number of applications processed), quality (the number of errors in processing applications), time (the time it takes to process and approve an application), and cost (for processing each application).

Soft data are needed on training programs that focus on developing such "soft" skills as communication. Typically, soft data—such as employee absenteeism and turnover—are subjective because they have to do with behavior. They're difficult to measure and convert to monetary values. And when compared with hard data, soft data are usually found to be less credible as a performance measure.

This third—and final—article in the series on training ROI shows how to convert program results to monetary benefits. It's easier than you think.

BY JACK J. PHILLIPS

The conversion

Here are five steps for converting either hard or soft data to monetary values.

Step 1: Focus on a single unit. For hard data, identify a particular unit of improvement in output (such as products, services, and sales), quality (often measured in terms of errors, rework, and product defects or rejects), or time (to complete a project or respond to a customer order). A single unit of soft data can be one employee grievance, one case of employee turnover, or a one-point change in the customer-service index.

Step 2: Determine a value for each unit. Place a value on the unit identified in step 1. That's easy for measures of production, quality, time, and cost. Most organizations record the value of one unit of production or the cost of a product defect. But the cost of one employee absence, for example, is difficult to pinpoint.

Step 3: Calculate the change in performance. Determine the performance change after factoring out other potential influences on the training results. This change is the output performance, measured as hard or soft data, that is directly attributable to training.

Step 4: Obtain an annual amount. The industry standard for an annual-performance change is equal to the total change in performance data during one year. Actual benefits may vary over the course of a year or extend past one year.

Step 5: Determine the annual value. The annual value of improvement equals the annual performance change, multiplied by the unit value. Compare the product of this equation to the cost of the program, using this formula: ROI = net annual value of improvement - program cost.

There are several other ways to convert data to monetary values. Some are appropriate for a specific type of data or data category; others are appropriate for any type of data. Here are some options.

Converting output to contribution. When a training program has produced a change in output, the value of the increased output can be determined from accounting or operational records. In for-profit organizations, this value reflects the "profit contribution" of an additional unit of product or service. In not-for-profit organizations, the contribution or value may show in the savings from producing an additional unit of output for the same input.

The calculations for measuring such contributions depend on the organization and its records. Most monitor performance output. If such data aren't available, managers may use marginal-cost statements and sensitivity analyses to pinpoint the values associated with changes in output.

For example, a bank's sales seminar for consumer-loan officers resulted in an increase in the volume of loans (output). To measure the training's return-on-investment, it was necessary to determine the value (profit contribution) of one additional consumer loan—an easy item to calculate from the bank's records.

> ■ *Perhaps the highest cost of poor quality is customer dissatisfaction, which is difficult to quantify* ■

The first step was determining the yield, also available from bank records. The next step was calculating the average spread between the cost of funds and the yield received on a loan. For example, the bank could obtain funds from depositors at 5.5 percent on average, minus the cost of making a loan, including advertising and employees' salaries.

Calculating the cost of quality. The cost of quality is an important measure in most manufacturing and service firms. Because many training programs are designed to improve quality, the HR department must place a tangible value on quality improvement. For some quality measures, that's easy. For example, if quality is measured as a product-defect rate, the value of an improvement is shown in eliminating the cost to repair or replace a defective product.

The most obvious cost of poor quality is the waste generated by mistakes: defective products, spoiled raw materials, and discarded paperwork. Such waste is translatable to monetary values. In addition, employee errors can cause expensive rework. The most costly rework is when a product is delivered to a customer and returned for repair. Maintaining a staff to perform rework is added overhead. In most manufacturing plants, the cost of rework is from 15 to 70 percent of productivity. In most banks, about 35 percent of operating costs are due to rework.

Perhaps the highest cost of poor quality is customer dissatisfaction. It can lead to lost business. Customer dissatisfaction is difficult to quantify. Typically, sales-and-marketing managers and marketing surveys are the best sources for measuring the effects of customer dissatisfaction.

Converting employees' time. Many training programs focus on reducing employees' work time. Employee time is money, including wages and benefits. A training program may enable a team to perform tasks in less time or with fewer members; time management can help individual employees save time. The value of the time saved is an important measure of a program's success, and conversion is relatively easy. The most obvious time savings is the reduced labor costs of performing work. The monetary savings equal the hours saved, multiplied by the per-hour labor cost.

For example, after attending a time-management training program, participants estimated that they now save an average of 74 minutes per day, worth $31.25 per day or $7,500 per year in labor. This time savings is based on the participants' average salary, plus benefits.

Generally, the average wage (with a percent added for employee benefits) is sufficient for most ROI calculations. But some employees' time is worth more. Some experts recommend that "employee maintenance" costs other than employee benefits be figured into the average labor cost per employee, including such items as office space, furniture, telephone, utilities, computers, calculators, and administrative support. Then, the average wage rate may rise. The most conservative approach is to use salary plus employee benefits.

In addition to a reduced labor cost, other benefits can result from a time savings, including improved service, avoided penalties, and added opportunities for profit.

A word of caution: Time savings are realized only when the amount of time saved translates to a cost reduction or profit contribution. The time saved must be used productively.

Using historic costs. Sometimes a company's records will show the cost and value of one unit of improvement. It's necessary to identify the appropriate records and tabulate the actual cost of items in question. Historic data are usually available for hard data and some selected soft data.

For example, a training program for improving safety performance used various measures for all safety-related items, including the accident-frequency rate and the total cost of workers' compensation. By examining the company's records and using a year of data, the HR department was able to calculate the average cost of each safety measure.

Using internal and external experts. When converting soft data without historic records, it's recommended to consider input from experts on the processes involved. They can provide the cost (or value) of one unit of improvement. They tend to be close to the situation and to have earned management's respect. When internal experts aren't available, external experts can fill in the gap. Most experts use their own approaches, so it's best to explain specifically what's needed. They should understand the processes and be willing to provide estimates, with explanations.

In one organization, a training program for reducing the number of employee grievances ended in soft data, to be monitored by the organization. Except for one instance of reimbursed back pay, the organization had no records on the costs of grievances (such as, the cost of external assistance or the time involved in working with a complainant). An expert had to estimate—in this case, the manager of the labor-relations department. He based his estimate on his perception of the average settlement

HARD AND SOFT DATA

Here are some examples of hard and soft data.

HARD

Output
- units produced
- items assembled or sold
- forms processed
- tasks completed

Quality
- scrap
- waste
- rework
- product defects or rejects

Time
- equipment downtime
- employee overtime
- time to complete projects
- training time

Cost
- overhead
- variable costs
- accident costs
- sales expenses

SOFT

Work Habits
- employee absenteeism
- tardiness
- visits to the dispensary
- safety-rule violations

Work Climate
- employee grievances
- employee turnover
- discrimination charges
- job satisfaction

Attitudes
- employee loyalty
- employees' self-confidence
- employees' perceptions of job responsibilities
- perceived changes in performance

New Skills
- decisions made
- problems solved
- conflicts avoided
- frequency in use of new skills

Development and Advancement
- number of promotions or pay increases
- number of training programs attended
- requests for transfer
- performance-appraisal ratings

Initiative
- implementation of new ideas
- successful completion of projects
- number of employee suggestions

when a grievance is lost, including such costs as arbitration and legal fees. He also factored in an estimated amount of time spent by supervisors, staff, and employees associated with the grievance. This internal estimate, though imprecise, was appropriate for the analysis. And management found it credible.

Using data from external studies. For some soft data, it may be appropriate to use research to estimate the value. It's fortunate that many databases contain studies on the costs of various items related to training, in-

cluding employee turnover, absenteeism, and grievances, as well as safety and customer satisfaction. Ideally, the data should come from a similar setting in the same industry.

For example, the evaluation of an HR program for reducing the turnover of branch managers in a financial-services company included the cost of employee turnover, including recruitment, orientation, and training for a new manager, as well as the costs of severance and unemployment pay for an exiting manager. Many HR practition-

■ *Time savings are realized only when used productively* ■

ers don't want to calculate the cost of turnover, particularly when it's needed just for a one-time event, such as a training evaluation. In the example, the cost was determined (based on industry standards) to be about one-and-a-half times the average annual salary of an employee, adjusted for the average base salary of a branch manager.

Using participants' estimates. Sometimes, the people closest to an improvement can provide the most reliable estimates on its value. Training participants can estimate the value of a soft-data improvement they've made by applying the skills they learned in training.

For example, to calculate ROI on a supervisory training program on lowering the rate of employee absenteeism, it was necessary to determine the average value of one absence, without the benefit of historic records. During the training, participants estimated the cost of an absence, based on their personal experience. Next, supervisors were asked to estimate the average cost of an absence in their work units, based on how an employee's absence is compensated. Then, all of the estimated values were averaged.

Using supervisors' estimates. Participants' supervisors are another source for determining the value of a unit of improvement due to training. For example, after completion of a training program for managers at Yellow Freight Systems, participants estimated the value of the improvements directly related to the training. Their managers also provided estimates after reviewing the process by which the participants had created their estimates. Then, the managers either confirmed or adjusted participants' values.

Using senior managers' estimates. Senior managers can place a value on

an improvement, based on their perception of its worth, when it's too difficult to calculate the value or when other sources for estimates are unavailable or unreliable.

Using HR's estimates. This approach may be perceived as biased. After all, the HR department will determine the basis for its claim for improvements due to training. For example, in a training program for dispatchers at an oil company, the HR department estimated the cost of one employee absence to be $200. Then, it used that value to calculate the savings due to training on reducing the absenteeism rate.

Raising credibility

The conversion approaches assume that the data items can be converted to monetary values. Highly subjective soft data—such as a change in employee morale—are difficult to convert. The key question is: "Would I be comfortable presenting these results to senior management?" If the results don't meet this test, they shouldn't be converted to dollars and cents. Instead, they should be presented as intangible benefits.

When reporting training results, credibility is always an issue. It's crucial that the data be accurate and that the conversion process be believable. Many HR practitioners are more comfortable reporting that training resulted in a reduction in employee absenteeism from 6 percent to 4 percent, without placing a monetary value on

> ■ *It's crucial that the data be accurate and the conversion process be believable* ■

the improvement. They assume that the people receiving the information will assign their own values. Unfortunately, those people may know little about the cost of absenteeism. Or, they may underestimate the actual value. That's why accurate ROI is important.

Less-than-precise estimates, assumptions, forecasts, and external data may make some HR practitioners hesitant to conduct conversion. But they can raise credibility by following these guidelines:

▶ Take a conservative approach when making estimates and assumptions.
▶ Use the most credible and reliable sources for estimates.
▶ Explain the approaches and assumptions used in the conversion.
▶ When results appear overstated, consider adjusting the numbers to achieve more realistic values.
▶ Use hard data whenever possible.

With soft data, senior management may adjust the results so that they're more linear and concrete. Or, they may adjust the results to reflect the time value of money because most investments in training are made at one time and the return is realized at a later time. Such adjustments are usually negligible compared with the benefits.

Many organizations are trying to become more aggressive in determining the monetary benefits of training. They're no longer satisfied just to report business results. Instead, they're converting business results to monetary values and comparing them with the cost of training to obtain the true return-on-investment—and the financial contributions of HR.

Jack Phillips *is a principal of Performance Resources Organization, Box 1969, Murfreesboro, TN 37111-1969. Phone 615/896-7694; fax 615/896-7181.*

Return on Investment: Accounting for Training

By Anthony P. Carnevale and Eric R. Schulz

Economic Accountability for Training: Demands and Responses—growing demand for accounting for training...advances in accounting for training... plans for moving forward.

Strategic Accounting—what's wrong with traditional management accounting systems...why and how they should be changed for costing human resources.

The Consensus Accounting Model—a new structure for accounting for training.

Evaluation Framework, Design, and Reports—the Kirkpatrick Model...front-end analyses...tying learning and performance objectives to organizational objectives...marketing training to management... evaluation steps and phases...evaluation design and data-collection tools.

Evaluation Practices—overview of trends...profiles of current practices in major U.S. organizations.

Advisors.

References.

Economic Accountability for Training: Demands and Responses

Accounting for the positive economic influence of training and development is the most critical issue in the training profession today. Business leaders realize how large the investment in training and development has become. So inevitably, questions about economic value arise: which training and development activities work? And at what cost?

Pressed to address those questions, many training and development professionals find themselves struggling to meet the expectations of managers and employees who want more training—and proof that particular training programs are worthwhile.

Despite the growing demand for accountability, financial accounting for training shows only a slight increase. As a rule, although training and development are undergoing more financial analysis, they are accounted for less than any other major corporate investment. For instance, a 1988 ASTD poll of organizations that led in training evaluation found that only 20 percent evaluated in terms of training's economic effect on the organization. In other words, when it comes to investments in training and development, subjective decisions prevail.

Many training professionals contend that accounting for training (through measurement and evaluation) takes too much time or is too costly. But, ASTD research has revealed organizations that account for training in flexible, practical ways, using relatively simple and inexpensive means. The training professionals in those organizations understand that, in the current business climate, accounting for training is essential to success and, sometimes, to survival.

At times economic conditions demand measurement and evaluation of all business functions, or top management may require financial justification for a training department's budget. But more often, and most important, are the routine judgments that top managers make about training's worth, using whatever information is available. Whether they say so or not, top managers constantly evaluate training efforts and assign a value to them.

If realistic information isn't at hand, these decision makers may draw arbitrary, inaccurate conclusions. Their view of training's worth may be based on its word-of-mouth reputation for past efforts or on their perceptions about training personnel. In the past, many training departments thrived on the basis of excellent reputa-

Examples of Corporate Training Expenditures

Corporation	Total Training Expenditure*	Percentage of Payroll Costs
IBM	$750,000,000	5.0%
General Electric	$260,000,000	2.0%
Xerox	$257,000,000	4.0%
Texas Instruments	$ 45,000,000	3.5%
Motorola	$ 42,000,000	2.6%
Honeywell	$ 30,000,000	3.0%

* Not including training participants' salaries. If salaries were included, the figures would roughly double.

Source: American Society for Training and Development, 1987.

tions. But during the economic downturns and downsizing of the 1980s, a reputation unsubstantiated by data often proved inadequate evidence of a department's contribution and worth.

Human resource development (HRD) professionals reluctant to account for training need to reorient their thinking to face the business realities of the nineties. Instead of deciding *whether* to measure and evaluate training's results, they must decide *how* to determine its costs and benefits.

Advances in accounting for training

Many HRD professionals have discovered that accounting for training doesn't have to be cumbersome and doesn't necessarily lead to criticism. Trainers and training participants are usually more accepting of evaluation when its purpose is clear: evaluation information signals whether a training program is improving participant or organizational performance to an extent that's worth the investment.

Analysis of performance data may indicate that, to help improve participants' on-the-job performance and promote achievement of business goals, training program components need to be changed. Other aspects of performance management—personnel selection and compensation, tools and job aids, and so on—may also require adjustment.

These days, HRD professionals have more performance measurement information available than ever before. Fortunately, computer software has made information collection, analysis, and retrieval easier and more accurate.

Sound human resource management is embedded in

Carnevale *is chief economist and vice-president of national affairs for the American Society for Training and Development. He was chief investigator for the joint ASTD/Department of Labor research project, "Best Practices: What Works in Training and Development" (DOL Grant No. 99-6-0705-75-079-02).* **Schulz** *was the researcher for the accounting and evaluation portions of the project.*

an organization's strategic change process. When an organizational issue or need comes to light—such as changes in products, technologies, or competition—decision makers will consider whether and how human resources might help. And they will determine what support, possibly including training and development, people will require.

Basic considerations include the following: in what ways and to what extent can training and development help resolve this organizational concern? If training accomplishes as much as possible toward resolving this concern, can the "top dollar" worth of its contribution be estimated? What else, if anything, might substitute for all or part of proposed training? What is liable to happen if training isn't provided?

One justification for training and development is compelling need. Today, because line managers control resources and are well-positioned to perceive training needs, they initiate most employee training in most organizations. Trainers cooperate closely with line managers to identify training problems and propose solutions—because, although training has its costs, the costs of *not* training may be considerably higher.

At its best, evaluation is inherent in all the phases of an organization's instructional systems development (ISD) process. If training is subject to "continuous improvement" and refinement from front-end problem analysis onward, it has a powerful bias toward success. Ongoing evaluation and corresponding de-emphasis of after-the-fact evaluation suit the fast pace of today's business environment, where training needs can emerge—and opportunities pass—quickly.

Trainers cooperate closely with line managers to identify training problems and propose solutions because, although training has its costs, the costs of not training may be considerably higher.

In fact, use of state-of-the-art instructional design technologies, rather than post-training evaluation, is the best way to assure training's effectiveness. When appropriate techniques are employed, effectiveness can be built in as training proceeds. Good training begins with a needs analysis that measures performance gaps and tailors learning objectives to performance objectives. The more precise and clear the needs analysis, the more likely the training will be appropriate and achieve valuable outcomes.

Understandably, different organizations don't place equal emphasis on measurement and evaluation. Small and medium-size organizations with informal training programs tend to devote less time and fewer resources

to those activities than large organizations do. But whatever its size, an organization is more likely to account for training programs that affect the organization broadly and require significant investment.

The most precise means for evaluating human resource events descend from the Planning-Programming-Budgeting Evaluation System (PPBS) introduced in the Department of Defense and other federal agencies in the 1960s. But few training departments ever require such sophisticated and expensive accounting and evaluation. Technically precise, entirely objective evaluation simply isn't feasible for many training programs. And elaborate evaluation isn't cost-effective when it's obvious that positive changes resulted from training, when a program's backers have little interest in an evaluation, or when the cost of evaluation would clearly outstrip possible benefits.

Yet more and more organizational leaders want and demand measures that clearly show training's contributions to accomplishment of business objectives. So the watchword of modern accounting for training is *appropriate*—rather than technically best—measurement and evaluation. Up-to-date organizations use only as much and as complex measurement and evaluation as is necessary. And their training and development efforts are planned and assessed in the broader context of human resource and general business strategies.

Advances in accounting for training were long stymied by the belief that it was imperative to use only quantitative data. Today most organizational leaders also consider qualitative data useful. Choices about data collection now are seen as depending on the organization's information needs, purpose for evaluation, and available resources.

Current practices

■ Polaroid's management has decided to undertake training only if it will affect the bottom line and to evaluate training programs using the same standards and units that local managers use as performance standards.

■ Upjohn is pioneering a concept of measuring return on investment in human resources based on a total performance-management system that considers business strategies, organization structure, and job design as well as training.

■ Vulcan Materials draws up detailed cost proposals for top management on all training projects.

■ Arthur Andersen bases many of its training investment decisions on a quantitative analysis and needs-determination process that precedes all course development.

■ AT&T's individual business units often initiate training and usually contract corporate trainers to work with them to identify needs.

■ Chase Manhattan Bank's training staff works directly with line and staff managers; virtually all training investment decisions are shared.

■ Aetna's line management allocates about 90 percent of training money to the continuation of successful training programs and planned company training strategies.

Moving forward

Wider dissemination of knowledge is needed. Most

Economic Accountability

HRD practitioners agree that the basic know-how for accounting for training exists, although its application falls considerably short of potential. Many well-publicized measurement and evaluation methods often are prohibitively expensive and time consuming. Meanwhile, the growing number of more convenient, exemplary evaluation practices have, for the most part, remained unknown outside the organizations that use them.

This *Journal* supplement brings attention to some of these practices. It also presents a new accounting model that will allow organizations to begin setting financial benchmarks for training within their institutions.

Accounting for training would benefit from more comparative data about training costs. Rules of thumb exist, but more guidelines and benchmarks are needed; research and analysis hold the promise of discovering or creating them. HRD professionals need to share and keep up with future advances in accounting for training.

Strategic Accounting

Human resources are an increasingly important factor in the economy. Investments in human capital are already key to improvements in productivity, wages, and national income. And the role of human capital will continue to expand as the economy becomes dominated by service-oriented jobs requiring extensive knowledge and training. At the same time, jobs in manufacturing will be highly skilled and vital for maintaining the operating efficiency of manufacturing technology.

HRD's heightened importance must be viewed in light of the labor pool preparing to enter the workforce: millions of potential employees are unequipped with basic workplace skills. For these people to be constructively assimilated in the economy, employers will have to intervene and provide training. What's more, the current workforce will require continual skills upgrading to keep pace with technological advances. But, partly because of traditional tax incentives and managerial accounting systems, organizational expenditures for human capital lag behind investments in physical and equipment capital.

A strategic element that today's managerial accounting statements lack, but that decision makers need, is financial information about human resources. Conventional accounting systems don't provide adequate data for decision making and planning about human resource use. And they don't provide feedback to permit evaluation of organizational effectiveness in using human resources. Yet accounting information is what's reported to upper management about an organization's overall performance.

An organization's management accounting system acts as a two-way communications device for upper and middle management because it lists important organizational and departmental goals. Economic indicators measured by the accounting system also serve as the basis for promotion of middle managers. So information that appears on management accounting reports has a strong influence on management behavior.

When accounting systems don't feature performance reports on effectiveness in managing people, it's only to be expected that managers will concentrate on the aspects of their jobs for which they *are* held accountable. This encourages managers to reduce or eliminate training expenditures and sacrifice long-term profit gains in favor short-term cost cutting. Under current management accounting standards, the economic impact of such mismanagement is not assessed.

An economy based largely on the knowledge and skills of human capital has important implications for the role of HRD professionals in organizations. They have come under pressure to become full business partners who make money for their organizations. This new role requires HRD professionals to think, speak, and operate more in economic and financial terms.

At present, training activities typically are evaluated in other terms—such as participants' reactions to training or supervisors' observation of participants' post-training, on-the-job behavior. But it has become necessary to account for and evaluate training activities in terms that assess the *value* of investment in employees.

Most American organizations have published statements that extol the importance of human resources. But where managerial accounting systems fail to look at the tradeoffs of training costs and benefits, HRD is likely to be treated as a secondary activity.

Antiquation of management accounting systems

Corporate management accounting systems are antiquated for the modern work environment. Despite the evolution of product and process technologies, management accounting systems have remained essentially unchanged for more than 50 years.

In part, this stagnation results from the integration of organizational balance sheets and income statements. A balance sheet represents the organization's total assets, debts, and net worth. Nobel prize-winning economist Paul Samuelson has likened a balance sheet to a snapshot of water at the end of a tub; it shows how much is there at the moment, but not whether or how much is flowing in or out. An income statement represents the result of operations (profit/loss) for a specific accounting period.

Integrating these two reports requires their foundation on the same financial transactions. If they were not integrated with income statements, balance sheets wouldn't emphasize short-term profit goals as much. As matters stand, investments in product development or human capital are discouraged because, as a rule, their benefits flow into the organization over longer periods than monthly, quarterly, or annual accounting reports consider.

Wall Street's emphasis on short-term earning targets and other corporate pressures have lead management cost accounting systems to focus narrowly on monthly earning reports. These reports—based largely on the distribution of manufacturing costs between goods sold and inventory in stock—don't represent the actual increase or decrease in an organization's economic value during the accounting period.

For example, modern just-in-time inventory systems significantly reduce inventory. Organizations that implement just-in-time manufacturing without upgrading their accounting systems leave managers without timely information for measuring product costs and promoting operating efficiency. Besides, even if the time lag in reporting operating costs were overcome, most reports now have too many important cost components hidden in summary figures to be of much benefit to production supervisors.

Because cash outlays for training are long-term investments, they distort an organization's profit measurements over short-term periods. So short-term reports obscure a manager's view of true value-creating activities, such as investment in human capital. Managers may be reassured by an incomplete picture and not realize that accounting systems aren't providing appropriate measures of operational growth or decline.

As B. Charles Ames, chairman and CEO of Uniroyal Goodrich Tire Company, and James D. Hlavacek, training consultant and professor of management at Wake Forest University, have noted, managers may "tighten the belt in the wrong way in the wrong places" if they simply inform their decisions with data "from accounting systems designed primarily to meet outside financial reporting requirements." In depicting a "cycle of competitive decay," Ames and Hlavacek show training as a competitive factor that may suffer inadequate investment because of inadequate accounting systems.

In many organizations, recent strategies—such as automation, quality improvements, reduced inventory, more efficient production processes—for replacing people with machines and minimizing waste have done almost all they can to reduce costs. The conventional savings strategy of reducing direct labor costs (wages, salaries, mandated employer contributions to Social Security, and so forth) by cutting back on employees no longer works as well.

Consider a manufacturing unit that once consisted of four employees who were replaced by a programmable controller and a robotic "eye" and arm.

The unit now has no direct labor costs, but people are still needed—to decide when to change what the unit makes, design its products, program and reprogram the controller, maintain the controller and robots, market the products, review legal documents such as service agreements for the unit's machinery, keep records of product sales, and train and retrain maintenance technicians and others—in support of the unit's operation. But now the costs associated with human resources are categorized as *indirect* costs (also known as overhead; see page S-10 for definitions).

The actual processing of services and products is increasingly a function of such overhead human resource activities, as the Manufacturing Studies Board of the National Research Council noted in a 1986 study.

Office automation has the same effect. One administrative assistant using a personal computer may do work that used to be done by four people: an office manager, secretary, clerk typist, and bookkeeper. But the administrative assistant needs the support of training on how to use software, repair computers, and so on.

Traditional managerial accounting systems' emphasis on direct labor costs is outdated. At the height of the manufacturing economy, direct labor costs far exceeded indirect costs. Overhead costs were distributed (allocated) throughout organizations by requiring managers to multiply their department or division's direct labor costs by a percentage. Accounting systems' use of direct labor costs as the means of distributing overhead costs to products, services, and departments reflected direct labor costs' predominance then.

But workplace automation has escalated overhead costs while greatly diminishing direct labor as a percentage of total costs. A June 1988 *Business Week* article stated that, in automated factories, direct labor typically represents 8 to 12 percent of production costs. In the electronics industry, the percentages are halved. Accounting systems haven't adapted to this major change.

Considerable management time is still devoted to recording and reducing direct labor costs, although these are a small fraction of overhead costs. According to H.T. Johnson of Pacific Lutheran University and R.S. Kaplan of Harvard Business School and Carnegie-Mellon University, overhead burden rates on direct labor ranged from 400 to 1,000 percent in the late 1980s. Obviously, any

Strategic Accounting

activity involving large amounts of direct labor costs appears expensive—and saving on direct labor costs has significant impact on cost records, if not actual costs.

Allocation of overhead costs to departments and products by direct labor also distorts product costs. Products made with low labor content have their overhead costs placed on products with high direct labor hours. Customized, infrequently produced products incur few direct labor hours, but create significant overhead costs for specialized design, engineering, and marketing. So these products appear less costly in comparison with high-volume mature, stable products. In short, in a direct labor cost-allocation system, mature products subsidize customized products.

A direct labor cost-allocation system also promotes decisions to "buy" rather than "make" labor. Managers can reduce direct labor costs by finding suppliers of cheaper labor. So corporate accounting systems favor subcontracting work to people outside the organization ("buy decisions") over assigning work to people in-house ("make decisions"). Buy decisions may defeat the purpose of reducing organizational costs, though, because overhead costs tend to rise as subcontracting does. For example, subcontracting places demands on the departments (such as purchasing, scheduling, and training) that specify product or service requirements for the subcontractor. Yet these overhead costs aren't traced to the practice of subcontracting because it has zero direct labor content.

Ability to assess, with reasonable accuracy, the overhead human resource costs of a product or service would bring a new order of management of human resource investments.

A focus on direct labor costs prevents organizations from getting a good financial management grasp on human resource costs. Overhead costs are the most rapidly increasing human resource costs in organizations. Because the impact of overhead costs is often underrated, few managements understand the economic impact of human resource elements on the profitability of their products or organization.

Management accounting systems must be altered to reflect the growing importance of overhead and equipment costs, and the diminished importance of direct labor costs.

Potentially most productive now are structural organizational changes such as more efficient communication systems and better worker management. Intangible benefits that stem from structural change, advanced technology, and training—such as design and process

flexibility and more knowledgeable and skilled employees to speed turnaround time—have become crucial to organizational competitiveness. But current cost accounting systems don't deal with intangible benefits, so they are rarely measured or estimated and factored into cost management. Managerial accounting systems must begin to consider such factors.

The National Association of Accountants, Harvard and Stanford business school representatives, several of the nation's largest accounting firms, and dozens of corporate sponsors have joined in a cost-management task force to recommend changes to help accounting "catch up" with computer-aided manufacturing.

The task force's first report concluded that, for sound investment decisions, qualitative factors (such as quality, flexibility, and timeliness) are more important than quantitative factors, although those should be measured. Having hammered out a new philosophy of accounting, the task force moves into the 1990s with plans to release new accounting software in keeping with the thinking behind the first report.

These days, as *Business Week* has noted, time is the "most precious commodity." This has many implications and effects. For instance, one Cleveland manufacturer no longer measures an employee's "pieces per hour." Now "throughput" (time to turn material into product) is the emphasis, so the company calculates how long each subprocess (including those in the overhead category) takes and how much each adds to product cost.

Ability to assess, with reasonable accuracy, the overhead human resource costs of a product or service would bring a new order of management of human resource investments. Return on investments in human capital could be improved through training and other personnel interventions. The human resource component of operational finances is poorly managed now because it isn't counted or measured well enough to allow for its true control.

Global competition and accounting systems

The obsolescence of management accounting systems is particularly damaging to American organizations competing in the global arena. In some cases, a foreign manufacturer may produce products—at significantly lower prices—for direct competition with an American organization's high-volume mature products.

Meanwhile, the American organization's cost accounting system leads its decision makers to conclude that their organization can't make money if it matches the foreign competitor's lower prices. This conclusion has driven many American companies to abandon product lines or move production of mature, stable products to low wage countries.

The December 25/January 1, 1990, issue of *U.S. News & World Report* reported that the United States, after a decade of restructuring, "now has an average cost advantage of about 20 percent." But Martin Starr of Columbia University, in a study that compared American-owned companies with foreign-owned U.S. operations, found that "Japanese and European managers spend three to five times as much on worker training."

Computer-integrated manufacturing has led the revolution of improvements in quality, inventory reduction, reduced set-up time, and product customization. The new technologies of computer-integrated manufacturing allow factories to change rapidly from one product to another, driving down economies of scale for production processes. That is, it may cost the same (or almost the same) to produce a few widgets as it does to make thousands.

Product life cycles are also shrinking rapidly, especially in high-tech markets, where a generation of technology may become outmoded in three years. Traditional management accounting systems also lose relevance as more costs—for research and development, physical investment, and training—must be incurred before production begins.

In response to the fast-paced competitive environment, many organizations have hastened to increase the number of products and services they offer, making it harder to attach inputs of resources (costs) to outputs (products and services).

Changes required in management accounting systems

New manufacturing and office technologies call for new cost accounting procedures to deal with such matters as measurement and justification of investments in employees. Data about such managerial considerations shouldn't necessarily be used for external financial reports, but it's critical that they be accounted for in internal management reports.

Organizations must understand the full costs of acquiring and developing resources: technologies, equipment, materials, and people. Organizations must also be aware of the long-term costs of translating those resources into final products or services. Management accounting systems that fail to provide measures of and warning signals about the efficiency and profitability of products and services undermine managers' ability to guide their organizations.

For operational control in the contemporary work environment, managers need accounting systems that provide information on important resources *during* an accounting period. And, to assess progress toward long-term profitability goals, greater use of nonfinancial indicators (such as more complimentary or fewer complaint letters from customers) is required.

Traditional accounting methods treat people only as expenses, so funds used to train people are computed as expenses when an organization's net income is figured. Accordingly, managers tend to regard human resources as expenses to be minimized instead of assets to be optimized.

Human resource management accounting is the next step for organizations progressively adopting a human resource management perspective. Human resource accounting would enable organizations to quantify the worth of people as organizational assets. Human resource accounting systems would strengthen the human resource professional's role as advisor to senior management on the human resource implications of business strategies. By measuring the expected worth of proposed investments in human capital, human resource

accounting also would facilitate management decisions about training.

Costing human resources

HRD has expense and asset components. For a human resource expenditure to be treated as an asset, it must return benefits to the organization in future accounting periods. If the benefits of training or development all take place during the current accounting period, the expenditure is treated as an expense.

There are no generally accepted accounting procedures for valuation of human assets—employees. Valuation of employees differs from valuation of things because people are not owned. But, like other assets, people have future usefulness that adds value to an organization.

The first attempt at implementing employee valuation came from the R.G. Barry Corporation. The aim was to improve planning, management, and investments in human resources. Training and development costs were accumulated in individual subsidiary accounts. Costs were amortized (written off gradually) over a person's expected term of employment or over the time a training program's effects were expected to have worth. If an employee left the organization before the end of the expected working-life estimate, unamortized costs were written off during the quarterly earnings period of the employee's departure. Quarterly accounting reports monitored managers' investments in employees and motivated managers to view human resources as valuable assets.

The Barry system employed historical costs (that is, original expenses incurred) for employee valuation. This method follows an asset model of accounting that measures the costs organizations sacrifice to develop people. The historical cost accounting approach has the advantage of helping managers understand that investments in human resources are parallel to investments in other organizational resources such as equipment.

One difficulty with using this approach for human resource accounting is that writing off unamortized costs based on turnover involves a great deal of subjectivity. It's also difficult to pinpoint to what extent organizational investments in an employee should be attributed to and written off for recruitment costs versus orientation costs or training and development costs. And, this approach only accounts for costs, not for an employee's worth to the organization.

An alternative to the historical costs method is to measure the cost of replacing employees. Replacement costs refer to the expenditure of organizational resources that would go to replacing current employees. Replacement costs include recruitment and training costs for new employees, and income not gained because newcomers are in training rather than producing on the job.

The major drawback of both the historical and replacement cost models of human resource accounting is their limited focus. They highlight investments in human resources, but ignore human resource effectiveness. They fail to gather and assess information about the economic effects of employees' behavior.

A better approach is to tie dollar estimates to positive

Strategic Accounting

changes in employee behavior that were produced by training interventions. This expense model measures the economic consequence of training programs in dollar-expense terms. The idea behind this way of "costing" human resource behavior is to measure the contribution of employees to overall organizational efficiency—while recognizing that an employee's contribution isn't dependent on the size of organizational investments in him or her, but relates to how effective and efficient the employee's on-the-job performance is.

This cost accounting strategy is different because it quantifies the benefits that training and development programs bring to employee performance. As in all frameworks for financial analysis, anticipated cost-benefit ratios are determined and applied. Calculating training programs' costs and benefits requires an understanding of how accountants categorize them.

Direct costs are expenses associated with costs that can be traced directly to specific projects or activities. Out-of-pocket direct costs are expenses for which money is paid on a specific project. In terms of training, these include travel fees and daily expense allowances (per diems), costs for purchased learning materials, contracted consultants, training room rental, and food service.

Out-of-pocket direct costs are the most obvious and easily tracked costs associated with training. But, according to Lyle M. Spencer, Jr., of McBer and Company, these expenses rarely equal more than 10 percent of a training program's total costs. The major costs of training activities relate to people's time—to salary costs for people conducting or participating in a specific training program.

Indirect costs are expenses that can't be directly associated with a specific project or activity but which are necessary for the organization to function. Sometimes the term *overhead* is used to describe all the indirect costs of doing business.

Examples include costs for interest on organizational debt, general building maintenance and repair, lights, heat, office equipment, and administrative salaries and expenses (for example, for a main receptionist or a legal staff). Some organizations subdivide such costs into overhead and general and administrative expenses (G&A) categories, and some calculate overhead on bases other than direct labor.

Fringe benefits are overhead costs related to time for which employees are paid but don't work (vacations, sick leave, and holidays) plus employer payments for health insurance, pensions, and other indirect compensation. Spencer states that, in American industry in the late 1980s, fringe benefits averaged 35 percent of direct salary costs. And in professional service firms, overhead averaged around 115 percent of direct salary plus fringe expenses.

Full costs are the total of direct costs plus indirect costs. Full costs are the best measure of how much it actually costs an organization to deliver a training service. In particular, recognition of the full cost of people's time is the basis for understanding the total costs of training programs.

It's useful to track training's full costs according to eight phases: administration, research and development, analysis, design, development, delivery, evaluation, and marketing.

Costs for each training phase can be subdivided into:
- Personnel costs—for people involved in a training project including in-house subject-matter experts and outside personnel's fees and expenses
- Outside purchase of goods and services—for materials and supplies bought from an outside provider for a specific training program
- Facilities costs—for the use of rental facilities such as classrooms, research and development laboratories, or production shops
- Incidental expenses—for travel and daily expense allowances during a training program
- General and administrative costs—for costs that, although associated with maintaining the training department, can't be directly traced to a particular training program. Such costs include general supplies and materials, equipment, facilities, and administrative and staff support salaries, wages, and fringe benefits.

Benefits of training programs:
- Increased revenue. By affecting quantity of output or sales per unit of time, training-based improvements can increase revenue. Increased output or sales can be documented and training's share in the increase claimed.
- Decreased or avoided expenses. A frequent benefit of training programs is the reduction (saving) or avoidance of costs. By ensuring employees' skills, training can help improve the quality of a product or service. Measurement of the related organizational benefit relate to reduction of scrap, absenteeism, inaccuracy, grievances, accidents, and wasted time or materials.
- Intangible benefits. Intangible benefits are activities, qualities, or conditions that have value but are extremely difficult or impossible to quantify. For instance, employee flexibility benefits an organization, but its worth is difficult to quantify. To keep investment in these benefits in perspective, decision makers should consider the potential risk of not investing in them and should estimate how substantial intangible benefits might possibly be. And, a brief narrative about anticipated intangible benefits (and indicators of them) may add meaning to the "hard numbers" of internal financial reports.

The Consensus Accounting Model

Accounting measures the economic track record of organizational activities and functions. Managers have long recognized that a standardized accounting model for training would facilitate decision making and enhance training department effectiveness. But until now, no standardized model for accounting for training has been widely accepted.

Now there's a new standardized accounting model for training. The model represents the consensus of training and accounting experts (see page S-30) who contributed ideas to a research project underwritten by a grant from the U.S. Department of Labor and conducted under the auspices of ASTD.

This consensus accounting model ties the procedures of existing accounting practices to the desired outcomes sought by management in a specific organization. The model's four steps support strategic accounting. For example, using the model could help a manager determine the percentage of a department's resources spent for training, which departments or individuals require training, and important training considerations that should influence future budgets.

Step 1: Establish an organization-specific definition of training

Accounting for training begins when decision makers in an organization reach agreement about what training is. For purposes of this discussion, training will be defined as "a structured program with identified objectives and learning plans to improve the knowledge, skills, and attitudes of trainees for use in their current and future job assignments." According to that definition, a consultant-provided program in which new employees learn how to use an organization's computer software is training. But an executive meeting convened to introduce a new corporate product or a performance appraisal to set employee work objectives is not.

This definition of training encompasses the following activities:
■ formal training courses offered by the organization or by outside training providers;
■ structured on-the-job training conducted by an employee's immediate supervisor or a qualified substitute and supplemented by written learning objectives and schedules;
■ satellite broadcasts, job rotation assignments, and assessment center activities—if their primary purpose is employee development.

But this definition of training does not include activities such as these:
■ conferences, seminars, meetings, and performance appraisals—unless employee development is their primary purpose;
■ self-development that an employee carries out on non-work time or using personal resources.

Step 2: Determine all training cost categories

It isn't easy to establish a uniform accounting system

Consensus Accounting Model

Step 1: Establish an organization-specific definition of training

Step 2: Determine all training cost categories

Direct Costs
■ personnel
■ outside goods and services
■ facilities
■ travel, per diems, accommodations, and incidental expenses.

Indirect Costs
■ overhead
■ facilities
■ general and administrative expenses
■ miscellaneous costs.

Step 3: Calculate training costs

Direct Costs
■ personnel
■ travel, per diems, accommodations, and incidental expenses
■ outside goods and services
■ facilities.

Indirect Costs
■ overhead (and general and administrative)
■ facilities
■ equipment.

Targeted Costs
■ training populations
■ subject matter
■ training providers
■ training phases.

Step 4: Code costs

---

within an organization or across organizations. But identifying and defining training costs leads to a clear understanding of where training monies go. In reviewing expenses for a training program, an organization must explore direct, indirect, and miscellaneous costs.

Direct costs

Personnel. Personnel costs include total costs for people involved in training:
■ Salaries and employee benefits of supervisory and non-supervisory training department directly engaged in developing, delivering, evaluating, and supporting training programs; for example, instructors, program designers, needs analysts, in-house evaluators of training, and clerical staff;
■ Salaries and employee benefits of other company personnel who assist training staff by serving as resources for developing or delivering training; for example, subject-matter experts and line managers;

The Consensus Accounting Model

Why Collect Training Cost Information?

To know costs. An organization's leaders should know approximately how much money is spent on training. Total expenditures often go beyond the training department budget to such costs as participants' salaries, their travel expenses, and fees for temporary help to replace them. Yet few organizations calculate their total HRD expenditures, and even fewer compare their expenditures to those of other organizations.

To compare costs and make choices. Admittedly, exact comparisons are difficult. To begin with, organizations use different bases for cost calculations. For example, although some programs at IBM account for student salaries, programs at AT&T never account for them. Also, organizations don't publish much training cost data. Still, general comparisons are possible. It's helpful to compare specific training program charge-back measurements (such as cost per hour or day) with those of other programs or organizations. Wide differences may indicate a problem.

Cost data are essential for making choices about alternatives to or among proposed and existing training: whether to invest in physical or human capital; whether to invest in training or other personnel interventions (such as job aids or new staff); what training programs to fund. For instance, cost data are needed to determine whether it's more cost-beneficial for the organization to design and develop a program in-house or to "contract out."

To monitor costs. Besides allowing comparison of resource allocations to particular programs and projects, monitoring costs allows managers to evaluate the proportion of investment in specific training populations, subject areas, and training providers. Costs may also be monitored according to their association with training administration, research and development, analysis, design, development, delivery, evaluation, or marketing.

To control costs. This is a major management responsibility from which training managers are not exempt. Accurate, current, well-organized cost information helps managers plan, arrange, control, price, and evaluate training programs.

■ Salaries and employee benefits of training participants;
■ Fees and expenses of people from outside the organization who render training department services; for example, temporary clerical staff, training consultants, and outside evaluators of training.

Outside goods and services. Outside goods and services include costs for design, development, reproduction, distribution, or review of training materials purchased from an outside provider. To be a direct cost,

these goods and services must be for, and used up during, a specific training program. So equipment purchased or rented for one program are in this category. But equipment or materials used during a training program are to become part of the organizational stockpile, they are consider indirect costs. Outside goods and services costs may be subdivided into:

■ Program materials and supplies. Materials and supplies purchased from an outside provider and for a specific training program; for example, off-the-shelf program materials, standardized tests, artwork, and audiotapes.
■ Outside printing and reproduction costs.
■ Equipment rental or lease.
■ Equipment purchase.

Facilities. Facilities costs include those incurred from a training program in a rented classroom, learning center, laboratory, or workshop.

Travel. Travel includes per diems (daily expense allowances), accommodations, and incidental expenses. This category includes total costs, but training personnel and participants' meals, travel, accommodations, and other expenses are accounted for separately within it.

Indirect costs

Indirect costs can't be traced back and directly tied to a specific training program. Although indirect costs for training are less visible than direct costs, they are substantial. Sometimes all indirect cost are termed overhead, but indirect costs usually are accounted for by sorting them into categories called "overhead" and "general and administrative" (G&A) costs.

When indirect costs are categorized that way, overhead costs relate to *things*—such as a training department's share of organizational materials, equipment, and facilities. G&A costs relate to *people*—such as a general administrator, main receptionist, or payroll clerk. The overhead and G&A costs and categories listed here are common, but they vary by organization. It's a good idea to enlist the organization's comptroller (or representative) to assist with selection of specific methods for capturing indirect costs related to the following categories.

Overhead costs. These include the following:
■ Materials. General office supplies and related expenses; for example, each training program will absorb a share of the expenses for general training department stationery, subscriptions, postage, photocopying, and telephones;
■ Equipment. A training program's fair share of expenses associated with equipment purchased by the organization and used by numerous training programs. Overhead equipment costs include equipment capitalization allocation (portion of original cost allocated to a particular training program) and equipment operation and maintenance costs;
■ Facilities. A training program's share of expenses for use of general office space in an organization's facilities.

General and administrative. One method for determining the G&A costs for a particular training program is to compare the program's length and expenses to those of other organizational activities. G&A costs are divided into these personnel-related categories:

■ Travel and expenses not directly billed to one program or client; for example, each training program must incur a share of executive staff travel;

■ Training department management and staff salaries, wages, and fringe benefits that can't be tied to a particular training program;

■ Administrative and staff support salaries, wages, and fringe benefits; for example, for legal and accounting department personnel.

Hidden costs. The direct and indirect costs described above constitute the information necessary to begin accounting for a training program. But there may be "hidden" cost information. To look for it, Glenn E. Head suggests that a training manager consider the following:

■ number of training program participants

■ average annual salaries of training participants

■ annual employee fringe benefits percentage for the organization

■ average travel and per diem expenses for training programs from the prior year

■ number of training instructors

■ number of subject-matter experts from other departments who help conduct a training program

■ number of times the training program will run each year

■ need to run pilot versions of the program

■ expected life of the program

■ location of the program's training facilities

■ equipment necessary to conduct the program

■ organizational method for allocating overhead and G&A costs.

Patterns or surprises that emerge may indicate a need to make changes in how training is managed or administered.

Step 3: Calculate training costs

After determining the basic cost categories of a training program, a training manager is ready to begin calculating the costs of training. A checklist (see page S-13)—and, possibly, consultation with an in-house accountant—can help a manager identify the costs to account for and those not to.

This decision is organization-specific: one organization may account for subject-matter experts' salaries and travel expenses, but not account for participants' salaries or travel; another organization may do the reverse. To calculate a training program's costs, a training manager can apply the simple formulas that follow.

Direct costs
Personnel.

■ Training participant costs. An estimate of the average salary or wage for training participants plus the organizational overhead rate gives the basis for participant costs. To estimate people's yearly earnings according to job classifications, consult payroll/compensation department data, supervisors, and other organizations that employ people in the same job classifications.

If a training program is for people in several job classifications, estimate the typical participant's earnings by looking at the participant roster, noting which jobs are represented and the number of participants that occupy each job, and factoring in each job's salary average. The median (the point that half the salary values are above and half are below) represents the participant salary or wage to use in subsequent calculations.

Next determine the organization's percentage costs to cover fringe benefits: health insurance, pensions, time when employees are paid but don't work (sick leave, vacation, holidays, and personal days), and educational opportunities.

The participant daily cost is based on the annual number of working days per employee. Subtracting the number of paid vacation days, holidays, and leave days from 260 (the number of weekdays in a year) gives this number. For example, if each employee gets 10 vacation days, 10 holidays, and 10 leave days a year, the equation would be: $260 - 30 = 230$ potentially productive days.

Multiplying the participant annual salary or wage by the organization's fringe benefit rate finds the total personnel costs for an employee. Personnel or payroll departments can often provide the current fringe benefit rate. For example, if the median participant salary was $30,000 and the fringe benefit rate was 30 percent, the total loaded personnel cost (with fringe benefits added) is 1.3 times the salary $(1 + .30 = 1.30)$. So, the total loaded annual personnel costs per training participant would be $39,000 ($30,000 x 1.3 = $39,000).

Dividing that total by the number of productive days determines the participant cost per day. Continuing the example, the average annual loaded personnel costs were $39,000, and the annual number of productive days was 230. So the average daily participant wage or salary was $169.57 ($39,000 ÷ 230 = $169.565). For an estimate of an hourly rate, the average participant salary is divided by the number of hours in a workday. In this example, employees work 8-hour days, so their average hourly cost is $21.20 ($169.57 ÷ 8 = $21.196).

Final participant costs may also include the average costs of meals, travel, and accommodations. Those are direct costs if they are used by the end of the training program (see "Travel . . ." below).

■ Training personnel costs. These are determined in the same manner as participant costs. The average yearly salary or wage is multiplied by the fringe rate to find the total loaded personnel costs of a training staffer. This figure, when divided by the annual number of productive days, yields the average cost per day for training personnel. The average travel and per diem costs can be derived from personnel/payroll records or by asking people directly (see "Travel . . ." below).

Training personnel costs include more than the time each instructor spends on a training program's preparation and delivery. Cost calculations should include the days or hours spent by task analysts, program designers and developers, and clerical staff in support of instruction.

The category also includes the time of any internal training program evaluators.

■ Other in-house personnel costs. Frequently, training departments rely on other employees in the organization

The Consensus Accounting Model

to design, deliver, or support a training program. For example, in-house subject-matter experts may advise on program design or may do stand-up training. The methods used for capturing the daily or hourly costs of training participants and staff also work for determining the costs of these other employees.

■ Direct outside personnel costs. Outside personnel may design or lead training, or temporary personnel may be used to do the work of employees in training. Multiplying those outside people's costs per day by the number (or fraction) of days they worked determines the total outside personnel costs. For example, if an outside evaluator whose fee is $200 a day takes four and a half days to evaluate a training program, that costs the organization $900 (200 x 4.5 = $900).

Travel, per diems, accommodations, and incidental expenses.

Multiplying the average travel costs per person by the number of travelers gives the total travel cost. Payroll or personnel records may offer information useful for determining average travel expenses. Similarly, analysis of records can reveal the average for per diems (daily expense allowances) for meals, local transportation, and so on.

If payroll or personnel records aren't available, a training manager can estimate travel and per diem costs by surveying participants and training personnel and then averaging the estimates from each group separately.

Outside goods and services.

To find the total cost for these services simply add the subsidiary costs that make up this category. Some subsidiary cost totals are the result of multiplying a per participant cost by the number of participants, while others already are a per program total. For example, if 10 participants each received a $5 purchased workbook and if the one piece of demonstration equipment was rented for $75, the cumulative total is $125 ($5 x 10 = $50; $50 + $75 = $125).

Facilities.

If the rent for a facility isn't a flat fee, calculate the total by multiplying its daily or weekly fee by the number of days or weeks of rental. For example, if a three-day workshop is to be held at a conference center that charges $1,000 a day, the total facility cost is $3,000 ($1,000 x 3 = $3,000).

Indirect costs

Although indirect costs often equal or exceed the direct costs of a training program, they are frequently overlooked in accounting for training programs. The various methods for determining indirect costs vary in their precision, in the amount of information they require from an organization, and in whether general and administrative costs are separated from other indirect expenses.

Overhead costs.

The simplest method for estimating overhead costs is to establish a base percentage rate of indirect costs for all training programs. In this approach, a training program's estimated indirect cost is estimated by multiplying the base percentage indirect cost rate by training's total direct personnel costs.

For example, if the direct personnel costs of a training program were $10,000 and the base percentage rate of indirect costs was 45 percent, the indirect program cost equalled $4,500 ($10,000 x .45 = $4,500). This approach to indirect costs requires an estimate by the comptroller of the typical base percentage rate of indirect program costs, but it lacks precision and may seriously underestimate training programs' typical costs.

A more precise method for capturing indirect training costs relies on total training department budget information. Total training budget costs include loaded employee salaries (costs for salaries and fringe benefits), facilities costs, equipment depreciation, and a fair share of administrative and executive costs.

All costs—except for loaded salaries of employes who make a direct contribution to training programs—are added. The total is divided by those loaded salaries. The numerator (training budget costs less those loaded costs) of the calculation is the base rate of indirect costs for all training programs.

For example, an organizations's total training budget was $500,000. Five employees, each of whom received $40,000 in loaded salaries (5 x $40,000 = $200,000), contributed directly to training programs. So the base percentage rate is 110, and the estimated overhead costs for all training programs equals $220,000 ($500,000 - $200,000 = $300,000; then, $300,000 - $200,000 = 1.1 or 110 percent; finally, $200,000 x 1.1 = $220,000).

Facilities costs.

These costs should be accounted for separately from other indirect costs. Indirect facilities costs usually are relatively small costs and hard to determine, so it's not worth spending too much time trying to measure them precisely. But it could be worth spending somewhat more time if major new construction or renovation is underway and causing an increase in these costs.

In some organizations, the accounting department will have the figures (costs for mortgage, electricity, maintenance, and building administration) that, added together, equal total facilities costs. For leased buildings, the total is found by multiplying the cost per square foot by the square footage for the facility in which a training program is held, and then dividing the total by the number of annual productive working days to determine a per day cost. This cost is multiplied by the number of days the facility is used for the training program.

For example, an organization leased its building for $10 a square foot. So, the average annual cost of its training room's 1,000 square feet was $10,000 ($10 x 1,000 = $10,000). Its per workday cost was $38.46 ($10,000 ÷ 260 = $38.46). A particular training program occupies the room five days each year, so its facilities cost was $192.30 ($38.46 x 5 = $192.30).

Equipment costs.

Equipment purchase and maintenance costs are divided by the equipment's useful life to find its annual cost. The annual cost is distributed evenly to all training programs. For example, a training department purchased a $500 videocassette recorder with an estimated life of five years. The VCR is used for 10 training programs. So the indirect cost to each program for this piece of equipment was $10 ($500 ÷ 5 = $100 per year; $100 ÷ 10 = $10 per program).

If a program uses several pieces of equipment, their costs are added together.

Targeted costs.

An organization may require specialized accounting beyond the standard categories used to break down total training costs into direct or indirect costs. Targeting particular cost areas for scrutiny can improve an organization's ability to determine how many dollars should be allocated to training—and where—for improved management of HRD.

For example, managers may have an interest in determining how much the training department is spending on:

Specific training populations.

It may be useful to look at training requirements and costs for groups of job classifications such as: executive, administrative, and managerial occupations; management support occupations (such as accounting); technical occupations (engineers and technicians); marketing and sales occupations; and administrative support populations (clerical and administrative assistants).

Comparisons with the time and dollars that other organizations spend on these populations can provide clues about whether an organization is devoting enough training attention to these populations. Comparative information is sometimes available through professional associations and publications for human resource specialists or for trainees' occupations.

Subject matter.

It also may be useful to track training dollars spent on entry, mid-level, and upper-level career programs, or to track the dollars spent on levels of a particular topic or on generic courses.

Training providers.

It may be useful to collect information on internal and outside providers. Information on external providers should distinguish between regularly presented customized programs and one-time programs.

Training phases.

It may be useful to treat training program phases as a classification system for direct and indirect costs. Typically:
■ Analysis costs relate to analysis of needs, resources, or constraints and to selection of training participants.
■ Design costs relate to the choice of learning objectives, preparation of a program proposal, and broad curriculum planning.
■ Development costs relate to such materials as participant workbooks, instructor guides, slides, tapes, tests, and computer software.
■ Delivery costs relate to personnel, outside goods and services, and facilities.
■ Evaluation costs relate to training tests, observations, interviews, and discussions.

Checklist of Training Cost-Account Classifications

Direct Costs

Personnel
- 100 —Salaries and benefits of training personnel
- 101 —Salaries and benefits of other company personnel (when assisting training)
- 102 —Salaries and benefits of training participants
- 103 —Outside personnel services

Outside Goods and Services
- 201 —Purchased program materials and supplies
- 202 —Outside printing and reproduction costs
- 203 —Equipment rental or lease
- 204 —Equipment purchase costs

Facilities
- 300 —Facilities rental or lease

Travel, Per Diems, Accommodations, and Incidental Expenses
- 400 —Training staff travel
- 401 —Training staff per diems
- 402 —Training staff accommodations
- 403 —Other training staff expenses
- 404 —Participant travel
- 405 —Participant per diems
- 406 —Participant accommodations
- 407 —Other participant expenses

Indirect Costs

Overhead
- 500 —Office supplies and materials
- 501 —Equipment capitalization
- 502 —Equipment maintenance
- 503 —Equipment repair
- 504 —Facilities

General and Administrative
- 600 —Travel and expenses charged to overhead
- 601 —Department management and staff salaries
- 602 —Administration and support staff salaries

Source: Adapted from Phillips, 1983.

■ Administration costs relate to course scheduling, activity coordination, and report drafting.
■ Research and development costs relate to exploration of new training techniques and strategies.
■ Marketing costs relate to advertising training internally and externally (for example, for brochures promoting a program).

The Consensus Accounting Model

Step 4: Code costs

Whatever accounting categories an organization uses, coding subsidiary cost components facilitates record keeping. Then it's important for the training manager to train the trainers in how to use whatever coding system is adopted.

For an overview of where costs fall, a training manager might create and fill in a cost classification matrix based on a checklist of training cost-account classifications and, for example, the phases of a training program (see pages S-13 and S-14). Then, for instance, if a piece of leased equipment were used in several phases, its cost would be allocated proportionately among those phases.

Cost Classification Matrix

	Analysis	Design	Development	Delivery	Evaluation	Administration	Research & Development	Marketing
Direct Costs								
Personnel								
100—Salaries and benefits of training personnel	■	■	■	■	■	■	■	■
101— Salaries and benefits of other company personnel (when assisting training)	■	■	■	■	■	■	■	■
102—Salaries and benefits of training participants	■	■	■	■	■	■	■	■
103—Outside personnel services	■	■	■	■	■	■	■	■
Outside Goods and Services								
201—Purchased program materials and supplies	■	■	■	■	■	■	■	■
202—Outside printing and reproduction costs	■	■	■	■	■	■	■	■
203—Equipment rental or lease	■	■	■	■	■	■	■	■
204—Equipment purchase costs	■	■	■	■	■	■	■	■
Facilities								
300—Facilities rental or lease	■	■	■	■	■	■	■	■
Travel, Per Diems, Accommodations, and Incidental Expenses								
400—Training staff travel	■	■	■	■	■	■	■	■
401—Training staff per diems	■	■	■	■	■	■	■	■
402—Training staff accommodations	■	■	■	■	■	■	■	■
403—Other training staff expenses	■	■	■	■	■	■	■	■
404—Participant travel	■	■	■	■	■	■	■	■
405—Participant per diems	■	■	■	■	■	■	■	■
406—Participant accommodations	■	■	■	■	■	■	■	■
407—Other participant expenses	■	■	■	■	■	■	■	■
Indirect Costs								
Overhead								
500—Office supplies and materials	■	■	■	■	■	■	■	■
501—Equipment capitalization	■	■	■	■	■	■	■	■
502—Equipment maintenance	■	■	■	■	■	■	■	■
503—Equipment repair	■	■	■	■	■	■	■	■
504—Facilities	■	■	■	■	■	■	■	■
General and Administrative								
600—Travel and expenses charged to overhead	■	■	■	■	■	■	■	■
601—Department management and staff salaries	■	■	■	■	■	■	■	■
602—Administration and support staff salaries	■	■	■	■	■	■	■	■

Source: Adapted from Phillips, 1983, p.118.

Evaluation Framework, Design, and Reports

A training program is a success if it achieves timely results consistent with pre-established participant performance objectives related to wider organizational goals.

Much the same is true of evaluation methods. Evaluation only needs to provide sufficient information to assure that a training program is meeting its objectives—and that those objectives further attainment of organizational goals and objectives. Evaluation methods must provide results in time to inform decision makers as they consider choices for current and future training. After all, the purpose of training is to improve performance, and the purpose of evaluation is to improve training's effectiveness and efficiency.

Historical failures or inabilities to evaluate both training's costs and benefits have rendered training particularly vulnerable to cost-cutting pressures, and have inhibited its use as a lever for effecting strategic change. The fact that fewer than half of America's training programs are formally evaluated indicates implicit managerial trust that, somehow or other, training facilitates attainment of organizational goals. Yet, as one trainer put it, "The worst thing that ever happened to training is that it was taken on faith that it was good." As cost pressures increase, trainers must demonstrate training's value in more substantive ways if it is to gain its rightful place among investment alternatives.

Admittedly, measurement can never completely ascertain a training program's effectiveness or its efficiency in achieving beneficial effects. What worked at one time at one training location with a unique group of participants can't necessarily be transferred to another time, setting, and group and be expected to work as well. Still, evaluations build a case of support for training by providing an approximation of its value.

The Kirkpatrick Model

The evaluation framework that most training practitioners use is the Kirkpatrick Model. Although this model doesn't accommodate all the evaluation methodologies that training managers employ, it's the most widely known evaluation model and illustrates a commonly used set of levels or rigors of evaluation.

Almost universally, organizations evaluate their training programs by emphasizing one or more of the model's four levels. In summary, these levels are as follows:

■ **Reaction.** How well did training participants like the program?

■ **Learning.** What knowledge (principles, facts, and techniques) did participants gain from the program?

■ **Behavior.** What positive changes in participants' job behaviors stemmed from the training program?

■ **Results.** What were the training program's organizational effects in terms of reduced costs, improved quality of work, increased quantity of work, and so forth?

Participant reactions are easy to collect, but provide little substantive information about training's worth. At the other end of the scale, results-level information is more difficult to collect, but provides data to analyze for assessment of training's organizational impact. In general, the more data sources used to evaluate a training program, the more complete is the picture of its effectiveness.

The appropriate levels of evaluation data to gather and analyze depend on the evaluation "clients." As a rule, line managers have more interest in performance change and organizational results than in participant reaction and learning. A training department, on the other hand, would have an interest in collecting reaction and learning data to determine what components of training could be improved.

Currently, most employee training is evaluated at the reaction level. Evaluation at this level is associated with the terms "smile test" or "happiness test," because reaction information usually is obtained through a participant questionnaire adminstered near or at the end of a training program.

The fact that fewer than half of America's training programs are formally evaluated indicates implicit managerial trust that, somehow or another, training facilitates attainment.

Because reaction information doesn't reveal what participants have learned or whether what they've learned will transfer to their jobs, it isn't indicative of training's return on investment. This lack and frequent misapplications of reaction data have caused some evaluators to deride its collection. But most training practitioners believe that participants' favorable reactions are critical to training program success—because people learn better when they accept training willingly and react positively to the form it takes. The clients of training most likely to have an interest in reaction data are participants themselves, training department staff, and course instructors.

At the learning level of evaluation, tests are used to measure the knowledge, skills, or attitudes that participants acquired during training. These tests should reflect each training program's particular objectives. For instance, for an introductory skills course, a participant may be considered successful if able to accomplish a given task at all, or the participant may be required to perform at a speed consistent with on-the-job requirements.

Measures of learning changes may be taken during a training program or at its conclusion. This level may indicate that a program's instructional methods are effective, but it doesn't show whether or how participants new learning will be applied on the job. The clients with the greatest interest in examining learning changes through training are participants, training department staff, and course instructors.

The third evaluation level deals with behavior or performance changes on the job. When learning doesn't transfer to the job, the two most likely reasons are that the work environment doesn't support the learned behavior or that a participant thinks the training was irrelevant. A participant's supervisor or colleagues (or the physical work setting) may discourage newly learned behavior, or the participant may believe that the new behavior, although encouraged by management, won't lead to personal benefit. That's why participants' managers should be consulted about training program development and should tell participants how the training will help them maintain or improve their positions.

Even with supervisory support and participants' commitment to transfer new learning to the job, there's no guarantee of increased productivity or ability to meet intended organizational economic objectives—because the changed performance may not be in keeping with desired organizational outcomes. That sorry situation occurs if a training program's objectives aren't properly established when the program is developed.

The ultimate measure of training program success relates results to organizational objectives. The results level of evaluation is the highest level of rigor and usually is expressed in organizational terms such as reduced costs. Organizational analysis, needs-identification forecasting, and needs analysis reveal areas where training can make a contribution to achievement of key organizational objectives.

It's difficult to isolate the beneficial organizational results of training. Nonetheless, training specialists need to work from an understanding of organization-level goals and objectives. In many cases, it is possible and feasible to link training contributions to organizational improvements. Doing so doesn't require absolute isolation of training's contributions. Rather, it requires indicators that demonstrate training's valuable role within the organization's systems

A systems view

Taking a systems approach means looking at the interconnection of organizational parts and the relationship of the whole organization to its environment. Organizations are composed of functionally related components arranged in a hierarchy of subsystems, systems, and supersystems. Organizing subsystems into larger systems helps coordinate activities and processes in order to fulfill the overall mission of facilitating organizational goals. The translation of organizational goals into identified current objectives provides context, meaning, and direction for the entire organization.

Training is one operating system within an organization, and training should be evaluated as a system in support of other systems. In other words, training must contribute to achievement of the goals of the departments it serves and, through these, contribute to organizational objectives. This view of training necessitates the use of systematic means for performing training and linking its role to the goals of higher systems. Evaluation of training is the main method used to assess whether training is accomplishing desired effects of sufficient value.

Evaluation steps

According to Ratzlaff, there are three major evaluation steps:
- setting an evaluation's purpose
- selecting evaluation methods and design
- reporting evaluation findings.

Setting an evaluation's purpose. First, an evaluator must determine whether an evaluation will be formative or summative. Formative evaluations, sometimes called improvement-oriented evaluations, are used to decide whether a program should be modified. Summative evaluations are used to decide whether a training program should be maintained, expanded, contracted out, or eliminated. Formative evaluations are carried out before and during the first running of a training program. Summative evaluations collect information about a training program's value after it has run at least once.

An evaluation's purpose and scope are delineated by the questions or issues its clients want addressed. If clients aren't directive about what an evaluation should accomplish, it may take some discussion between a client and an evaluator to set evaluation objectives.

Continuing client involvement in planning provides opportunities to reassess a training program's objectives and, consequently, to reassess aspects of evaluation. Talking about relationships between training and intended outcomes can reveal whether training activities logically flow from objectives that support goals, and whether objectives reflect realistic expectations.

An evaluator needs interpersonal skills to build rapport with clients and establish a climate of trust and candor. A free flow of information will be promoted by an evaluator's flexibility in considering possibilities for an evaluation's focus and design.

Laying the groundwork for a common understanding of a training program means asking clients questions. An evaluator should paraphrase and restate each client's views about a training program. That helps ensure that the evaluator has interpreted the client's views correctly; if not, it points out what needs clarification.

An evaluator must learn how the primary client—the evaluation's financial sponsor—expects to use evaluation information, because the answer affects what data will be collected and how findings will be presented.

The evaluator must also decide to what extent an evaluation can accommodate the questions of other stakeholders in the program—such clients as instructional designers, instructors, training participants, and their supervisors. It may not be possible to deal with all the issues raised.

Evaluation Framework, Design, and Reports

The evaluator next develops tentative evaluation objectives and plans that lead to new questions. For a formative evaluation, there will be questions about how to staff the program and how to select participants (for instance, by mandatory or voluntary attendance, or on the basis of current or future needs). A summative evaluation might seek to characterize participants who benefited most from a program (for example, those who stayed for optional lab work or those who had a designated learning partner).

Selecting evaluation methods and design. As soon as the training program's nature (formative or summative) and tentative objectives are established, the evaluator begins to develop data-collection plans.

Data-collection strategy relates to how participants will be grouped for evaluation and when evaluation measures will be taken. An evaluator should always use more than one data source when assessing an aspect of a training program, because multiple sources give a rounded view and build stronger support for evaluation findings.

Once evaluation questions are outlined, the evaluator begins listing quantitative or qualitative data-collection techniques to use to answer each question. Whenever possible, to be economical, participants' learning and application of learning should be measured through data-collection techniques and tools already in use. Constructing measurement procedures and instruments is time-consuming and expensive, and existing procedures and standardized instruments for assessing participants' reactions or learning gains are often adaptable for many training programs.

Still, new data-collection procedures and instruments should be developed for a training program if the evaluation addresses unique questions, if no suitable standardized data-collection tool or technique is available, or if it's critical that program outcomes be measured precisely.

Evaluators should determine whether the primary client disapproves of a particular data-collection technique. For instance, if a client regards observation of training participants as obtrusive, then this technique should not be used. An evaluator must also select techniques that can answer evaluation questions within timeframes set by management. To do so, the evaluator must allow for time to train any others who'll be responsible for data collection. For example, line managers may need a training session or two if they'll be responsible for observing and assessing participants' learning.

Before proceeding, an evaluator may want discuss with the client the tradeoffs between implementing a practical but less rigorous evaluation design versus a more credible—but also more expensive—design. If necessary, the client may be persuaded to provide more time or funding or to narrow evaluation objectives.

At this point, an evaluator should review background information about the training program from the following items:

■ Documents. For a formative evaluation these include front-end analyses prompting training and previous training project proposals, goal statements, budgets, materials, publicity, evaluations, memoranda by program staff. For summative evluation, the evaluator will survey documents for the current program.

■ Discussions with participant and training staff— instructors, instructional designers, previous evaluators, and training administrators. Consultation with program staff gains their insights and promotes their comfort and cooperation with the evaluation.

■ Observations. After examining a variety of sources, an evaluator should write a detailed description of training program objectives and activities designed to achieve them. Once the client and evaluator reach agreement about the description, it should be distributed to program staff and training administrators.

In keeping with the program description, an evaluator should establish a schedule of evaluation activities to determine the order, length and, possibly, dates of evaluation activities. The schedule begins with a list of each major evaluation effort and its subactivities to each of which the evaluator adds starting and completion dates.

Throughout evaluation, an evaluator should note and record variances between expected and actual time devoted to each evaluation activity. This information facilitates the development of realistic scheduling for future evaluations or, in some cases, may be useful for specifying to the current client when and where unexpected obstacles are being encountered.

Reporting evaluation findings. An effective communication strategy disseminates information to the client and program staff through informal conversations, discussions, and meetings; formal meetings and presentations; and memoranda and written reports.

Depending on the scope, complexity, and formality of an evaluation and on clients' expectations, an evaluator may incorporate the evaluation schedule into a formal evaluation contract that states the following:
■ the evaluation's purposes
■ aspects of the training program to be examined
■ data-collection techniques to be used
■ a schedule of evaluation activities
■ formats for reports of evaluation findings.

Even if there's no formal evaluation contract, timing and formats for the release of evaluation information should be part of the strategic planning for evaluation. Especially for a formative evaluation, frequent piecemeal release of findings to client and program staff is preferable to a lengthy final report because frequent communication allows for adjustments to the program or its evaluation.

Again, an evaluator's interpersonal skills are important because an evaluator needs to be tactful, informative, and non-threatening when communicating that a training program needs improvement.

The key evaluation document is an evaluation report of final evaluation findings that will affect clients' future actions. These reports tend to be longer for summative evaluations because, by nature, formative evaluations call for numerous interim reports. In any case, to be valuable, an evaluation report must be timely and clearly reported in four sections:

■ *Executive summary.* This one-page synopsis of the report may be all that some clients read, so it should be particularly well reported.

■ *Introduction.* This describes the training program's

Evaluation Framework, Design, and Reports

objectives, which training program aspects were assessed, and what the evaluation's timeframe and budget were.

■ *Data-collection techniques and evaluation design.* This describes overall evaluation design, how training participants were selected, each data-collection tool or instrument, the evaluation questions each addressed, and the participants to whom each was applied or administered and the number of times administered.

■ *Evaluation findings.* This reiterates which training program aspects were investigated and provides the evaluation's major findings. This section should also describe constraints on the evaluation—because such major restrictions as time and budget can influence an evaluation's credibility.

If the evaluation is to recommend a program's cancellation or continuance, that recommendation appears here. For continuing programs, suggestions for revisions are offered, focusing on strengthening the programs for attaining better results.

Phases of evaluation

As already noted, training program evaluation doesn't just follow a training program's completion. Evaluation has three, roughly sequential phases:
■ It determines whether a training program as planned is likely to meet organizational goals.
■ It monitors training in process to ensure that training is being conducted as planned or undergoes necessary correction.
■ It determines whether a program as implemented met expectations.

Planned evaluation

Planned evaluations analyze the environment in which training is developed. This involves analysis of the organization's goals, resources, and people. Planned evaluations assess discrepancies between expected and actual performance, current employee skills, and training deficiencies.

Planned evaluation begins with the fundamental question: Will the organization benefit from providing training? The major components of a planned evaluation are organizational analysis, needs-identification forecasting, needs analysis, and marketing of training.

Organizational analysis

Trainers must be proactive rather than reactive in considering an organization's need for training. Instead of waiting to respond to training needs and opportunities that become clearly evident, trainers must anticipate performance problems and opportunities.

To be proactive, training departments translate organizational objectives into training objectives, then plan, deliver, and evaluate programs in terms of contribution to objectives. This means that preparation of training programs must begin with analysis of data collected from the entire organization.

It's impractical to conduct a full-scale organizational analysis at the beginning of each training program. But periodic organizational analysis gives guidance and a realistic context to training decisions. An annual analysis sponsored by the training department can:

■ help determine the state of the organization
■ help training department staff understand the organization and organizational systems they serve
■ serve as a base for future training programs by revealing training needs.

In the end, organizational analysis should:
■ describe immediate and future objectives of the organization and of its particular subdivisions
■ identify as realistically as possible the organization's economic, social, and political environment
■ discuss the organization's structure and resources, including equipment and facilities, monetary resources, and current human resource skills.

This analysis allows trainers to design training programs that meet organizational goals in the face of economic and political pressures and in the context of available resources.

Organizational analysis is often overlooked because of immediate pressure to conduct a particular training program for which the need is regarded as obvious. But without such analysis, trainers may believe a training program will fill a fundamental organizational need when, in fact, it would fill a only short-term or minimal need or treat only one symptom of a larger problem. If so, program staff may mistake their roles, misunderstand organizational direction, and actually impede accomplishment of goals. The result is frustration, waste, and trivialization of the training function.

So an organizational analysis should be as thorough as possible and based on information from a variety of sources. First, analysts should examine written organizational mission statements, strategic plans, statements of philosophy, and statements of objectives.

Consultations with executives or their managerial representatives can be an excellent source of information about the organization's immediate and long-term objectives. But it may be difficult for an analyst to gain access to upper managers or, in the short time available for consultation, it may be hard to arrive at focused objectives with a clear bearing on training.

Training managers can overcome these difficulties through the following activities:
■ devising and distributing short forms that allow managers to explain themselves easily and whenever their schedules permit
■ requesting appointments with selected managers and sending interview instruments to them in advance so they see what kind of information is needed and have time to consider their responses.

Needs-identification forecasting

A needs-identification forecast begins with scrutiny of key business indicators to determine whether an organization is achieving its goals or falling short of them.

As in organizational analysis, that involves examination of the organization's stated objectives, strategies for achieving those objectives, and measures of its success in attaining goals. But for needs-identification forecasting, analytical activity is directly aimed at discovering how organizational objectives may translate into training needs. So special attention should be given to human resource plans incorporated into goals, strategies, and objectives in order to analyze whether the plans

- are feasible in terms of dedication of organizational resources to support them
- relate to training programs that have run in the past or are likely to require development of new training programs.

If organizational goals and objectives aren't being met despite dedication of seemingly sufficient resources, human performance must be analyzed in terms of current and future business goals and objectives.

The data for this needs-identification forecast may be gathered from the following sources:

Organizational performance records related to

- productivity measures quantifying output per units produced, tons manufactured, items assembled, money collected, items sold, forms processed, loans approved, inventory turnover, patients visited, applications processed, students graduated, tasks completed, output per hour, work backlog, number of incentive bonuses earned, or shipments made
- quality measures such as scrap level, error rates, amount of waste, rework time, shortages, deviation from standards, product failures, inventory adjustments, and customer compliments or complaints
- workforce performance and behavior indicators such as measures of absenteeism, tardiness, injuries, number of promotions or pay increases, training program attendance, requests for transfer, performance appraisals, and turnover rate
- Safety and regulatory measures that count the number of employee accidents and injuries, Occupational Safety and Health Administration (OSHA) litigations, and reported safety violations.

Questionnaires for or interviews with managers and employees.

"Self-reports" of training deficiencies are the most obvious clues to training needs. If training is institutionalized as a benefit, employees will be more willing to report training deficiencies and seek assistance. But the analyst must determine whether a reported performance deficiency calls for additional training or for an alternative human resource management solution.

Observations of employees performing tasks.

Observers can form impressions by listening to employees talk about their organization, jobs, supervisors, and working environment. Observers should note any obvious interpersonal difficulties or communication breakdowns among supervisors and employees, employees' complaints or inattentiveness to work, or supervisors' poor communication of direction. If such problems appear, observers should seek the underlying reasons for these difficulties. Observers should also collect data on employees' overt performance and skills.

Plans for new business strategies.

Organizational changes—such as plans to expand or reduce the number of employees, products, or services, to merge or reorganize the company, or to introduce new operating methods or technology—often demand new or improved human resource capabilities.

If the completed needs-identification forecast shows

variance between actual and expected performance, it may be possible to document dollar losses caused by performance deficiencies. With evidence of dollar losses, a training department can take a proactive approach. Dollar-loss estimates are the part of cost-benefit analysis that communicates to managers in the way they understand best.

Training needs analysis

If organizational analysis and needs-identification forecasting reveal a performance deficiency, that doesn't necessarily mean that a training problem exists. Training is an appropriate intervention only if employees lack the vocational preparation needed to perform their job tasks adequately.

As a guideline, Mager and Pipe suggest that trainers consider this scenario: if an employee is at gunpoint and cannot find the skill and motivation to perform a task effectively, the performance deficiency is likely to be a training problem. While Mager and Pipe don't literally recommend this drastic test, it dramatizes the importance of motivation. Employees may know how to perform a task well but, for various reasons, may not demonstrate it on the job.

Training needs analysis has two parts:
- job or task analysis to describe the tasks performed on a job regardless of who performs them
- personal analysis to describe the specific knowledge, skills, and attitudes that a particular person needs to develop in order to perform the job adequately.

Many methods are available for gathering information for training needs analysis. They vary in the time and resources they require and in their suitability for particular purposes. For instance, work participation helps an analyst take the novice's point of view, which suits analysis of entry-level training. But that method isn't appropriate for analyzing complex high-level tasks.

Frequently, more than one method is used because the strengths of one can offset the weakness of another. For example, to avoid disrupting an observation, an analyst may withhold questions while employees complete a series of tasks, and then conduct a brief group interview.

The most common information-gathering methods for needs analysis:
- questionnaires and checklists, possibly open-ended
- individual or group interviews with job holders, supervisors, or subject-matter experts
- observations, perhaps including "critical incident" observation to determine what behaviors lead to excellent or unsatisfactory (as opposed to adequate or average) performance
- work participation by the analyst.

Tying learning and performance objectives to organizational objectives

Learning objectives are evaluated by measures of training's impact on participants' knowledge or skills as evidenced in the training setting. Performance objectives, on the other hand, are evaluated by assessment of training's impact on participants' on-the-job performance and by organizational measures.

For example, if observation indicates that warehouse

employees frequently misplace inventory because of inability to read shipping manifests and crate labels, a training program's learning objectives will be measures of improved reading skills. Success in achieving the performance objective is evaluated by subsequent assessment of workers' performance in correctly unloading and storing inventory.

When learning and performances objectives are both achieved, a training program's development process is validated. If neither set of objectives is achieved or if learning objectives are met but performance objectives are not, training's design and development are suspect.

In structuring learning and performance objectives, a trainer should set goals for employee improvements attainable through training. Learning and performance objectives should clearly set out competencies and performance expected. In most cases, training shouldn't begin until most employees agree that learning and performance goals are reasonable and attainable.

Finally, the objective-setting process links learning objectives and performance objectives to organizational goals and objectives. Because organizational goals are broad, it may be difficult to connect them directly to specific job knowledge or skill. As a rule, they can be linked to team or departmental objectives and, through them, connected to institutional goals. That is useful because it encourages integrity in training design and development and because participants may find additional motivation if they understand how their immediate efforts fit into overall organizational success.

Marketing training to management

If preliminary organizational analysis, needs-identification forecasting, and needs analysis lead a training manager to conclude that training is necessary to counteract current or anticipated deficiencies, then the manager will prepare a recommendation to management proposing and advocating development of a training program.

To build support, a training advocate must present a strategy and action plan. This plan is the centerpiece of an in-house marketing effort aimed at internal decision makers. The presentation should be well thought out and comprehensive but concise. The manager should give a copy of the written plan to each person whose approval of the proposal is needed.

The plan should include the following:
■ a one-to-two page summary of the most compelling data from the analysis effort
■ a list of the positive effects expected if the training is conducted and a list of problems or risks associated with not providing the training
■ a recommendation for training program development, including a general description of the proposed program's content, timeframe, design, staff responsibilities, resource constraints, and estimated cost.

There are colloquial expressions that categorize training plans depending on the resources they require. Mager and Pipe identified these expressions in their "shoulda, oughta, wanna" (should, ought, want to) principle:
■ "Shoulda" refers to a training plan that will cover essential needs.

■ "Oughta" refers to a plan for a program that would go beyond the bare basics and would be better to have if enough money and other resources are available.
■ "Wanna" is the comprehensive, sophisticated training program the organization would have in ideal circumstances.

In a few organizations, upper management might easily afford funding for the "wanna" option and might even see it as essential. But, in many organizations, recommending the "oughta" option necessitates a "hard sell" to unsympathetic managers. Recommending a "shoulda" option usually looks good in short-term costs and may be easier to achieve, but it could prove more costly in the long run if supplementary training is required later.

For the initial presentation, a training manager probably won't have enough information to prepare a complete operational budget, but an estimated cost figure should be possible. Before implementation becomes a reality, decision makers need to see an operational budget with firm dollar figures and an exact time schedule. So, once enough information has been gathered, a training manager should present (in person or in writing) a second, more detailed training plan.

A later evaluation step is re-examination of training's preliminary cost-benefit analysis for comparison of actual and projected returns. Demonstration of positive results will increase the likelihood that training will be seen and used as a strategic tool for managing performance.

Training as a Performance Management Tool

Training is one among numerous tools for performance management. Gherson and Moore analyzed training a tool for strategic intervention and found that it has these advantages:
■ Comparatively low cost. In general, the costs of training resources are spread over time and are less expensive than those of other interventions.
■ Discretion. When training is part of competitive repositioning, training activity usually isn't visible to outside competitors.
■ Ease of adjustment. Training can be altered relatively quickly if evaluations reveal that it's not meeting organizational needs.
■ Potential for improving internal relations. Training is an investment in present staff capabilities. Recognition of that can raise employee morale.
Training also has disadvantages:
■ Time. Training takes time. In technical jobs, for instance, training may be a relatively slow-response intervention compared to hiring new, already trained staff.
■ Lost production. Employees aren't producing while they're in training or traveling to and from a training site.
■ Potential for ineffectiveness. Employees may respond negatively to training they perceive as ineffective or unconnected to rewards or career development.

Just as involving managers in program design and evaluation plans encourages their sense of ownership, providing written and oral reports about potential economic improvement from training will encourage managerial appreciation of training's value.

Such a report should describe the following:
■ performance problems to be alleviated or eliminated
■ accounting model that will be used
■ method used to estimate the dollar value of training benefits
■ method used to estimate training's return on investment (ROI)
■ potential benefits to the overall organization.

Process and implementation evaluation

Process evaluations monitor training programs to ensure that they follow a rigorous design, development, and implementation process. They gather information to answer whether a program will follow (or is following) a structured format.

If a program isn't proceeding exactly as planned, there may be good reasons (for example, a better workbook came on the market at the last minute) or bad ones (for example, most participants are finding the sequence of material confusing).

Process evaluations of training designs are future-oriented. They represent crucial questions that training designers must ask after they have decided to initiate training but before and as they actually do it. Process evaluations involve deciding how to produce the necessary learning and behavior changes in participants. Of course, training designs are compromises between theory and practicality, and require some degree of alteration as they are installed and operated. On-the-spot control and revisions adapt programs to the environment and make them work.

Most training programs have many discrepancies between design and implementation. As they are noticed and considered, modifications may be made. Changes often consist of implementing different or additional controls (such as simpler instructions, repeated learning activities, more frequent reviews, smaller task divisions, or more focused materials). In some cases, the design itself or expectations for participant performance must change.

The goal is to note and assess program progress, look for discrepancies, make revisions, try the revised version, and then reobserve and reassess to see whether acceptable progress is now being made.

Evaluations after implemention may be formative or summative. Summative evaluations stress that training programs that don't "pay off" may be abandoned or replaced by other human resource interventions. Spencer suggests that training managers turn this into an advantage. For example, if a training manager uses evaluation to avoid continuing a program past its prime, the manager can also make a case for transferring the old program's funds to a desirable program previously considered unaffordable.

Rigor and practicality in evaluation design

Evaluation designs differ in rigor and practicality.

Rigor relates to the quality and quantity of information the evaluation produces and to how well the information traces participant and organizational changes to the training program. Practicality relates to lessening the time and expense or increasing ease of conducting and reporting an evaluation.

Unlimited resources would permit an evaluator to answer all questions that a client might have about a training program. An absence of constraints also would permit an evaluator to select the most rigorous evaluation design to ensure the validity of evaluation conclusions. Instead, the only certainty of an evaluation is that constraints, both economic and political, will restrict the ability to answer clients' questions.

Constraints also affect the rigor of evaluation design. An evaluator is responsible for selecting the most rigorous evaluation design possible under existing constraints and for lobbying management to lift undue, organizationally self-imposed constraints.

Rigorous evaluation designs have the following characteristics:
■ They collect data from many, perhaps all, participants.
■ They collect data more than once, possibly many times.
■ They evaluate more at the organizational results level.
■ They employ quantitative data-collection methods.
■ They are more expensive.
■ They are more time-consuming.
■ They yield formal reports.
■ They are used for making decisions about program continuation or cutback (that is, they are summative).
■ They are used when training's success is critical for safety or strategic business purposes.

Less rigorous practical evaluation designs, in contrast, have these traits:
■ They collect detailed data from a small sample of participants.
■ They are usually conducted to determine participants' reactions and learning gains.
■ They use qualitative data-collection methods.
■ They are less expensive.
■ They take less time.
■ They use informal reporting procedures.
■ They are used to identify program strengths and weaknesses and to recommend areas for program improvement (that is, they are formative).
■ They are used when training success is desirable, but not critical.
■ They are used when a rigorous evaluation design is unjustified or impossible.

It might seem that all training programs should use rigorous evaluation designs, but Brandenburg and Smith cite an evaluation study that dispels that notion.

New England Telephone's management commissioned a $40,000 evaluation to determine the effectiveness of a training program for technicians. The evaluation used a rigorous evaluation design and multiple measures of performance. The evaluation design permitted the evaluator to make strong recommendations in favor of the training program after a year. But the

Evaluation Framework, Design, and Reports

report arrived too late for the client to make a decision based on program findings. By then, everyone familiar with the evaluation had left the department that had financed it. Despite the evaluation's rigor, resources were wasted because practical considerations about reporting the findings were neglected.

Two evaluation design concepts are often referred to in discussions of specific evaluation designs:
- pre- and post-testing
- control groups.

Pre- and post-testing. Measures collected before the implementation are called pre-tests; those collected afterward are called post-tests. Tests are commonly thought of as written examinations, but virtually any measure that's meaningful for an organization can serve as a test. Test data can be collected from training participants directly or from organizational data.

Pre-tests are usually taken once. It's advisable, though, to pre-test at least twice. Here's why: people are often placed in a training program because of extremely high or low pre-test data. For example, management trainees may be slated for an accelerated training program on the basis of their high performance-appraisal ratings. Conversely, people may be placed in a remedial training program on the basis of their poor records.

The danger of selecting training participants on the basis of one test is that extreme results tend to drift toward the average in subsequent tests, whether there's intervention or not. This is because an individual's test results often can't go any higher or lower. This tendency for extremes to drift back toward the middle can make accelerated programs look unduly unsuccessful and remedial programs look unduly successful in post-tests.

Control groups. Pre- and post-measurement tests alone don't always prove that training is responsible for positive change in participants, because outside factors may be responsible. Using a control group improves an evaluator's ability to determine whether changes are attributable to training. A control group consists of employees who receive the same treatment as the training group except that they don't receive training. Control groups can range from a few people to a department or division.

Evaluation Designs

These descriptions of evaluation designs are ordered roughly by the degree of experimental control or scientific rigor that they provide.

Pre- and post-test control group design. The evaluator assigns employees to either an instructional or a control group. Only the instructional group receives training. Data are gathered from both groups through pre- and post-tests. If, upon completion of the program, the instructional group shows greater post-test performance gains than the control group, training is held responsible.

But this design is effective only when the instructional and control groups share nearly equivalent characteristics. With an instructional group of experienced workers and a control group of entry-level workers, for example, it would be difficult to determine whether training or job experience was the cause of change. Random selection of employees for each group usually results in groups with similar characteristics. But in practice, it's almost impossible to assign employees randomly to training programs.

Many HRD specialists commend this design's ability to quantify and neutralize the effects of other factors. Others criticize it as an inappropriate attempt to inject scientific rigor despite the impossibility of keeping outside factors (such as management styles and peer pressure) constant. Another problem is that organizations may be unwilling to withhold training from some employees just because the evaluator needs a control group.

Multiple baseline design. This design may be used if a training program is to be introduced to different parts of the organization at different times. Data are collected on each group of participants before and after the training. The second group serves as a control group for the first group, the third is the control group for the second, and so on. If each group shows marked improvement in performance as compared to its control group, success is indicated.

This design is more cost-effective than the pre- and post-test control group design, because it eliminates the need for a rigidly maintained control group. But more time is required to introduce the training at different places. Besides, training is often location-specific, so a training program may have to be altered somewhat to meet the specific needs of each group.

Time series design. This design examines effectiveness through repeated measures of performance before and after program implementation. If the measures show improvement after training, training's effectiveness is supported. In this design, the training group serves as its own control group.

The time series design is more practical and cost-effective than, but not as rigorous as, the control group or multiple baseline designs because it provides less experimental control. It can't prove that outside events (such as a new product line or marketing program) occurring simultaneously with training aren't partly or entirely responsible for improvements. Still, if several measures are taken and performance improves only after the introduction of a training program, a case is built for the effectiveness of the program.

Single group pre- and post-test design. This design is widely used because it offers a measure of comparison and is inexpensive, but the absence of a control group makes it difficult to attribute changes to training. This design is criticized for its lack of rigor, but it can be useful.

For example, Aetna Life Insurance Company instituted a claims processing training program for new employees and used the single group pre- and post-test design to evaluate its effectiveness. A pre-test revealed that new employees had little knowledge of claims processing. A post-test given shortly after the training program revealed significant learning gains, and it was unlikely that anything other than training was responsible for performance improvement.

One-time case study. This design trains participants without any pre-test measure, but a post-training measure is taken during the program or shortly afterward. When this design is used, there usually has been little thought given to evaluation design.

Still, this is the most convenient evaluation design, because it requires only one measure. It's also the least rigorous, because it makes no comparisons. The best that can be said for this design is that it's more informative than taking no measure. This design is useful when constraints permit no preliminary data collection, if the primary evaluation client simply wants data collected to confirm or invalidate perceptions about a training program, or when participants have no prior background in the subject matter.

For example, when IBM offered a program on beginning Japanese to their international sales representatives, no pre-test was given because of their lack of exposure to the language. In such a case, it's appropriate to take performance measures only after participants' initial exposure to training content.

Data-collection tools

Data-collection tools can be categorized as either quantitative or qualitative instruments.

Quantitative instruments:
■ performance records and tests
■ standardized questionnaires and survey instruments
■ personnel assessment instruments.

The candor and accuracy of questionnaire and survey responses can be strengthened by assuring respondents of anonymity

Quantitative data have the following characteristics:
■ relatively easy to measure and assign dollar values to
■ objectively based
■ use a common measure of performance
■ credible to management

Qualitative instruments:
■ interviews
■ observations
■ focus group meetings
■ case studies

These instruments may be supplemented by other tools. For example, an evaluator may use a checklist of behaviors to guide an observation or may videotape a focus group meeting for additional review.

Qualitative data have the following characteristics:
■ difficulty in standardizing
■ subjectivity
■ behaviorally oriented
■ less credible to management.

At first, quantitative data appear superior to qualitative data. Yet quantitative data are more influenced by outside factors than qualitative data. Also, the appropriateness of quantitative or qualitative data depends on an evaluation's purposes, evaluation clients' questions, and overall evaluation design.

Quantitative data collection is more suitable when the following circumstances hold:
■ Evaluation is to determine whether a training program should be continued or expanded.
■ Evaluation's purpose is to identify a training program's economic impact on the organization.
■ A rigorous evaluation design is used.
■ Standardized data about a training program are needed.
■ The specific training is crucial to strategic businesses goals or safety.
■ Formal evaluation reports are required.

Qualitative data is more suitable under these circumstances:
■ The focus of the evaluation is to improve the program, to discover unanticipated consequences of the evaluation itself or to determine how a training program's success varies at different sites or among categories of participants.
■ Quantitative information needs to be augmented to provide depth and detail about a program's success.
■ Quantitative data are unavailable (for example, employee agreements may prohibit the collection of certain quantitative data).

To select the best design and data-collection methods for use in the real world of politics, personalities, and methodological imperfections, evaluators must match appropriate data-collection methods with evaluation purposes. So, in practice, quantitative and qualitative data are often gathered together.

The collection of information from multiple data sources or through several methods is called triangulation. Although more expensive and time-consuming, triangulation increases the probability of confirming training's responsibility for changes in employee performance and organizational measures.

Evaluation Practices

ASTD's research revealed that the actual practice of evaluation doesn't often follow the strict recommendations of evaluation literature. This is largely explained by the fact that many training practitioners haven't found the literature's advice applicable or useful for their organizations.

But, as well-known author and management consultant Thomas J. Peters has said, "What gets measured gets done. . . . Even imperfect measures provide an accurate strategic indication of progress, or lack thereof." So practitioners have employed various practical evaluations.

Here's an overview of current evaluation practices among organizational leaders in training, telling how and why they subscribe to their various practices. The evaluation techniques and practices explored don't meet traditional academic notions of rigor, but do provide valuable information, are reproducible, and can be quickly and easily conducted. Most of the training managers that par-

Evaluation Practices

ticipated in ASTD's research effort believe that there's value in a concerted effort to increase the practice of employee training evaluation along these lines.

All the organizations represented in this study evaluate some aspect of their training programs. In terms of the four-level Kirkpatrick model (see page S-15), 75 to 100 percent of them evaluated training programs at the participant reaction level. Virtually all of them also evaluated participants' knowledge gains in some of their training programs. Twenty-five percent of their training programs were evaluated at this, the learning level.

Behavior change on the job was the least measured: among companies surveyed, only about 10 percent evaluated training at this level. Employee training was only evaluated at the organizational results level about 25 percent of the time, despite new pressures on training practitioners to assess the economic worth of HRD activities.

Sixty-six percent of the training managers reported that HRD professionals are under increasing pressure to show that programs are producing favorable bottom-line results. These managers had a strong track record in training evaluation or had high management acceptance of training as a way to meet real operational needs. So, in their experience, increased pressure did not mean that upper management doubted that training could be beneficial.

The reason usually given for closer scrutiny by management was that employee training is being recognized as a significantly large expenditure. But greater management attention coupled with movement toward cost reduction can be particularly injurious to expenditures (such as those for employee training) that have hard-to-isolate or long-term payoffs.

Although most training programs are evaluated at the reaction and learning levels, these levels aren't always consistent with the reasons for evaluation. Research suggests that evaluation conducted for the proper reasons helps determine training's impact on job performance and economics within an organization.

Most organizations evaluate training programs to meet the following demands:

■ **Training department demands.** For quality assurance, trainers gather information to direct their efforts to improve training effectiveness. Trainers also want to demonstrate training programs' worth to top or operating management. And, training managers want to build databases for future planning and analysis.

■ **Employee demands.** After a decade of downsizing and flattening of organizational pyramids, employees are seeking training that's timely and useful in meeting their new job responsibilities.

■ **Management demands.** Managers are scrutinizing training as a tool for gaining a competitive advantage. Many managers believe that having the best trained work force increases competitiveness.

In an economically competitive environment, however, it's necessary to justify training expenditures to ensure adequate returns on training investments. Many organizations are only willing to pay for training that seems relevant and efficient. But interviews with training managers reveal that most training decisions aren't the result of data derived from assessment of training's worth. Rather, decisions about training still tend to be based on management's perceptions about training programs' worth.

Consequently, most training investment decisions are a matter of tradition—and continue training programs already in place. For example, supervisory programs usually are conducted year after year but undergo occasional adjustments to meet changing company needs.

Some organizations, such as Aetna and Johnson & Johnson, allocate as much as 85 percent of their training budgets for the continuation of existing programs. That necessitates program monitoring to make sure that the programs continue to address line managers' training needs. Frequent communication between training department staff and line managers promotes trainers' understanding of business needs.

Of couse, some trainers avoid evaluating training programs for fear that hard results may not justify the expense of progams. That sentiment is particularly likely in an evaluator who had direct responsibility for part of a program. To avoid having evaluators with a vested interest in evaluation results, some organizations prefer to use outside evaluators of in-house training and in-house evalutors of contracted training. But outside evaluators' lack of in-depth familiarity with the organization carries its own problems. At least one company, Digital Equipment Corporation, has established an internal department for training quality assurance to resolve the inside/outside evaluator dilemma.

The driving force behind most employee training investment decisions is the line management structure. Line managers usually recognize employees' needs for new job knowledge and skills, because managers are closest to the problem. Generally, training investment decisions are made when line management reports a performance problem likely to require a training solution. The training department performs a front-end analysis to determine whether the difficulty is a problem that has a training solution. In most organizations, line or operating management initiates 75 to 95 percent of employee training.

But top management usually is responsible for initiating programs that are company-wide, related to overall company policy, start-ups of broad new activities (such as comprehensive quality assurance programs), or of special interest to executives. And, it's often top management that approves or disapproves a specific dollar amount for training investments.

Many decision makers assume that once behaviors are corrected, organizational economic indicators will improve automatically. Some organizations require no tangible evidence from training to demonstrate its worth other than its connection to strategic business goals. That's the straightforward reason some training managers give for not accounting for training through measurement and evaluation: upper management doesn't require it. But without accounting for training, it's hard to say whether training contributes enough organizational value or might be worthy of greater resources.

Who initiates training investments?

The extent of line management's involvement in initiating training was reflected in ASTD's investigation of training practices in pace-setting organizations. The figure shows the percentages of training investment decisions made by line management (alone or in conjunction with the training department), top management, or the training department.

As the percentages shown in the box indicate, line management increasingly controls the investments an organization makes in training. Almost all of the organizations listed noted that a market-driven approach—repeated assessment of what training the organization needs rather than unquestioned continuance of existing training in products or programs—is the key to training department success. Therefore, training departments must learn which programs line management needs to function effectively.

Specific questions that training managers should ask line managers include:

■ What are departmental objectives for the fiscal year?
■ What problems have impinged on employees' assigned responsibilities, and do these responsibilities have training implications?
■ What are the career goals of key people in the department?

Communication between training department staff and line managers to focus on organizational problems, issues, plans, and strategies will identify the knowledge and skill gaps that impede performance. For example, Chase Manhattan Bank encouraged constant communication between line management and the training department by eliminating their training course catalogue. Chase believed that having a course catalogue discouraged the training department from rigorously assessing the training needs of each business unit.

Who Initiates Training Investments?

Company	Line Management	Top Management	Training Department
Polaroid	75%	—	25%
Johnson & Johnson	75%	15%	10%
Motorola	75%	10%	15%
Upjohn	95%	5%	—
DEC	95%	5%	—
Arthur Andersen	75%	—	25%
Aetna	90%	—	10%
AT&T	80%	5%	15%
Chase Manhattan	95%	5%	—
Vulcan Materials	50%	50%	—
New England Telephone	90%	10%	—
Xerox Customer & Marketing Education	75%	—	25%

Source: American Society for Training and Development, 1988.

Current practices

Research shows that most companies evaluate all their training programs in some fashion. All the companies that ASTD investigated build evaluation design into overall program design when program objectives are established. Sophisticated evaluation methods aren't applied to training programs if a program is a continuation of previously successful efforts, when less rigorous techniques are adequate for answering the majority of questions decision makers have about a program, or when practical considerations prohibit using such techniques.

Even in organizations relatively advanced in training evaluation, sophisticated statistical methods or controls are rarely employed. This may be for the good reasons above or, as Brandenburg and Smith have pointed out, because of trainers' lack of evaluation expertise (including uncertainty about when to measure for change) and training departments' concern that evaluation is tantamount to criticism.

Whatever methods they use, evaluators are increasingly using multiple data sources—combinations of quantitative and qualitative data.

The organizations that ASTD investigated hold conflicting views about the need to demonstrate training's connection to the bottom line. These views affect how evaluations are conducted within the organizations and how information about training programs is communicated to management.

One of the most popular purposes for evaluating training is to demonstrate training's worth to top or operating management. The organizations investigated showed the following approaches to training and evaluation:
■ Top management is greatly supportive of investments in training. Management perceptions of training's worth is the critical factor.
■ Line management views training as a strategic lever for achieving key business objectives. Evaluation focuses on whether key business objectives improve after training. If so, training receives at least a share of the credit.
■ An evaluation must demonstrate how a training program has made specific contributions to business objectives and, to the extent possible, must discount alternative explanations for improvements.

Organizational profiles: economic results

Vulcan Materials. Because of increasing industrial com-

Evaluation Practices

petitiveness, Vulcan Materials is evaluating training at the economic-results level. One-third of Vulcan's training investment decisions are based on cost-benefit analysis conducted before training implementation. Training department proposals include analysis of projected costs and benefits; evaluation design is built into program development. In order of preference, the priority of evaluation measures is output, quality, cost, timeliness, behavior (including that associated with attitudes), and observations.

As noted earlier, few organizations use statistical methods and controls for training evaluation, but management now places greater emphasis on demonstrating training's contribution to organizational economic goals. Accordingly, Vulcan's former human resources manager developed a simple, practical method for estimating training's economic results.

Before a training program's implementation, middle managers are asked to estimate the savings they expect to result for their departments. Middle managers also rate their confidence (on a 0 to 100 percent scale) that the training program itself (as opposed to other factors such as managers' on-the-job reinforcement of practices) will be responsible for the savings. Projected savings are multiplied by the "confidence" percentage to yield a forecast of total cost savings from training.

Once a training program is complete, evaluators examine actual cost savings or revenue increases that the organization has incurred. The training department then asks line managers to estimate (again, through a percentage) how responsible they now believe the training was for these improvements. The actual savings/revenue amount is multiplied by this percentage to provide an estimate of the total cost savings the training program has provided to the organization; that figure is then compared to the forecast.

An example of that practice is shown in a course to improve first-line supervisors' supervisory skills. En-hancement of supervisory skills was expected to reduce production worker turnover. Middle managers of the production workers' supervisors were asked to project the savings they expected to make as a result of reduced turnover and to indicate their confidence rating of training's responsibility for this savings.

The supervisor training program was estimated to cost $10,000. The overall confidence level of middle managers that the program would be responsible for reducing turnover among production workers was 50 percent. Middle managers estimated that the benefits from reduced turnover would equal $200,000. That expected total benefit of reduced turnover was multiplied by the 50 percent confidence level to estimate the training program's contribution to reduced turnover at $100,000—10 dollars of benefit for each dollar spent on training.

After the training department made a proposal to top management outlining the expected savings from reduced turnover, the training program was implemented. A study six months later revealed that actual total savings from reduced turnover amounted to $100,000. Middle managers again expressed a 50 percent confidence level that training was responsible for the savings. The actual total benefits of reduced turnover was multiplied by 50 percent to determine training's actual responsibility for improvements in turnover.

The result showed that the training program's contribution to reduced turnover was estimated at $50,000 or five dollars of benefits for each dollar spent on training. This success story of reduced turnover was publicized with some credit attributed to the training department.

Arthur Andersen regularly evaluates training at the economic results level, and evaluation is invariably designed into the original program. Anticipation of economic downturns when training may need more justification causes increasing pressure to show cost-benefit

---

Savings Forecast and Actual Savings

Savings forecast

| $200,000 (expected benfits) | x | .50 (managers' confidence level of training responsibility) | − $100,000 |

| $100,000 (expected savings) | − | $10,000 (training costs) | = $90,000 savings forecast |

Post-training evaluation

| $100,000 (expected benefits) | x | .50 (manager's confidence level of training responsibility) | = $50,000 |

Actual savings

$$\$50,000 - \$10,000 = \$40,000$$

Source: interview, Phillips, 1988.

for training. Top management is generally receptive and supportive of evaluation, and eager to assess training's contribution to achieve strategic business goals.

That management perspective is reflected in evaluation policies. Arthur Andersen is unusual in that its evaluators use a wide range of statistical methods, including experimental designs as well as more common qualitative and quantitative evaluation designs.

For example, using the control group method and a pre-test and post-test evaluation design, Arthur Andersen compared accelerated learning techniques to traditional instructional design and delivery methods to determine which was more economical.

Thirty-eight participants were selected based on similar work and personal characteristics. Participants were randomly assigned to either an experimental (accelerated learning approach) or control (traditional approach) group. The traditional learning approach was evaluated as being superior to the accelerated learning approach in terms of participants' knowledge gains and reactions at the end of training. But a study four months later revealed that behavior improvements on the job were higher for the accelerated learning group than for the control group. That led to further research by Arthur Andersen on the economic value of accelerated learning techniques.

Johnson & Johnson, an organization with an excellent training reputation, evaluates all training programs. Although the primary evaluation emphasis is on behavioral change on the job, evaluators also pay attention to advancing operational initiatives.

For example, a training program that addresses a key operational initiative shares credit if improvements occur after training. Management accepts this claim of credit without requiring the support of statistical controls.

More than 80 percent of training decisions are based on qualitative data obtained from worldwide employee surveys. Actual decision making about where to invest in employee development usually isn't based on hard data or rigorous financial analysis to determine potential return on investment.

Recently, Johnson & Johnson's corporate headquarters mandated a strong training policy that exhibits top management's greatly increased support for training. Management is looking more intensely at employee training as a tool for gaining competitive advantage. But that hasn't lead to increased management demands for an economic-results level of evaluation. Instead, in response to the strong increase in management support and funding for training, there is self-generated pressure from the training department to demonstrate training's effectiveness.

Motorola is another organization where management's perspective of training influences evaluation of training programs at the results level. Top management is strongly supportive of training, to the point of having set mandatory minimums for training investments in all divisions. The chairman of Motorola has publicly singled out training as a critical means of supporting strategic business goals.

At Motorola, training plays a key role in achieving strategic goals and operating initiatives. Motorola's training and education center is charged with operating an extensive in-house training and education effort to raise employee skill and knowledge levels. Motorola's managers are committed to training as a means of improving productivity, performance, and profitability through the development and expansion of workforce skills.

Motorola's managers are also committed to the principle of participative management. They believe that only by developing and expanding employee skills will effective employee contributions—and consequent productivity improvements—be maximized.

Because sophisticated evaluation methodology is costly, it isn't used to evaluate training at the results level. In fact, Motorola's trainers believe that if they were to be required to evaluate training at the results level it would mean they had lost the demonstrated confidence of management. Almost all Motorola training programs are evaluated using quantitative and qualitative measures in order to improve training effectiveness and, at times, to demonstrate training's effectiveness.

There's no direct effort to demonstrate training's contribution to Motorola's economic objectives. But indirectly, by designing training programs to address five of Motorola's key operational initiatives, the training department can make a legitimate claim of contributing to those objectives' achievements. The operational initiatives are having zero production defects, reducing total cycle time, integrating production and manufacturing, becoming a customer-driven company, and developing a participative management culture.

Although training programs can share in the credit for improvements in measures of these initiatives that occur after training, the absence of statistical controls means training can't eliminate rival explanations for improvement. Still, Motorola's management has great confidence in training's contribution to economic goals.

That is partly because the training population is predominately production personnel. Result measures are usually more precise for them because their work is highly task-specific. For managers and staff personnel, units of measure are much less obvious, and there are many more non-training influences on work output.

Polaroid. Since the 1980s, Polaroid has demonstrated that training affects the bottom line. Usually, the link between training and an improved bottom line hasn't been a matter of hard data analysis.

The units of measurement for evaluating training are the same units that line managers use as performance standards (for example, production units per hour per employee). Projections about training's impact on the bottom line are made without the support of statistical controls. Sometimes, benefits are based on projections of past, similar experiences.

A few years ago management showed strong endorsement of training's worth at Polaroid: training expenditures were doubled although the company was going through an overall downsizing of 30 percent. Training was perceived as a vital economic force although hard data were not available to substantiate that belief.

Evaluation Practices

Evaluating training at the behavioral level

In some organizations, evaluation looks at training's on-the-job application. ASTD's research indicates that the most popular reasons for evaluation are to gather information that helps decision makers improve the training process and to facilitate participants' job performance. So evaluation is done to ascertain whether participants meet program learning objectives and transfer learning to the job setting.

The organizations that ASTD investigated regarded training and line management collaboration in evaluation design as critical to accurate identification of employee behaviors necessary to contribute to organizational goals. The organizations interviewed use a variety of practical qualitative and quantitative measures of individual performance to demonstrate positive transfer of what was learned in training to the employees' jobs. Virtually all the organization's representatives stressed the importance of waiting a reasonable period of time before assessment of on-the-job behavioral changes, typically six months.

Organizational profiles: behavioral results

Johnson & Johnson acknowledges training's direct support of strategic business goals. One reflection of that is the recent addition of a state-of-the art management training center at Johnson & Johnson's corporate headquarters. To ensure quality training, Johnson & Johnson's training director has directed that new emphasis be placed on measurement to augment the company's existing evaluation strategy.

Currently, Johnson & Johnson uses a six-months' post-training self-report form to evaluate training participants' on-the-job changes and management potential and practices. Subordinates of managerial and supervisory trainees are surveyed annually about their perceptions of mangerial performance changes.

Johnson & Johnson recently developed a first-line supervisors' competency model to identify critical supervisory work skills. Training now is being used as the strategic lever to help 350 first-line supervisors achieve these competencies. This program's evaluation has three parts:
■ Participants define their perceptions of change.
■ Participants' supervisors fill out questionnaires.
■ Five to eight subordinates of participants respond to a survey.

The Travelers Companies. Travelers stresses evaluation of training at the behavioral level with few attempts at translating evaluation results into dollars earned or profit. Travelers believes that it's more relevant to evaluate a training program's effect on management behavior, supervisory tactics, and strategic planning efforts.

Travelers regards evaluation as critical because of training's perceived importance in facilitating management competencies.

Travelers has identified management competencies for supervisors, managers, and directors and has developed a continuum of 27 training programs for each managerial level. The critical managerial competencies were identified through surveys of 700 Travelers' managerial personnel nationwide. Fifty Travelers managers identified as extremely high-potential performers were also surveyed to help isolate the traits and skills most important for Travelers' managers. Analysis of those traits and skills shaped training and evaluation plans.

Trainers at Travelers would prefer to use more statistical evaluation methods, but current staff capacity and lack of management demand for precision discourage the use of more rigorous evaluation methodology.

At present, Travelers uses a variety of qualitative and quantitative methods. Quality assurance checks are performed to assess program validity and relevance during training implementation. Qualitative data indicate whether management training programs need revision. Information about participant reactions is gathered immediately after a training program and also through evaluator observations and post-course group discussions with participants. The useful evaluation information that management has been receiving may eventually pave the way for more rigorous evaluations.

Travelers trains instructors in evaluation techniques (such as observation and group discussion), so they can act as self-evaluators. It's not economical for Travelers to use an evaluator/observer in every training program, especially those that have been operating for some time. For such programs, evaluation is primarily a maintenance function. But, a professional evaluator is usually called in to assess new training programs.

Evaluation data are used to discern whether participants view training as valuable and transferable to the job. As a rule, if more than 25 percent of a training program's participants express dissatisfaction with an aspect of the course as it relates to the job, revisions will be made. When revisions are to be made in a course, evaluation data are reviewed by a committee of line managers, instructional designers, an evaluator, and an instructor.

New England Telephone conducts evaluations primarily to determine whether training meets the expectations of line managers once workers are back on the job. New England Telephone doesn't use statistical methods but prefers to build a case for their training programs' value through reaction sheets and focus group meetings.

New England Telephone uses three reaction forms.

The first, a questionnaire, is administered to participants immediately after a course. It asks participants to evaluate the instructor and facilities and to assess the transferability of the training curriculum to the job.

The evaluator administers a second reaction form to participants three months after training. This form asks participants to reconsider training's effectiveness for their overall jobs and to assess why it was or wasn't valuable.

The training department sends a third questionnaire to line managers in order to assess a program's effectiveness in serving their mission.

Focus group meetings are held several months after training and may last an entire working day. These meetings reinforce the role of training in facilitating corporate objectives by tying training closer to strategic business goals. By examining discrepancies between training expectations and actual accomplishments, they also enhance credibility.

Focus group members include an evaluator, middle managers, instructors, participants, and instructional designers. The evaluator moderates the meeting and is responsible for condensing participant reaction sheet data and qualitative data gathered from interviews with participants and the instructor. Discussions focus on necessary changes. Major questions that the focus groups address:

■ Does the course content require revision to meet existing or changing business requirements?
■ Could the program be redesigned to be more effective?

When developing a training program, New England Telephone holds focus group meetings for the client and training department. The client is closely involved in establishing training program learning objectives. The after-training focus group discusses whether these learning objectives were adequately realized in the program.

The training department claims that by understanding and serving the strategic business goals of the various business groups in the organization, training can share in the responsibility for economic improvements. Consistent with this approach, New England Telephone's evaluation unit doesn't serve as a judge but as an internal consultant charged with improving the relevance and quality of training.

Xerox Corporation Customer and Marketing Education.

Customer and Marketing Education's primary evaluation goal is to determine whether the skills, knowledge, and attitudes presented in training have been transferred to particpants' job performance. Line management and the training department communicate before training program goals and objectives are established. After a training program, an evaluator determines whether the learning objectives set by the client have had a positive effect on job performance.

Questions that evaluations seek to answer:
■ Do participants use what they learned in the training program?
■ Which training objectives did and did not transfer to participants' jobs?
■ If transfer to jobs didn't occur, what factors are responsible?

Evaluation begins in the pilot stage of program development, when a program is in draft form and hasn't been delivered on a regular basis. This first evaluation step helps identify and correct confusing or irrelevant material before a program is fully implemented.

Pilot-test evaluation data are gathered from several sources: tests of the curriculum, classroom observations, reaction forms from participants, and debriefings with participants, instructional designers, and line managers.

The second evaluation step takes place during training. Training programs are monitored to determine whether participants achieve program learning objectives and to obtain participants' reactions. Data are collected over a period of time—from tests, participant reaction sheets, and after-course discussions with participants.

This approach permits the evaluator to examine trends and patterns. For instance, the evaluator looks to see whether specific types of participants are reacting similarly to a training program. In this step, evaluation seeks to answer whether participants achieve program objectives and whether the program meets participants' needs.

The final evaluation step takes place after training and centers on whether programs learning objectives have transferred to the job setting. This evaluation steps seeks answers to these questions:
■ Are people using what they learned in training on the job?
■ Which learning objectives transferred to the job?
■ What factors were responsible for training not transferring to the job?

For this step, data-collection methods include the following:
■ observations of former participants in their job settings
■ roundtable discussions with participants after training
■ interviews with participants
■ questionnaires
■ telephone interviews
■ examinations of job records
■ interviews with former participants' managers.

Unless a client requests quantitative measures, Xerox finds that using a variety of qualitative methods is more appropriate for evaluation than a strictly quantitative or statistical approach.

Evaluators observe former participants in their work settings in order to verify data gathered from interviews, round table discussions, or questionnaires. These observations also help determine whether any barriers in the working environment inhibit the transfer of learning to the job setting. Barriers might include conflicting organizational policies and practices or managerial unwillingness to allow participants to use the new skills. Barriers also may be uncovered through group discussions (among an evaluator, managers, instructors, and instructional designers) that focus on learning transfer.

AT&T uses a combination certification-evaluation program that measures whether trainees can do a job following training. Very few AT&T training investment decisions are based on hard productivity data, but decisions are invariably based on analysis of competency deficiencies.

AT&T enables an individual business unit to certify employee performance on the job through a needs analysis. The analysis identifies the knowledge, skills, and abilities that workers must have to perform adequately.

Evaluation design is always a part of program design. And AT&T evaluates to ensure that design standards are valid for producing effective training. AT&T regularly evaluates all training with the primary aim of assuring that the training is up to date and serves its intended purpose. Rather than using criterion-based testing, AT&T uses line managers and expert judgment to determine whether behavior has improved on the job. AT&T also uses a five-point self-report form to measure participants' reactions to a course, their perceptions about whether training has enabled them to perform their jobs better, and their assessment of training's contribution to their career development. ■

Advisors

Advisors

Accounting and Evaluation Advisory Panel

Robert L. Craig
Vice-President (ret.)
American Society for Training and
    Development

Henry L. Dahl
Manager
Training and Development
UpJohn Corporation

Jim Donohue
Associate Director
Training and Development
Champion International Corporation

Robert S. Fenn
National Director of Training
The Travelers Insurance Company

Allen R. Gilberti
Manager Systems Training and Development
Niagra Mohawk Power

Gerald Gundersen
Chief, Research
U.S. Department of Labor

Kenneth H. Hansen
Manager
Corporate Education and Training
Xerox Corporation

John Hurley
Vice-President
Corporate Training and Educational
    Resources
The Chase Manhattan Bank

Charles E. LaPier
Personnel Director
Polaroid Corporation

William Luithle, Jr.
Division Manager
Corporate Training Support Group
American Telephone & Telegraph
    Technologies

Mike Lyden
Senior Consultant for Evaluation
Aetna Institute for Corporate Education

Jeff Oberlin
Manager
Planning and Evaluation
Motorola, Incorporated

David R. Schwandt
Director
Organization and Human Development
U.S. General Accounting Office

Don Tracey
Manager
Education Support
International Business Machines

Laurence Weinstein
Professor of Business
Sacred Heart University

Accounting and Evaluation Network

Richard Anderson
Columbia University Teachers College

Robert M. Anderson
General Electric Company

Sandi Auerbach
International Business Machines

Ronald Bacon
CIGNA

Doug Barney
Northern Telecom

Frank Biso
Texas Instruments

Christina Caron
British Embassy

Neal E. Chalofsky
George Washington University

Ivan Charner
National Institute for Work and Learning

Maurice E. Coleman
Arthur Andersen & Company

Donald K. Conover
American Telephone & Telegraph

Robert L. Craig
Vice-President (ret.)
American Society for Training and
    Development

Oliver Cummings
Arthur Anderson & Company

Henry Dahl
Upjohn Corporation

Chester Delaney
The Chase Manhattan Bank

James DeVito
Johnson & Johnson

James Donohue
Champion International Corporation

C. David Esch
United States Agency for International
    Development

Michael Emmott
Manpower Services Commission

Robert S. Fenn
The Travelers Insurance Company

John Fox
Los Alamos National Laboratory

Donald Fronzaglia
Polaroid Corporation

Allen R. Gilberti
Niagra Mohawk Power

Margaret Goodwin
Dayton Hudson Department Store Company

Gerald Gunderson
U.S. Department of Labor

Kenneth H. Hansen
Xerox Corporation

Jack Head
Batesville Casket Company

David Hobbs
Frito Lay

John Hurley
The Chase Manhattan Bank

Elizabeth Kasl
Columbia University Teachers College

Linda Kemp
Dayton Hudson Department Store Company

Donald L. Kirkpatrick
University of Wisconsin

Rosslyn S. Kleeman
United States General Accounting Office

Robert Lake
Xerox Corporation

Harry Litchfield
Deere and Company

Judy Long
Liggett and Myers Tobacco

William Luithle, Jr.
American Telephone & Telegraph

Mike Lyden
Aetna Institute of Corporate Education

Leslie May
Digital Equipment

Dorothy Ellen McNutt
College of the Mainland

Walter J. Michalski
General Telephone and Electronics

Mike Negrelli
International Business Machines

Dianne O'Connell
Harris Corrporation

Jeff Oberlin
Motorola Training and Education Center

Susan H. Ogle
People's Bank

Tom Orton
Integrated Semiconductor

Jane Perkins
New England Telephone

Jack J. Phillips
Alabama Federal

Herb Quinn
The Chase Manahattan Bank

Jane Rifkin
American Telephone & Telegraph

Geary A. Rummler
Warren, New Jersey

Kathleen N. Ryerson
St. Elizabeth Medical Center

David R. Schwandt
U.S. General Accounting Office

Mick Sheppick
Honeywell

Martin E. Smith
New England Telephone

Gene Sullivan
American Council on Education

Richard A. Swanson
University of Minnesota

Peter Tatrzewski
International Business Machines

Don Tracey
International Business Machines

Mary Kay Ulleman
Bank One, Dayton, N.A.

Meredith Ward
Squibb Corporation

Laurence Weinstein
Sacred Heart University

Audre Wenzler
Laventhol & Horwath

Other contributors

Sue Griffith
The Travelers Insurance Company

Jay Orlin
Northern Telecom

Dave Ford
Southland Corporation

Bob Shaw
International Business Machines

Valerie Beer
Xerox Corporation

AT&T. *Developing Training Tests*. Trainer's Library series. Reading, Mass.: Addison-Wesley, 1987.

AT&T. *Measurement and Evaluation*. Trainer's Library series. Reading, Mass.: Addison-Wesley, 1987.

Ames, B.C., and Hlavacek, J.D. "Vital Truths About Managing Your Costs." *Harvard Business Review*, January–February 1990, 140–147.

Anderson, R.E. "Cost Accounting for Adult Education and Training." In R.E. Anderson and E.S. Kasl (eds.), *The Costs and Financing of Adult Education*. Lexington, Mass.: Lexington, 1982.

Brandenburg, D., and Smith, M.E. *Evaluation of Corporate Training Programs*. Time Report 91. Princeton, N.J.: ERIC Clearinghouse on Tests, Measurement, and Evaluation, Educational Testing Service, 1986.

Brethower, K.S., and Rummler, G.A. "Evaluating Training." *Training & Development Journal*, May 1979, 14–22.

Brinkerhoff, R.O. *Achieving Results From Training*. San Francisco: Jossey-Bass, 1987.

Business Week Staff. "How the New Math of Productivity Adds Up." *Business Week*, June 6, 1988, 102–115.

Business Week Staff. "Human Capital: The Decline of America's Work Force." *Business Week*, September 19, 1988, 100–141.

Carnevale, A.P. "The Learning Enterprise." *Training & Development Journal*, January 1986, 18–26.

Carnevale, A.P. *Training America: Strategies for the Nation*. Alexandria, Va.: American Society for Training and Development/National Center on Education and the Economy, 1989.

Cascio, W.F. *Applied Psychology in Personnel Management*. (2nd ed.) Reston, Va.: Reston Publishing, 1982.

Cascio, W.F. *Costing Human Resources: The Financial Impact of Behavior in Organizations*. (2nd ed.) Boston: PWS-Kent, 1987.

Cetron, M.J., Rocha, W., Luckins, R. "Into the 21st Century: Long-Term Trends Affecting the United States." *The Futurist*, July-August 1988, 29–40.

Dennison, E.F., *Accounting for United States Economic Growth 1929#1969*. Washington, D.C.: The Brookings Institution, 1974.

Flamholtz, E.G. *Human Resource Accounting*. (2d ed.) San Francisco: Jossey-Bass, 1986.

Gherson, D.J., and Moore, C.A. "The Role of Training in Implementing Strategic Change." In L.S. May, C.A. Moore, and S.J. Zammit (eds.), *Evaluating Business and Industry Training*. Boston: Kluwer Academic, 1987.

Gilbert, T.F. *Human Competence Engineering: Worthy Performance*. New York: McGraw-Hill, 1978.

Goldstein, I.I. *Training: Program Development and Evaluation*. Monterey, Calif.: Brooks/Cole, 1974.

Head, G.E. *Training Cost Analysis: A Practical Guide*. Washington, D.C.: Marlin Press, 1985.

Herman, J.L., Morris, L.L., and Fitz-Gibbon, C.T. *Evaluator's Handbook*. Beverly Hills, Calif.: Sage, 1987.

Horngren, C., and Foster, G. *Cost Accounting: A Managerial Emphasis*. (6th ed.) Englewood Cliffs, N.J.: Prentice-Hall, 1987.

Johnson, H.T., and Kaplan, R.S. *Relevance Lost: The Rise and Fall of Management Accounting*. Boston: Harvard Business School Press, 1987.

Johnston, W.B. and Packer, A.E. *Workforce 2000: Work and Workers for the Twenty-First Century*. Indianapolis, Ind.: Hudson Institute, 1987.

Kaplan, R.S. "Yesterday's Accounting Undermines Production." *Harvard Business Review*, July–August 1984, 95–101.

Kaplan, R.S. "One Cost Accounting System is not Enough." *Harvard Business Review*, January–February 1988, 23–31.

Kearsley, G. *Costs, Benefits and Productivity in Training Systems*. Reading, Mass.: Addison-Wesley, 1982.

Kirkpatrick, D.L. "Techniques for Evaluating Training Programs." In D.L. Kirkpatrick (ed.), *Evaluating Training Programs*. Alexandria, Va.: American Society for Training and Development, 1975.

Kirrane, D.E. "Cost Accounting Today." *Training & Development Journal*, September 1986, 24–27.

Lawler, E.E. *Pay and Organizational Development*. Reading, Mass.: Addison-Wesley, 1981.

Lombardo, C.A. "Do the Benefits of Training Justify the Costs?" *Training & Development Journal*, December 1989, 60–64.

Lombardo, C.A. "Cost/Benefit Analysis of Training: An Introduction." In H. Birnbrauer (ed.), *The ASTD Handbook for Technical and Skills Training*. Alexandria, Va.: American Society for Training and Development, 1986.

Mager, R.F. *Preparing Instructional Objectives*. Belmont, Calif.: Fearon, 1962.

Mager, R.F., and Pipe, P. *Analyzing Performance Problems; Or You Really Oughta Wanna*. Belmont, Calif.: Fearon, 1970.

Manufacturing Studies Board Commission on Engineering and Technical Systems, National Academy of Science. *Toward a New Era in U.S. Manufacturing: The Need for a National Vision*. Washington, D.C.: National Academy Press, 1986.

May, L.S. "Applying Quality Management Concepts and Techniques to Training Evaluation." In L.S. May, C.A. Moore, and S.J. Zammit (eds.), *Evaluating Business and Industry Training*. Boston: Kluwer Academic, 1987.

Mosier, N.R. *Financial Analysis: A Review of the Methods and Their Applications to Employee Training*. Project No. 9. St. Paul: Training and Development Research Center, University of Minnesota, 1986.

References

Nasar, Sylvia. "Outlook 1990: It's Gloves Off Time." *U.S. News & World Report*, Dec. 25/Jan. 1, 1990, 40–42.

Odiorne, G.S. "The Need for an Economic Approach to Training." *Journal of the American Society of Training Directors*, March 1964, 3–12.

Parker, B.L. *Summative Evaluation in Training and Development: A Review and Critique of the Literature, 1980 through 1983*. Project Number 4. St. Paul: Training and Development Research Center, University of Minnesota, 1984.

Patton, M.Q. *How to Use Qualitative Methods in Evaluation*. Beverly Hills, Calif.: Sage, 1987.

Peters, T.J. *Thriving on Chaos: Handbook for a Management Revolution*. New York: Knopf, 1988.

Phillips, J.J. *Handbook of Training Evaluation and Measurement Methods*. Houston: Gulf, 1983.

Porter, M.E. *Competitive Advantage: Creating and Sustaining Superior Performance*. New York: Free Press, 1985.

Ratzlaff, L.A. (ed.). *The Education Evaluator's Workbook: How to Assess Education Programs*. Vol.1. Alexandria, Va.: Education Research Group, Capitol Publications, 1987.

Robinson, D.G. and Robinson, J. *Training for Impact*. San Francisco: Jossey-Bass, 1989.

Rosenberg, M.J. "Evaluating Training Programs for Decision Making." In L.S. May, C.A. Moore, and S.J. Zammit (eds.), *Evaluating Business and Industry Training*. Boston: Kluwer Academic, 1987.

Ruth, G. *A Systems Approach to Cost-Benefit Analysis*. Training and Development Research Report. Columbus: Graduate Program in Training and Development, Department of Educational Policy and Leadership, Ohio State University, 1985.

Samuelson, P.A. *Economics*. (11th ed.) New York: McGraw-Hill, 1980.

Schein, E.H. *Organizational Psychology*. (3rd ed.) Englewood Cliffs, N. J.: Prentice-Hall, 1980.

Smith, M.E. "Measuring Results." In R. L. Craig (ed.), *Training and Development Handbook: A Guide to Human Resource Development* (3d ed.) New York: McGraw-Hill, 1987.

Spencer, L.M. "Calculating Human Resource Program Costs and Benefits." In W. R. Tracey (ed.), *Human Resources Management Development Handbook*. New York: AMACOM, 1985.

Spencer, L.M. *Calculating Human Resource Costs and Benefits: Cutting Costs and Improving Productivity*. New York: Wiley, 1986.

Stecher, B.M, and Davis, W.A. *How to Focus an Evaluation*. Beverly Hills, Calif.: Sage, 1987.

Stephans, E., Mills, G.E., Pace, R.W., and Ralphs, L. "HRD in the *Fortune* 500: A Survey." *Training & Development Journal*, January 1988, 26–32.

Steinmetz, A. "The Discrepancy Evaluation Model." In G.F. Madaus, M. Scriven, and D. Stufflebeam (eds.), *Evaluation Models: Viewpoints on Educational and Human Services Evaluation*. Boston: Kluwer Academic, 1983.

Swanson, R.A., and Geroy, G. *Forecasting the Economic Benefits of Training*. Project No. 1. St. Paul: Training and Development Research Center, University of Minnesota, 1984.

Swanson, R.A., and Gradous, D.B. *Forecasting Financial Benefits of Human Resource Development*. San Francisco: Jossey-Bass, 1988.

Toffler, A. *The Third Wave*. New York: William Morrow, 1980.

Work in America Institute. *Training for New Technology, Part III: Cost Effective Design and Delivery of Training Programs*. Scarsdale, N.Y.: Work in America Institute, 1985.

Credits

Editing: Diane E. Kirrane
Production: Judith Wojcik

About the
Author

Donald L. Kirkpatrick, Professor Emeritus of the University of Wisconsin, is an author, speaker, seminar leader, and consultant in the training and development field. He is a former training manager and personnel manager in industry. He holds a Ph.D. in Counseling and Adult Education from the University of Wisconsin where he wrote his dissertation "Evaluating a Human Relations Training Program for Supervisors." He is the author of six books, including *How to Plan and Conduct Productive Business Meetings, No-Nonsense Communication, How to Train and Develop Supervisors,* and *Evaluating Training Programs: The Four Levels*. His books *How to Improve Performance through Appraisal and Coaching* and *How to Manage Change Effectively* won ASPA's "Best Book of the Year" award in 1982 and 1985, respectively. He is also the author of seven Supervisory/Management Inventories on the topics of Communication; Human Relations; Managing Change; Time Management; Modern Management; Performance Appraisal and Coaching; and Leadership, Motivation, and Decision Making. He has conducted management seminars on evaluating training programs in many countries including England, France, Singapore, Australia, Greece, Venezuela, and Saudi Arabia. Some examples of seminar subjects he teaches are Leadership and Motivation, Effective Communication, Managing Change, Managing Time, Performance Appraisal and Coaching, Decision Making, and Team Building.

Kirkpatrick has organized and conducted management development programs for many organizations including Blockbuster, Ford, General Electric, ServiceMaster, Northern States Power, Continental Airlines, Abbott Laboratories, and State Farm Insurance. An honorary lifetime member of the American Society for Training & Development (ASTD), Kirkpatrick served as ASTD National President and Chairman of the Board of Directors in 1975. In 1997, Dr. Kirkpatrick was elected to *Training* magazine's HRD Hall of Fame.